Advance Pr

Mathematical Puzzles

Mathematical Puzzles

Peter Winkler

Illustrations by
Jess Johnson

CRC Press
Taylor & Francis Group
Boca Raton London New York

CRC Press is an imprint of the
Taylor & Francis Group, an **informa** business

AN A K PETERS BOOK

First edition published 2021
by CRC Press
6000 Broken Sound Parkway NW, Suite 300, Boca Raton, FL 33487-2742

and by CRC Press
4 Park Square, Milton Park, Abingdon, Oxon, OX14 4RN

Library of Congress Cataloging-in-Publication Data

Names: Winkler, P. (Peter), 1946- author.
Title: Mathematical puzzles / Peter Winkler ; Illustrations by Jess
  Johnson.
Description: First edition. | Boca Raton : AK Peters/CRC Press, 2020. |
  Includes bibliographical references and index.
Identifiers: LCCN 2020034978 (print) | LCCN 2020034979 (ebook) | ISBN
  9780367206925 (paperback) | ISBN 9780367206932 (hardback) | ISBN
  9780429262913 (ebook)
Subjects: LCSH: Mathematical recreations.
Classification: LCC QA95 .W645 2020 (print) | LCC QA95 (ebook) | DDC
  793.74--dc23
LC record available at https://lccn.loc.gov/2020034978
LC ebook record available at https://lccn.loc.gov/2020034979

ISBN: 978-0-36720-693-2 (hbk)
ISBN: 978-0-36720-692-5 (pbk)
ISBN: 978-0-42926-291-3 (ebk)

Typeset in CalistoMT
by KnowledgeWorks Global Ltd.

for Miles, Sage, and Beatrix

Contents

Acknowledgments   ix

Preface   xi

The Puzzles   xv

The Hints   lxxxix

 1. Out for the Count   1

 2. Achieving Parity   17

 3. Intermediate Math   29

 4. Graphography   41

 5. Algebra Too   55

 6. Safety in Numbers   67

 7. The Law of Small Numbers   77

 8. Weighs and Means   91

 9. The Power of Negative Thinking   101

10. In All Probability   109

11. Working for the System   137

12. The Pigeonhole Principle   151

13. Information, Please   163

14. Great Expectation   181

Contents

15. Brilliant Induction   205

16. Journey into Space   223

17. Nimbers and the Hamming Code   235

18. Unlimited Potentials   253

19. Hammer and Tongs   273

20. Let's Get Physical   295

21. Back from the Future   303

22. Seeing Is Believing   327

23. Infinite Choice   341

24. Startling Transformation   353

Notes and Sources   371

Bibliography   393

Author   399

Index   401

Acknowledgments

This volume owes its existence to two institutions and many individuals. Dartmouth College has provided inspiring colleagues, a stimulating environment, and steadfast encouragement; special thanks are due to the endowers of the William Morrill Professorship. The National Museum of Mathematics (with whom I spent the academic year 2019–2020), through the generosity of the Simons Foundation, provided a huge variety of outreach opportunities, with many audiences on whom to test puzzles.

Some of the individual contributors (witting and otherwise) are listed alphabetically below, with multiple contributors in italics and the two most prolific in bold. Specific contributions appear in *Notes & Sources*. Doubtless many deserving names have been left out owing to gaps in my knowledge or memory, but I hope to rectify some of those omissions as time goes on.

Dorit Aharonov · Arseniy Akopyan · David Aldous · Kasra Alishahi · *Noga Alon* · Dan Amir · A. V. Andjans · Titu Andreescu · Omer Angel · *Gary Antonick* · Steve Babbage · *Matt Baker* · Yuliy Baryshnikov · *Elwyn Berlekamp* · Béla Bollobás · Christian Borgs · Bancroft Brown · Daniel E. Brown · A. I. Bufetov · *Joe Buhler* · Caroline Calderbank · Neil Calkin · Teena Carroll · Amit Chakrabarti · Deeparnab Chakrabarty · Adam Chalcraft · A. S. Chebotarev · William Fitch Cheney, Jr. · Vladimir Chernov · N. L. Chernyatyev · Vašek Chvátal · *Barry Cipra* · Bob Connelly · **John H. Conway** · Tom Cover · Paul Cuff · Robert DeDomenico · Oskar van Deventer · Randy Dougherty · Peter Doyle · Ioana Dumitriu · Freeman Dyson · Todd Ebert · Sergi Elizalde · Noam Elkies · Rachel Esselstein · R. M. Fedorov · Steve Fisk · Gerald Folland · *B. R. Frenkin* · *Ehud Friedgut* · Alan Frieze · Anna Gal · *David Gale* · A. I. Galochkin · *Gregory Galperin* · **Martin Gardner** · Bill Gasarch · Dieter Gebhardt · *Giulio Genovese* · Carl Giffels · Sol Golomb · David Gontier ·

Acknowledgments

Bill Gosper · Ron Graham · Geoffrey Grimmett · Ori
Gurel-Gurevich · Olle Häggström · *Sergiu Hart · Bob Hen-
derson · Dick Hess* · Maya Bar Hillel · *Iwasawa Hirokazu* ·
Ander Holroyd · Ross Honsberger · Naoki Inaba · Svante
Janson · *A. Ya. Kanel-Belov* · Mark Kantrowitz · Howard
Karloff · Joseph Keane · Kiran Kedlaya · David Kempe
· Rick Kenyon · *Tanya Khovanova* · Guy Kindler · Mur-
ray Klamkin · Victor Klee · Danny Kleitman · V. A.
Kleptsyn · Niko Klewinghaus · Anton Klyachko · *Don
Knuth* · Maxim Kontsevich · Isaac Kornfeld · A. K. Ko-
valdzhi · Alex Krasnoshel'skii · Piotr Krason · Jeremy
Kun · Thomas Lafforgue · Michael Larsen · Andy Latto ·
V. N. Latyshev · Imre Leader · *Tamás Lengyel · Hendrik
Lenstra* · Anany Levitin · Jerome Lewis · Sol LeWitt ·
Michael Littman · Andy Liu · Po-Shen Loh · *László Lovász*
· Edouard Lucas · Russ Lyons · Marc Massar · David
McAllester · Peter Bro Miltersen · Grant Molnar · R. L.
Moore · Thierry Mora · Frank Morgan · Carl Morris ·
Lizz Moseman · Elchanon Mossel · *Frederick Mosteller* ·
Colm Mulcahy · *Pradeep Mutalik* · Muthu Muthukrish-
nan · Gerry Myerson · Assaf Naor · Girija Narlikar ·
Jeff Norman · Alon Orlitsky · Garth Payne · *Yuval Peres*
· Nick Pippenger · Dick Plotz · *V. V. Proizvolov · Jim
Propp* · Kevin Purbhoo · Anthony Quas · *Dana Randall*
· Lyle Ramshaw · Dieter Rautenbach · Sasha Razborov ·
Dan Romik · I. F. Sharygin · Bruce Shepherd · Raymond
Smullyan · Jeff Steif · Sara Robinson · Saul Rosenthal ·
Rustam Sadykov · Mahdi Saffari · Boris Schein · Christof
Schmalenbach · Markus Schmidmeier · Frederik Schuh ·
David Seal · Dexter Senft · *Alexander Shapovalov* · James
B. Shearer · S. A. Shestakov · *Senya Shlosman* · George
Sicherman · Vladas Sidoravicius · Sven Skyum · Laurie
Snell · *Pablo Soberón* · Emina Soljanin · *Joel Spencer* · A.
V. Spivak · Caleb Stanford · *Richard Stanley* · Einar Stein-
grimsson · Francis Su · Benny Sudakov · Andrzej Szulkin
· Karthik Tadinada · Bob Tarjan · Bridget Tenner · Prasad
Tetali · Mikkel Thorup · Mike Todd · Enrique Treviño ·
John Urschel · Felix Vardy · Tom Verhoeff · Balint Virag
· *Stan Wagon* · George Wang · Greg Warrington · *Johan
Wästlund* · Diana White · Avi Wigderson · Herb Wilf ·
Dana Williams · I. M. Yaglom · Steven Young · *Paul Zeitz*
· Leo Zhang · *Yan Zhang · Barukh Ziv*

Preface

What's in This Book?

This book contains 300 of the very best mathematical puzzles in the world. Some have appeared in previous books, including my own, or in contests; others are brand new. The book is organized as follows: First, all the puzzles are presented; then, for each puzzle, a hint together with the chapter number in which you can find the solution. (The *Hints* section lists puzzles in alphabetical order by name, so may be of use in locating a particular puzzle.)

The chapters themselves are classified by solution method, rather than by general puzzle topic. Each chapter introduces a technique, uses it to solve puzzles, and ends with an actual theorem of mathematics that uses the technique in its proof. Additional information on every puzzle is found in *Notes and Sources* at the end of the volume.

Who Is This Book For?

Puzzle lovers and math lovers. For background you'll need no more than high-school math. Calculus, abstract algebra, and higher college-level material will not be used. Some math that you might not find in every curriculum, like elementary graph theory and probability, will be introduced as needed.

This does not mean the puzzles are easy; some are, others will stump professional mathematicians. And, the existence of an easy solution does not mean that a solution is easy to find!

How Were the Puzzles Chosen?

To me a great puzzle offers entertainment and enlightenment. It should be fun to read and think about. It should make an interesting mathematical point, perhaps by confounding your intuition, or perhaps just by reminding you of the amazing things elementary mathematics can do.

Where Do the Puzzles Come From?

Some of the puzzles are original but most were contributed by puzzle composers and puzzle mavens all over the world—they are the real heroes of this book. Many of their names are found in the *Acknowledgments* and *Bibliography*. Problems from contests are not usually appropriate for this volume; they are designed to test, rather than to entertain, and often require more advanced mathematics. Nonetheless some irresistible gems have arisen in contests—especially, for some reason, the Moscow Mathematical Olympiads. Often a puzzle has multiple sources and rarely can its origin be identified, or even defined, precisely; my meager efforts at doing so are reported in *Notes and Sources*. Puzzles and solutions are presented in my own words, which in some cases will be similar to a source's, in others completely different.

How Should Puzzles Be Used—or Not Used?

Mathematical puzzles should not be used for triage! Degree of ability to solve puzzles does not tell you whether someone will make a good mathematician, or even be good at mathematics.

Why not? Don't mathematicians sit around all day solving puzzles?

Nothing could be farther from the truth. To begin with, most research mathematicians are not problem-solvers but theory-builders. They look for patterns, ask good questions, and put ideas together.

When mathematicians do solve problems, they take their time and usually work together, each contributing different skills. And they often change the problem as they work! Moreover, the problems they work on are not composed as clever traps, they arise naturally.

So it should not be a surprise that most mathematicians are not great puzzle solvers—and many non-mathematicians are.

On top of all this, it is useful to remember that the ability to solve puzzles has a lot to do with confidence in oneself—which in turn is inextricably tied to social norms and prejudice, and perhaps testosterone as well. The last thing we want to do is discourage women and minorities from pursuing mathematical careers.

So what are puzzles good for, if the ability to solve them does not tell us who should become a mathematician?

The key is that a good puzzle is a gem, a thing of beauty. I would argue that the ability to *appreciate* a good puzzle, and the willingness to put effort into solving it, tells us a great deal. (Smart employers know this and if you are given a puzzle in a job interview, they are interested in your attitude and approach, not whether you get the "right" answer.) Do you want to know what math can do? Or, equally important, what math *can't do?* Do you feel that you are desperate to know the answer? Then, maybe, math is for you.

And if math is for you, then learning the techniques needed to solve puzzles is a great way to dive in. As you will see in this volume, those same techniques show up in proofs of theorems; and as (I hope) you will also see, solving puzzles is eye-opening and a lot of fun.

Where Should You Start?

With the puzzles, of course. The puzzles are the stars of this show; the rest of the book is just support. If you simply read the puzzles, pick out the ones you like, and try to solve them, you are headed for days, weeks, or months of intrigue, frustration, joy, and enlightenment. Yes, the techniques proposed in the chapters are useful tools, but be warned that no two people will classify puzzle-solving techniques the same way or even assign puzzles the same way in a given classification. Indeed, my own assignments are often distressingly arbitrary.

Note also that for the harder puzzles, the solution given (which may be one of many) is sometimes presented in rather compact fashion. The heart symbol ($\heartsuit$) is used to signal the end of an argument or proof; if you reach one unedified, go back!

Whatever you do, have fun and keep an open mind. Everything in this volume will be free on line not more than $2\frac{1}{2}$ years after initial publication, and even before then all readers are encouraged to send me comments, corrections, complaints, additional solutions, and—always!—new puzzles.

The Puzzles

Bat and Ball

A bat costs $1 more than a ball; together they cost $1.10. How much does the bat cost?

No Twins Today

It was the first day of class and Mrs. O'Connor had two identical-looking pupils, Donald and Ronald Featheringstonehaugh (pronounced "Fanshaw"), sitting together in the first row.

"You two are twins, I take it?" she asked.

"No," they replied in unison.

But a check of their records showed that they had the same parents and were born on the same day. How could this be?

Portrait

A visitor points to a portrait on the wall and asks who it is. "Brothers and sisters have I none," says the host, "but that man's father is my father's son." Who is pictured?

Half Grown

At what age is the average child half the height that he or she will be as an adult?

Shoes, Socks, and Gloves

You need to pack for a midnight flight to Iceland but the power is out. In your closet are six pairs of shoes, six black socks, six gray socks, six pairs of brown gloves, and six pairs of tan gloves. Unfortunately, it's too dark to match shoes or to see any colors.

How many of each of these items do you need to take to be sure of getting a matched pair of shoes, two socks of the same color, and matching gloves?

Phone Call

A phone call in the continental United States is made from a west coast state to an east coast state, and it's the same time of day at both ends of the call. How is this possible?

Powers of Two

How many people are "two pairs of twins twice"?

Rotating Coin

While you hold a US 25-cent piece firmly to the tabletop with your left thumb, you rotate a second quarter with your right forefinger all the way around the first quarter. Since quarters are ridged, they will interlock like gears and the second will rotate as it moves around the first.

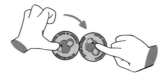

How many times will it rotate?

Squaring the Mountain State

Can West Virginia be inscribed in a square?

Natives in a Circle

An anthropologist is surrounded by a circle of natives, each of whom either always tells the truth or always lies. She asks each native whether the native to his right is a truth teller or a liar, and from their answers, she is able to deduce the fraction of liars in the circle.
   What fraction is it?

Winning at Wimbledon

As a result of temporary magical powers (which might need to be able to change your gender), you have made it to the women's singles tennis finals at Wimbledon and are playing Serena Williams for all the marbles. However, your powers cannot last the whole match. What score do you want it to be when they disappear, to maximize your chances of notching an upset win?

Polyhedron Faces

Prove that any convex polyhedron has two faces with the same number of edges.

Finding the Counterfeit

You have a balance scale and 12 coins, 11 of which are genuine and identical in weight; but one is counterfeit, and is either lighter or heavier than the others. Can you determine, in three weighings with a balance scale, which coin is counterfeit and whether it is heavy or light?

Signs in an Array

Suppose that you are given an $m \times n$ array of real numbers and permitted, at any time, to flip the signs of all the numbers in any row or column. Can you always arrange matters so that all the row sums and column sums are non-negative?

Birthday Match

You are on a cruise where you don't know anyone else. The ship announces a contest, the upshot of which is that if you can find someone who has the same birthday as yours, you (both) win a beef Wellington dinner.

How many people do you have to compare birthdays with in order to have a better than 50% chance of success?

Meeting the Ferry

Every day at noon GMT a ferry leaves New York and simultaneously another leaves Le Havre. Each trip takes seven days and seven nights, arriving before noon on the eighth day. How many of these cross-Atlantic ferries does one of them pass on its way across the pond?

Sinking 15

Carol and Desmond are playing pool with billiard balls number 1 through 9. They take turns sinking balls into pockets. The first to sink three numbers that sum to 15 wins. Does Carol (the first to play) have a winning strategy?

Monk on a Mountain

A monk begins an ascent of Mt. Fuji on Monday morning, reaching the summit by nightfall. He spends the night at the summit and starts down the mountain on the same path the following morning, reaching the bottom by dusk on Tuesday.

Prove that at some precise time of day, the monk was at exactly the same spot on the path on Tuesday as he was on Monday.

Mathematical Bookworm

The three volumes of Jacobson's *Lectures in Abstract Algebra* sit in order on your shelf. Each has $2''$ of pages and a front and back cover each $\frac{1}{4}''$, thus a total width of $2\frac{1}{2}''$.

A tiny bookworm bores its way straight through from page 1, Vol I to the last page of Vol III. How far does it travel?

Other Side of the Coin

A two-headed coin, a two-tailed coin, and an ordinary coin are placed in a bag. One of the coins is drawn at random and flipped; it comes up "heads." What is the probability that there is a head on the other side of this coin?

Slicing the Cube

Before you is a circular saw and a $3 \times 3 \times 3$ wooden cube that you must cut into twenty-seven $1 \times 1 \times 1$ cubelets. What's the smallest number of slices you must make in order to do this? You are allowed to stack pieces prior to running them through the saw.

Rolling Pencil

A pencil whose cross-section is a regular pentagon has the maker's logo imprinted on one of its five faces. If the pencil is rolled on the table, what is the probability that it stops with the logo facing up?

Watermelons

Yesterday a thousand pounds of watermelons lay in the watermelon patch. They were 99% water, but overnight they lost moisture to evaporation and now they are only 98% water. How much do they weigh now?

Boy Born on Tuesday

Mrs. Chance has two children of different ages. At least one of them is a boy born on a Tuesday. What is the probability that both of them are boys?

Air Routes in Aerostan

The country of Aerostan has three airlines which operate routes between various pairs of Aerostan's fifteen cities. What's the smallest possible total number of routes in Aerostan if bankruptcy of any of the three airlines still leaves a network that connects the cities?

Tipping the Scales

A balance scale sits on the desk of the science teacher, Ms. McGregor. There are weights on the scale, which is currently tipped to the right. On each weight is inscribed the name of at least one pupil.

On entering the classroom, each pupil moves every weight carrying his or her name to the opposite side of the scale. Must there be a set of pupils that Ms. McGregor can let in that will tip the scales to the left?

Chomp

Alice and Bob take turns biting off pieces of an $m \times n$ rectangular chocolate bar marked into unit squares. Each bite consists of selecting a square and biting off that square plus every remaining square above and/or to its right. Each player wishes to avoid getting stuck with the poisonous lower-left square.

Show that, assuming the bar contains more than one square, Alice (the first to play) has a winning strategy.

Whose Bullet?

Two marksmen, one of whom ("A") hits a certain small target 75% of the time and the other ("B") only 25%, aim simultaneously at that target. One bullet hits. What's the probability that it came from A?

Poker Quickie

What is the best full house?

(You may assume that you have five cards, and you have just one opponent who has five random other cards. There are no wild cards. As a result of the Goddess of Chance owing you a favor, you are entitled to a full house, and you get to choose whatever full house you want.)

Lines Through a Grid

If you want to cover all the vertices in a 10×10 square grid by lines none of which is parallel to a side of the square, how many lines are needed?

Bidding in the Dark

You have the opportunity to make one bid on a widget whose value to its owner is, as far as you know, uniformly random between $0 and $100. What you do know is that you are so much better at operating the widget than he is, that its value to *you* is 80% greater than its value to him.

If you offer more than the widget is worth to the owner, he will sell it. But you only get one shot. How much should you bid?

Easy Cake Division

Can you cut a cubical cake, iced on top and on the sides, into three pieces each of which contains the same amount of cake and the same amount of icing?

Crossing the River

In eighth century Europe, it was considered unseemly for a man to be in the presence of a married woman, even briefly, unless her husband was there as well. This posed problems for three married couples who wished to cross a river, the only means being a rowboat that could carry at most two people. Can they get to the other side without violating their social norms? If so, what's the minimum number of crossings needed?

Bacterial Reproduction

When two pixo-bacteria mate, a new bacterium results; if the parents are of different sexes the child is female, otherwise it is male. When food is scarce, matings are random and the parents die when the child is born.

It follows that if food remains scarce a colony of pixo-bacteria will eventually reduce to a single bacterium. If the colony originally had 10 males and 15 females, what is the probability that the ultimate pixo-bacterium will be female?

Sprinklers in a Field

Sprinklers in a large field are located at the vertices of a square grid. Each point of land is supposed to be watered by exactly the three closest sprinklers. What shape is covered by each sprinkler?

Bags of Marbles

You have 15 bags. How many marbles do you need so that you can have a different number of marbles in each bag?

Men with Sisters

On average, who has more sisters, men or women?

Dots and Boxes Variation

Suppose you are playing Dots and Boxes, but with each player having an *option* of whether to go again after making a box. Suppose the board has an odd number of cells (thus, someone must win). Show that the first player to make a box has a winning strategy.

Fair Play

How can you get a 50-50 decision by flipping a bent coin?

Salaries and Raises

Wendy, Monica, and Yancey were hired at the beginning of calendar 2020.

Wendy is paid weekly, $500 per week, but gets a $5 raise each week; Monica is paid monthly, $2500 per month, and gets a $50 raise each month; Yancey is paid yearly, $50,000 per year, and gets a $1500 raise each year.

Who will make the most money in 2030?

Two Runners

Two runners start together on a circular track, running at different constant speeds. If they head in opposite directions they meet after a minute; if they head in the same direction, they meet after an hour. What is the ratio of their speeds?

Broken ATM

George owns only $500, all in a bank account. He needs cash badly but his only option is a broken ATM that can process only withdrawals of exactly $300, and deposits of exactly $198. How much cash can George get out of his account?

Domino Task

An 8×8 chessboard is tiled arbitrarily with thirty-two 2×1 dominoes. A new square is added to the right-hand side of the board, making the top row length 9.

At any time you may move a domino from its current position to a new one, provided that after the domino is lifted, there are two adjacent empty squares to receive it.

Can you retile the augmented board so that all the dominoes are horizontal?

Big Pairs in a Matrix

The sum of the greatest two numbers in each row of a certain square matrix is r; the sum of the greatest two numbers in each column is c. Show that $r = c$.

Second Ace

What is the probability that a poker hand (5 cards dealt at random from a 52-card deck) contains at least two aces, given that it has at least one? What is the probability that it contains two aces, given that it has the Ace of Spades? If your answers are different, then: What's so special about the Ace of Spades?

Fewest Slopes

If you pick n random points in, say, a disk, the pairs of points among them will with probability 1 determine $n(n-1)/2$ distinct slopes. Suppose you get to pick the n points deliberately, subject to no three being collinear. What's the smallest number of distinct slopes they can determine?

Three Natives at the Crossroads

A logician is visiting the South Seas and as is usual for logicians in puzzles, she is at a fork, wanting to know which of two roads leads to the village. Present this time are three willing natives, one each from a tribe of invariable truth-tellers, a tribe of invariable liars, and a tribe of random answerers. Of course, the logician doesn't know which native is from which tribe. Moreover, she is permitted to ask only two yes-or-no questions, each question being directed to just one native. Can she get the information she needs? How about if she can ask only *one* yes-or-no question?

Rolling All the Numbers

On average, how many times do you need to roll a die before all six different numbers have turned up?

Uniform Unit Distances

Can it be that for any positive integer n, there's a configuration of points on the plane with the property that every point is at distance 1 from exactly n other points?

Life Is a Bowl of Cherries

In front of you and your friend Amit are 4 bowls of cherries, containing, respectively, 5, 6, 7, and 8 cherries. You and Amit will alternately pick a bowl and take one or more cherries from it. If you go first, and you want to be sure to get the last cherry, how many cherries should you take—and from which bowl?

Finding the Missing Number

All but one of the numbers from 1 to 100 are read to you, one every 10 seconds, but in no particular order. You have a good mind, but only a normal memory, and no means of recording information during the process. How can you ensure that you can determine afterward which number was not called out?

Three-Way Duel

Alice, Bob, and Carol arrange a three-way duel. Alice is a poor shot, hitting her target only 1/3 of the time on average. Bob is better, hitting his target 2/3 of the time. Carol is a sure shot.

They take turns shooting, first Alice, then Bob, then Carol, then back to Alice, and so on until only one is left. What is Alice's best course of action?

Splitting a Hexagon

Is there a hexagon that can be cut into four congruent triangles by a single line?

Expecting the Worst

Choose n numbers uniformly at random from the unit interval $[0,1]$. What is the expected value of their minimum?

Belt Around the Earth

Suppose the earth is a perfect sphere and a belt is tightened around the equator. Then the belt is loosened by adding 1 meter to its length, and it now sits the same amount off the ground all the way around. Is there enough room to slip a credit card underneath the belt?

Stopping After the Boy

In a certain country, a law is passed forbidding families to have another child after having a boy. Thus, families might consist of one boy, one girl and one boy, five girls and one boy... How will this law affect the ratio of boys to girls?

Attic Lamp

An old-fashioned incandescent lamp in the attic is controlled by one of three on-off switches downstairs—but which one? Your mission is to do something with the switches, then determine after *one* trip to the attic which switch is connected to the attic lamp.

Efficient Pizza-Cutting

What's the maximum number of pieces you can get by cutting a (round) pizza with 10 straight cuts?

Fourth Corner

Pegs occupy three corners of a square. At any time a peg can jump over another peg, landing an equal distance on the other side. Jumped pegs are not removed. Can you get a peg onto the fourth corner of the square?

Cutting the Necklace

Two thieves steal a necklace consisting of 10 red rubies and 14 pink diamonds, fixed in some arbitrary order on a loop of golden string. Show that they can cut the necklace in two places so that when each thief takes one of the resulting pieces, he gets half the rubies and half the diamonds.

Odd Run of Heads

On average, how many flips of a coin does it take to get a run of an odd number of heads, preceded and followed by tails?

Subsets with Constraints

What is the maximum number of numbers you can have between 1 and 30, such that no two have a product which is a perfect square? How about if (instead) no number divides another evenly? Or if no two have a factor (other than 1) in common?

Spinning Switches

Four identical, unlabeled switches are wired in series to a light bulb. The switches are simple buttons whose state cannot be directly observed, but can be changed by pushing; they are mounted on the corners of a rotatable square. At any point, you may push, simultaneously, any subset of the buttons, but then an adversary spins the square. Is there an algorithm that will enable you to turn on the bulb in at most a fixed number of spins?

Watches on the Table

Fifty accurate watches and a tiny diamond lie on a table. Prove that there exists a moment in time when the sum of the distances from the diamond to the ends of the minute hands exceeds the sum of the distances from the diamond to the centers of the watches.

Bugging a Disk

Elspeth, a new FBI recruit, has been asked to bug a circular room using seven ceiling microphones. If the room is 40 feet in diameter and the bugs are placed so as to minimize the maximum distance from any point in the room to the nearest bug, where should the bugs be located?

Points on a Circle

What is the probability that three uniformly random points on a circle will be contained in some semicircle?

Two Different Distances

Find all configurations of four points in the plane that determine only two different distances. (Note: there are more of these than you probably think!)

Same Sum Subsets

Amy asks Brad to pick 10 different numbers between 1 and 100, and to write them down secretly on a piece of paper. She now tells him she's willing to bet $100 to $1 that his numbers contain two nonempty disjoint subsets with the same sum! Is she nuts?

Players and Winners

Tristan and Isolde expect to be in a situation of severely limited communication, at which time Tristan will know which two of 16 basketball teams played a game, and Isolde will know who won. How many bits must be communicated between Tristan and Isolde in order for the former to find out who won?

Spaghetti Loops

The 100 ends of 50 strands of cooked spaghetti are paired at random and tied together. How many pasta loops should you expect to result from this process, on average?

Swapping Executives

The executives of Women in Action, Inc., are seated at a long table facing the stockholders. Unfortunately, according to the meeting organizer's chart, every one is in the wrong seat. The organizer can persuade two executives to switch seats, but only if they are adjacent and neither one is already in her correct seat.

Can the organizer organize the seat-switching so as to get everyone in her correct seat?

Life Isn't a Bowl of Cherries?

In front of you and your friend Amit are 4 bowls of cherries, containing, respectively, 5, 6, 7, and 8 cherries. You and Amit will alternately pick a bowl and take one or more cherries from it. If you go first, and you want to be sure Amit gets the last cherry, how many cherries should you take—and from which bowl?

Righting the Pancakes

An underchef of the great and persnickety Chef Bouillon has made a stack of pancakes, but, alas, some of them are upside-down—that is, according to the great Chef, they don't have their best side up. The underchef wants to fix the problem as follows: He finds a contiguous substack of pancakes (of size at least one) with the property that both the top pancake and the bottom pancake of the substack are upside-down. Then he removes the substack, flips it as a block and slips it back into the big stack in the same place.

Prove that this procedure, no matter which upside-down pancakes are chosen, will eventually result in all the pancakes being right side up.

Pie in the Sky

What fraction of the sky is occupied by a full moon?

Testing Ostrich Eggs

In preparation for an ad campaign, the Flightless Ostrich Farm needs to test its eggs for durability. The world standard for egg-hardness calls for rating an egg according to the highest floor of the Empire State Building from which the egg can be dropped without breaking.

Flightless' official tester, Oskar, realizes that if he takes only one egg along on his trip to New York, he'll need to drop it from (potentially) every one of the building's 102 floors, starting with the first, to determine its rating.

How many drops does he need in the worst case, if he takes *two* eggs?

Circular Shadows I

Suppose all three coordinate-plane projections of a convex solid are disks. Must the solid be a perfect ball?

Next Card Red

Paula shuffles a deck of cards thoroughly, then plays cards face up one at a time, from the top of the deck. At any time, Victor can interrupt Paula and bet $1 that the next card will be red. He bets once and only once; if he never interrupts, he's automatically betting on the last card.

What's Victor's best strategy? How much better than even can he do? (Assume there are 26 red and 26 black cards in the deck.)

Attention Paraskevidekatriaphobes

Is the 13th of the month more likely to be a Friday than any other day of the week, or does it just *seem* that way?

Unanimous Hats

One hundred prisoners are offered the chance to be freed if they can win the following game. In the dark, each will be fitted with a red or black hat according to a fair coinflip. When the lights are turned on, each will see the others' hat colors but not his own; no communication between prisoners will be permitted.

Each prisoner will be asked to write down his guess at his own hat color, and the prisoners will be freed if they all get it right.

The prisoners have a chance to conspire beforehand. Can you come up with a strategy for them that will maximize their probability of winning?

Raising Art Value

Fans of Gallery A like to say that last year, when Gallery A sold to Gallery B "Still Life with Kumquats," the average value of the pictures in each gallery went up. Assuming that statement is correct, and that the value of each of the 400 pictures in the galleries' combined holdings is an integral number of dollars, what is the least possible difference between the average picture values of Gallery A and Gallery B before the sale?

First Odd Number

What's the first odd number in the dictionary? More specifically, suppose that every whole number from 1 to, say, 10^{10} is written out in formal English (e.g., "two hundred eleven," "one thousand forty-two") and then listed in dictionary order, that is, alphabetical order with spaces and hyphens ignored. What's the first odd number in the list?

Lattice Points and Line Segments

How many lattice points (i.e., points with integer coordinates) can you have in 3-space with the property that no two of them have a third lattice point on the line segment between them?

Breaking a Chocolate Bar

You have a rectangular chocolate bar scored in a 6×4 grid of squares, and you wish to break up the bar into its constituent squares. At each step, you may pick up one piece and break it along any of its marked vertical or horizontal lines.

For example, you can break three times to form four rows of six squares each, then break each row five times into its constituent squares, accomplishing the desired task in $3 + 4 \times 5 = 23$ breaks. Can you do better?

Meet the Williams Sisters

Some tennis fans get excited when Venus and Serena Williams meet in a tournament. The likelihood of that happening normally depends on seeding and talent, so let's instead construct an idealized elimination tournament of 64 players, each as likely to win as to lose any given match, with bracketing chosen uniformly at random. What is the probability that the Williams sisters wind up playing each other?

Cards from Their Sum

The magician riffle-shuffled the cards and dealt five of them face down onto your outstretched hand. She asked you to pick any number of them secretly, then tell you their sum (jacks count as 11, queens 12, kings 13). You did that and the magician then told you exactly what cards you chose, including the suits!

After a while you figured out that her shuffles were defective— they didn't change the top five cards, so the magician had picked them in advance and knew what they were. But she still had to pick the ranks of those cards carefully (and memorize their suits), so that from the sum she could always recover the ranks.

In conclusion, if you want to do the trick yourself, you need to find five distinct numbers between 1 and 13 (inclusive), with the property that every subset has a different sum. Can you do it?

Waiting for Heads

If you flip a fair coin repeatedly, how long do you have to wait, on average, for a run of five heads?

Half-right Hats

One hundred prisoners are offered the chance to be freed if they can win the following game. In the dark, each will be fitted with a red or black hat according to a fair coinflip. When the lights are turned on, each will see the others' hat colors but not his own; no communication between prisoners will be permitted.

Each prisoner will be asked to write down his guess at his own hat color, and the prisoners will be freed if at least half get it right.

The prisoners have a chance to conspire beforehand. Can you come up with a strategy for them that will maximize their probability of winning?

Finding a Jack

In some poker games the right of first dealer is determined by dealing cards face-up (from a well-shuffled deck) until some player gets a jack. On average, how many cards are dealt during this procedure?

Measuring with Fuses

You have on hand two fuses (lengths of string), each of which will burn for exactly 1 minute. They do not, however, burn uniformly along their lengths. Can you nevertheless use them to measure 45 seconds?

Rating the Horses

You have 25 horses and can race them in groups of five, but having no stopwatch, you can only observe the order of finish. How many heats of five horses do you need to determine the fastest three of your 25?

Red and Blue Hats in a Line

There are n prisoners this time, and again each will be fitted with a red or black hat according to flips of a fair coin. The prisoners are to be arranged in a line, so that each prisoner can see only the colors of the hats in front of him. Each prisoner must guess the color of his own hat, and is executed if he is wrong; however, the guesses are made sequentially, from the back of the line toward the front. Thus, for example, the ith prisoner in line sees the hat colors of prisoners 1, 2, ..., $i-1$ and hears the guesses of prisoners n, $n-1$, ..., $i+1$ (but he isn't told which of those guesses were correct—the executions take place later).

The prisoners have a chance to collaborate beforehand on a strategy, with the object of guaranteeing as many survivors as possible. How many players can be saved in the worst case?

Three Sticks

You have three sticks that can't make a triangle; that is, one is longer than the sum of the lengths of the other two. You shorten the long one by an amount equal to the sum of the lengths of the other two, so you again have three sticks. If they also fail to make a triangle, you again shorten the longest stick by an amount equal to the sum of the lengths of the other two.

You repeat this operation until the sticks do make a triangle, or the long stick disappears entirely.

Can this process go on forever?

Spiders on a Cube

Three spiders are trying to catch an ant. All are constrained to the edges of a cube. Each spider can move one third as fast as the ant can. Can the spiders catch the ant?

Hopping and Skipping

A frog hops down a long line of lily pads; at each pad, he flips a coin to decide whether to hop two pads forward or one pad back. What fraction of the pads does he hit?

Divisibility Game

Alice chooses a number bigger than 100 and writes it down secretly. Bob now guesses a number greater than 1, say k; if k divides Alice's number, Bob wins. Otherwise k is subtracted from Alice's number and Bob tries again, but may not use a number he used before. This continues until either Bob succeeds by finding a number that divides Alice's, in which case Bob wins, or Alice's number becomes 0 or negative, in which case Alice wins.

Does Bob have a winning strategy for this game?

Candles on a Cake

It's Joanna's 18th birthday and her cake is cylindrical with 18 candles on its $18''$ circumference. The length of any arc (in inches) between two candles is greater than the number of candles on the arc, excluding the candles at the ends.

Prove that Joanna's cake can be cut into 18 equal wedges with a candle on each piece.

Sums of Two Squares

On average, how many ways are there to write a positive number n as the sum of two squares? In other words, suppose n is a random integer between 1 and a zillion. What is the expected number of ordered pairs (i, j) of integers such that $n = i^2 + j^2$?

Pairs at Maximum Distance

Let X be a finite set of points on the plane. Suppose X contains n points, and the maximum distance between any two of them is d. Prove that at most n pairs of points of X are at distance d.

Comparing Numbers, Version I

Paula (the perpetrator) takes two slips of paper and writes an integer on each. There are no restrictions on the two numbers except that they must be different. She then conceals one slip in each hand.

Victor (the victim) chooses one of Paula's hands, which Paula then opens, allowing Victor to see the number on that slip. Victor must now guess whether that number is the larger or the smaller of Paula's two numbers; if he guesses right, he wins $1, otherwise he loses $1.

Clearly, Victor can at least break even in this game, for example, by flipping a coin to decide whether to guess "larger" or "smaller"— or by always guessing "larger," but choosing the hand at random. The question is: Not knowing anything about Paula's psychology, is there any way he can do better than break even?

Comparing Numbers, Version II

Now let's make things more favorable to Victor: Instead of being chosen by Paula, the numbers are chosen independently at random from the uniform distribution on [0,1] (two outputs from a standard random number generator will do fine).

To compensate Paula, we allow her to examine the two random numbers and *to decide which one Victor will see*. Again, Victor must guess whether the number he sees is the larger or smaller of the two, with $1 at stake. Can he do better than break even? What are his and Paula's best (i.e., "equilibrium") strategies?

King's Salary

Democracy has come to the little kingdom of Zirconia, in which the king and each of the other 65 citizens has a salary of one zircon. The king cannot vote, but he has power to suggest changes—in particular, redistribution of salaries. Each person's salary must be a whole number of zircons, and the salaries must sum to 66. Each suggestion is voted on, and carried if there are more votes for than against. Each voter can be counted on to vote "yes" if his or her salary is to be increased, "no" if decreased, and otherwise not to bother voting.

The king is both selfish and clever. What is the maximum salary he can obtain for himself, and how many referenda does he need to get it?

Adding, Multiplying, and Grouping

Forty-two positive integers (not necessarily distinct) are written in a row. Show that you can put plus signs, times signs, and parentheses between the integers in such a way that the value of the resulting expression is evenly divisible by one million.

Missing Card

Yola and Zela have devised a clever card trick. While Yola is out of the room, audience members pull out five cards from a bridge deck and hand them to Zela. She looks them over, pulls one out, and calls Yola into the room. Yola is handed the four remaining cards and proceeds to guess correctly the identity of the pulled card.

How do they do it? And once you've figured that out, compute the size of the biggest deck of cards they could use and still perform the trick reliably.

Ping-Pong Progression

Alice and Bob play table tennis, with Bob's probability of winning any given point being 30%. They play until someone reaches a score of 21. What, approximately, is the expected number of points played?

Odd Light Flips

Suppose you are presented with a collection of light bulbs controlled by switches. Each switch flips the state of some subset of the bulbs, that is, turns on all the ones in the subset that were off, and turns off those that were on. You are told that for any nonempty set of bulbs, there is a switch that flips an odd number of bulbs *from that subset* (and perhaps other bulbs as well).

Show that, no matter what the initial state of the bulbs, you can use these switches to turn off all the bulbs.

Painting the Cubes

Can you paint 1000 unit cubes with 10 different colors in such a way that for any of the 10 colors, the cubelets can be arranged into a $10 \times 10 \times 10$ cube with only that color showing?

Red Points and Blue

Given n red points and n blue points on the plane, no three on a line, prove that there is a matching between them so that line segments from each red point to its corresponding blue point do not cross.

Identifying the Majority

A long list of names is read out, some names many times. Your object is to end up with a name that is guaranteed to be the name which was a called a majority of the time, if there is such a name.

However, you have only one counter, plus the ability to keep just one name at a time in your mind. Can you do it?

Returning Pool Shot

A ball is shot from a corner of a polygonal pool table (not necessarily convex) with right-angle corners; let's say all sides are aligned either exactly east-west or exactly north-south.

The starting corner is a convex corner, that is, has an interior angle of 90°. There are pockets at all corners, so that if the ball hits a corner exactly it falls in. Otherwise, the ball bounces true and without energy loss.

Can the ball ever return to the corner from which it began?

Pancake Stacks

At the table are two hungry students, Andrea and Bruce, and two stacks of pancakes, of height m and n. Each student, in turn, must eat from the larger stack a non-zero multiple of the number of pancakes in the smaller stack. Of course, the bottom pancake of each stack is soggy, so the player who first finishes a stack is the loser.

For which pairs (m, n) does Andrea (who plays first) have a winning strategy?

How about if the game's objective is reversed, so that the first player to finish a stack is the winner?

Trapped in Thickland

The inhabitants of Thickland, a world somewhere between Edwin Abbott's *Flatland* and our three-dimensional universe, are an infinite set of congruent convex polyhedra that live between two parallel planes. Up until now, they have been free to escape from their slab, but haven't wanted to. Now, however, they have been reproducing rapidly and thinking about colonizing other slabs. Their high priest is worried that conditions are so crowded, no inhabitant of Thickland can escape the slab unless others move first.

Is that even possible?

Hats and Infinity

Each of an infinite collection of prisoners, numbered $1, 2, \ldots$, is to be fitted with a red or black hat. At a prearranged signal, all the prisoners are revealed to one another, so that everyone gets to see all his fellow prisoners' hat colors—but no communication is permitted. Each prisoner is then taken aside and asked to guess the color of his own hat.

All the prisoners will be freed provided *only finitely many* guess wrongly. The prisoners have a chance to conspire beforehand; is there a strategy that will ensure freedom?

All Right or All Wrong

This time the circumstances are the same but the objective is different: The guesses must either be *all right* or *all wrong*. Is there a winning strategy?

Magnetic Dollars

One million magnetic "susans" (Susan B. Anthony dollar coins) are tossed into two urns in the following fashion: The urns begin with one coin in each, then the remaining 999,998 coins are thrown in the air one by one. If there are x coins in one urn and y in the other, magnetic attraction will cause the next coin to land in the first urn with probability $x/(x + y)$, and in the second with probability $y/(x + y)$.

How much should you be willing to pay, in advance, for the contents of the urn that ends up with fewer susans?

Shoelaces at the Airport

You are walking to your gate at O'Hare and need to tie your shoelace. Up ahead is a moving walkway that you plan to utilize. To minimize your time to get to the gate, should you stop and tie your shoelace now, or wait to tie it on the walkway?

Hazards of Electronic Coinflipping

You have been hired as an arbitrator and called upon to award a certain indivisible widget randomly to either Alice, Bob, or Charlie, each with probability $\frac{1}{3}$. Fortunately, you have an electronic coinflipping device with an analog dial that enables you to enter any desired probability p. Then you push a button and the device shows "Heads" with probability p, else "Tails."

Alas, your device is showing its "low battery" light and warning you that you may set the probability p only once, and then push the coinflip button at most 10 times. Can you still do your job?

Handshakes at a Party

Nicholas and Alexandra went to a reception with 10 other couples; each person there shook hands with everyone he or she didn't know. Later, Alexandra asked each of the other 21 partygoers how many people they shook hands with, and got a different answer every time.

How many people did Nicholas shake hands with?

Area—Perimeter Match

Find all integer-sided rectangles with equal area and perimeter.

Prime Test

Does $4^9 + 6^{10} + 3^{20}$ happen to be a prime number?

Lost Boarding-Pass

One hundred people line up to board a full jetliner, but the first has lost his boarding pass and takes a random seat instead. Each subsequent passenger takes his or her assigned seat if available, otherwise a random unoccupied seat.

What is the probability that the last passenger to board finds his seat unoccupied?

Lemming on a Chessboard

On each square in an $n \times n$ chessboard is an arrow pointing to one of its eight neighbors (or off the board, if it's an edge square). However, arrows in neighboring squares (diagonal neighbors included) may not differ in direction by more than 45 degrees.

A lemming begins in a center square, following the arrows from square to square. Is he doomed to fall off the board?

Packing Slashes

Given a 5 × 5 square grid, on how many of the squares can you draw diagonals (slashes or backslashes) in such a way that no two of the diagonals meet?

Peek Advantage

You are about to bet $100 on the color of the top card of a well-shuffled deck of cards. You get to pick the color; if you're right you win $100, otherwise you lose the same.

How much is it worth to you to get a peek at the *bottom* card of the deck? How much more for a peek at the bottom *two* cards?

Truly Even Split

Can you partition the integers from 1 to 16 into two sets of equal sizes so that each set has the same sum, the same sum of squares, and the same sum of cubes?

Line Up by Height

Yankees manager Casey Stengel famously once told his players to "line up alphabetically by height." Suppose 26 players, no two exactly the same height, are lined up alphabetically. Prove that there are at least six who are also in height order—either tallest to shortest, or shortest to tallest.

Curves on Potatoes

Given two potatoes, can you draw a closed curve on the surface of each so that the two curves are identical as curves in three-dimensional space?

Falling Ants

Twenty-four ants are placed randomly on a meter-long rod; each ant is facing east or west with equal probability. At a signal, they proceed to march forward (that is, in whatever direction they are facing) at 1 cm/sec; whenever two ants collide, they reverse directions. How long does it take before you can be certain that all the ants are off the rod?

Polygon Midpoints

Let n be an odd integer, and let a sequence of n distinct points be given in the plane. Find the vertices of a (possibly self-intersecting) n-gon that has the given points, in the given order, as midpoints of its sides.

Rows and Columns

Prove that if you sort each row of a matrix, then each column, the rows are still sorted!

Bugs on a Pyramid

Four bugs live on the four vertices of a triangular pyramid (tetrahedron). Each bug decides to go for a little walk on the surface of the tetrahedron. When they are done, two of them are back home, but the other two find that they have switched vertices.

Prove that there was an instant when all four bugs lay on the same plane.

Snake Game

Joan begins by marking any square of an $n \times n$ chessboard; Judy then marks an orthogonally adjacent square. Thereafter, Joan and Judy continue alternating, each marking a square adjacent to the last one marked, creating a snake on the board. The first player unable to play loses.

For which values of n does Joan have a winning strategy, and when she does, what square does she begin at?

Three Negatives

A set of 1000 integers has the property that every member of the set exceeds the sum of the rest. Show that the set includes at least three negative numbers.

Numbers on Foreheads

Each of 10 prisoners will have a digit between 0 and 9 painted on his forehead (they could be all 2's, for example). At the appointed time each will be exposed to all the others, then taken aside and asked to guess his own digit.

In order to avoid mass execution, at least one prisoner must guess correctly. The prisoners have an opportunity to conspire beforehand; find a scheme by means of which they can ensure success.

Bias Test

Before you are two coins; one is a fair coin, and the other is biased toward heads. You'd like to try to figure out which is which, and to do so you are permitted two flips. Should you flip each coin once, or one coin twice?

Non-Repeating String

Is there a finite string of letters from the Latin alphabet with the property that there is no pair of adjacent identical substrings, but the addition of any letter to either end would create one?

Three-way Election

Alison, Bonnie, and Clyde run for class president and finish in a three-way tie. To break it, they solicit their fellow students' second choices, but again there is a three-way tie. The election committee is stymied until Alison steps forward and points out that, since the number of voters is odd, they can make two-way decisions. She therefore proposes that the students choose between Bonnie and Clyde, and then the winner would face Alison in a runoff.

Bonnie complains that this is unfair because it gives Alison a better chance to win than either of the other two candidates. Is Bonnie right?

Skipping a Number

At the start of the 2019 season WNBA star Missy Overshoot's lifetime free-throw percentage was below 80%, but by the end of the season it was above 80%. Must there have been a moment in the season when Missy's free-throw percentage was exactly 80%?

Red and Black Sides

Bored in math class, George draws a square on a piece of paper and divides it into rectangles (with sides parallel to a side of the square), using a red pen after his black one runs out of ink. When he's done, he notices that every rectangular tile has at least one totally red side.

Prove that the total length of George's red lines is at least the length of a side of the big square.

Unbroken Lines

Can you re-arrange the 16 square tiles below into a 4 × 4 square—no rotation allowed!—in such a way that no line ends short of the boundary of the big square?

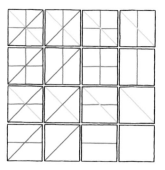

Baby Frog

To give a baby frog jumping practice, her four grandparents station themselves at the corners of a large square field. When a grandparent croaks, the baby leaps halfway to it. In the field is a small round clearing. Can the grandparents get the baby to that clearing, no matter where in the field she starts?

Sequencing the Digits

In how many ways can you write the numbers 0 through 9 in a row, such that each digit other than the leftmost is within one of some digit to the left of it?

Recovering the Polynomial

The Oracle at Delphi has in mind a certain polynomial p (in the variable x, say) with non-negative integer coefficients. You may query the Oracle with any integer x, and the Oracle will tell you the value of $p(x)$.

How many queries do you have to make to determine p?

Faulty Combination Lock

A combination lock with three dials, each numbered 1 through 8, is defective in that you only need to get two of the numbers right to open the lock. What is the minimum number of (three-number) combina-

tions you need to try in order to be sure of opening the lock?

Prisoners and Gloves

Each of 100 prisoners gets a different real number written on his forehead, a black glove, and a white glove. Seeing the numbers on the other prisoners' foreheads, he must put one glove on each hand in such a way that when the prisoners are later lined up in real-number order and hold hands, the gloves match. How can the prisoners, who are permitted to conspire beforehand, ensure success?

Gaming the Quilt

You come along just as a church lottery is closing; its prize is a quilt worth $100 to you, but they've only sold 25 tickets. At $1 a ticket, how many tickets should you buy?

Bracing the Grid

Suppose that you are given an $n \times n$ grid of unit-length rods, jointed at their ends. You may brace some subset S of the small squares with diagonal segments (of length $\sqrt{2}$).

Which choices of S suffice to make the grid rigid in the plane?

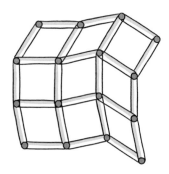

Rectangles Tall and Wide

It is easy to check that a 6×5 rectangle can be tiled with 3×2 rectangles, but only if some tiles are horizontally oriented and others are vertically oriented.

Show that this can't happen if the big rectangle is a square—in other words, if a square is tiled with congruent rectangles, then it can be re-tiled with the same rectangles all oriented the same way.

Guarding the Gallery

A certain museum room is shaped like a highly irregular, non-convex, 11-sided polygon. How many guard-posts are needed in the room, in the worst case, to ensure that every part of the room can be seen from at least one guard-post?

Flying Saucers

A fleet of saucers from planet Xylofon has been sent to bring back the inhabitants of a certain randomly selected house, for exhibition in the Xylofon Xoo. The house happens to contain five men and eight women, to be beamed up randomly one at a time.

Owing to the Xylofonians' strict sex separation policy, a saucer cannot bring back earthlings of both sexes. Thus, it beams people up until it gets a member of a second sex, at which point that one is beamed back down and the saucer takes off with whatever it has left. Another saucer then starts beaming people up, following the same rule, and so forth.

What is the probability that the last person beamed up is a woman?

Increasing Routes

On the Isle of Sporgesi, each segment of road (between one intersection and the next) has its own name. Let d be the average number of road segments meeting at an intersection. Show that you can take a drive on the Isle of Sporgesi that covers at least d road segments, and hits those segments in strict alphabetical order!

Curve on a Sphere

Prove that if a closed curve on the unit sphere has length less than 2π, then it is contained in some hemisphere.

Biased Betting

Alice and Bob each have $100 and a biased coin that comes up heads with probability 51%. At a signal, each begins flipping his or her coin once a minute and bets $1 (at even odds) on each outcome, against a bank with unlimited funds. Alice bets on heads, Bob on tails. Suppose both eventually go broke. Who is more likely to have gone broke first?

Suppose now that Alice and Bob are flipping the *same* coin, so that when the first one goes broke the second one's stack will be at $200. Same question: Given that they both go broke, who is more likely to have gone broke first?

Ascending and Descending

Fix a number n and call a list of the numbers from 1 to n "good" if there is no descending subsequence of length 10. Show that there are at most 81^n good lists.

Unbroken Curves

Can you re-arrange the 16 square tiles below into a 4×4 square—no rotation allowed!—in such a way that no curve ends short of the boundary of the big square?

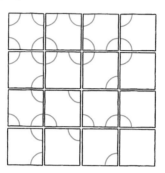

Dot-Town Exodus

Each resident of Dot-town carries a red or blue dot on his (or her) forehead, but if he ever thinks he knows what color it is he leaves town immediately and permanently. Each day the residents gather; one day a stranger comes and tells them something—*anything*—nontrivial about the number of blue dots. Prove that eventually Dot-town becomes a ghost town, *even if everyone can see that the stranger's statement is patently false.*

Drawing Socks

You have 60 red and 40 blue socks in a drawer, and you keep drawing a sock uniformly at random until you have drawn all the socks of one color. What is the expected number of socks left in the drawer?

Path Through the Cells

Cells in a certain cellular telephone network are assigned frequencies in such a way that no two adjacent cells use the same frequency. Show that if the number of frequencies used is as small as possible (subject to this condition), then it's possible to design a path that moves from cell to adjacent cell and hits each frequency exactly once—in ascending order of frequencies!

Painting the Polyhedron

Let P be a polyhedron with red and green faces such that every red face is surrounded by green ones, but the total red area exceeds the total green area. Prove that you can't inscribe a sphere in P.

Whim-Nim

You and a friend, bored with Nim and Nim Misére, decide to play a variation in which at any point, either player may declare "Nim" or "Misère" instead of removing chips. This happens at most once in a game, and then of course the game proceeds normally according to that variation of Nim. (Taking the whole single remaining stack in an undeclared game *loses*, as your opponent can then declare "Nim" as his last move.)

What's the correct strategy for this game, which its inventor, the late John Horton Conway, called "Whim"?

Bacteria on the Plane

Suppose the world begins with a single bacterium at the origin of the infinite plane grid. When it divides its two successors move one vertex north and one vertex east, so that there are now two bacteria, one at $(0,1)$ and one at $(1,0)$. Bacteria continue to divide, each time with one successor moving north and one east, provided both of those points are unoccupied.

Show that no matter how long this process continues, there's always a bacterium inside the circle of radius 3 about the origin.

Poorly Placed Dominoes

What's the smallest number of dominoes one can place on a chessboard (each covering two adjacent squares) so that no more fit?

Ants on the Circle

Twenty-four ants are randomly placed on a circular track of length 1 meter; each ant faces randomly clockwise or counterclockwise. At a signal, the ants begin marching at 1 cm/sec; when two ants collide they both reverse directions. What is the probability that after 100 seconds, every ant finds itself exactly where it began?

Chinese Nim

On the table are two piles of beans. Alex must either take some beans from one pile or the same number of beans from each pile; then Beth has the same options. They continue alternating until one wins the game by taking the last bean.

What's the correct strategy for this game? For example, if Alex is faced with piles of size 12,000 and 20,000, what should he do? How about 12,000 and 19,000?

Not Burning Brownies

When you bake a pan of brownies, those that share an edge with the edge of the pan will often burn. For example, if you bake 16 square brownies in a square pan, 12 are subject to burning. Design a pan and divide it into brownie shapes to make 16 identically shaped brownies of which as few as possible will burn. Can you get it down to four burned brownies? Great! How about three?

Double Cover by Lines

Let $\mathcal{L}_\theta$ be the set of all lines on the plane at angle θ to the horizontal. If θ and θ' are two different angles, the union of the sets $\mathcal{L}_\theta$ and $\mathcal{L}_{\theta'}$ constitutes a *double cover* of the plane, that is, every point belongs to exactly two lines.

Can this be done in any other way? That is, can you cover each point of the plane exactly twice using a set of lines that contains lines in more than two different directions?

Integral Rectangles

A rectangle in the plane is partitioned into smaller rectangles, each of which has either integer height or integer width (or both). Prove that the large rectangle itself has this property.

Losing at Dice

Visiting Las Vegas, you are offered the following game. Six dice are to be rolled, and the number of *different* numbers that appear will be counted. That could be any number from one to six, of course, but they are not equally likely.

If you get the number "four" this way you win $1, otherwise you lose $1. You decide you like this game, and plan to play it repeatedly until the $100 you came with is gone.

At one minute per game, how long, on the average, will it take before you are wiped out?

Even-Sum Billiards

You pick 10 times, with replacement (necessarily!), from an urn containing 9 billiard balls numbered 1 through 9. What is the probability that the sum of the numbers of the balls you picked is even?

Competing for Programmers

Two Silicon Valley startups are competing to hire programmers. Hiring on alternate days, each can begin by hiring anyone, but every subsequent hire by a company must be a friend of some previous hire—unless no such person exists, in which case that company can again hire anyone.

In the candidate pool are 10 geniuses; naturally each company would like to get as many of those as possible. Can it be that the company that hires first cannot prevent its rival from hiring nine of the geniuses?

Strength of Schedule

The 10 teams comprising the "Big 12" college football conference are all scheduled to play one another in the upcoming season, after which one will be declared conference champion. There are no ties, and each team scores a point for every team it beats.

Suppose that, worried about breaking ties, a member of the governing board of the conference suggests that each team receive in addition "strength of schedule points," calculated as the sum of the scores achieved by the teams it beat.

A second member asks: "What if all the teams end up with the same number of strength-of-schedule points?"

What, indeed? Can that even happen?

Locker Doors

Lockers numbered 1 to 100 stand in a row in the main hallway of Euclid Junior High School. The first student arrives and opens all the lockers. The second student then goes through and re-closes lockers numbered 2, 4, 6, etc.; the third student changes the state of every locker whose number is a multiple of 3, then the next every multiple of 4, etc., until the last student opens or closes only locker number 100.

After the 100 students have passed through, which lockers are open?

Gasoline Crisis

You need to make a long circular automobile trip during a gasoline crisis. Inquiries have ascertained that the gas stations along the route contain just enough fuel to make it all the way around. If you have an empty tank but can start at a station of your choice, can you complete a clockwise round trip?

Soldiers in a Field

An odd number of soldiers are stationed in a large field. No two soldiers are exactly the same distance apart as any other two soldiers. Their commanding officer radios instructions to each soldier to keep an eye on his nearest neighbor.

Is it possible that every soldier is being watched?

Home-field Advantage

Every year, the Elkton Earlies and the Linthicum Lates face off in a series of baseball games, the winner being the first to win four games. The teams are evenly matched but each has a small edge (say, a 51% chance of winning) when playing at home.

Every year, the first three games are played in Elkton, the rest in Linthicum.

Which team has the advantage?

Zeroes and Ones

Show that every natural number has a non-zero multiple which, when written out (base 10), contains only zeroes and ones. Even better, prove that your telephone number, if it ends with a 1, 3, 7 or 9, has a multiple whose decimal representation is all ones!

Random Intersection

Two unit-radius balls are randomly positioned subject to intersecting. What is the expected volume of their intersection? For that matter, what is the expected *surface area* of their intersection?

Profit and Loss

At a recent stockholders' meeting of Widget Industrials Inc., the Chief Financial Officer presented a chart of the month-by-month profits (or losses) since the last meeting. "Note," said she, "that we made a profit over every consecutive eight-month period."

"Maybe so," complained a shareholder, "but I see we *lost* money over every consecutive *five*-month period!"

What's the maximum number of months that could have passed since the last meeting?

Boarding the Manhole

An open manhole 4 meters in diameter has to be covered by boards of total width w. The boards are each more than 4 meters in length, so if $w \geq 4$ it's obvious that you can cover the manhole by laying the boards side by side (see figure below). If w is only 3.9, say, you still have plenty of wood and you are allowed to overlap the boards if you want. Can you still cover the manhole?

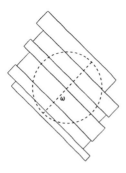

Pegs on the Half-Plane

Each grid point on the XY plane on or below the X-axis is occupied by a peg. At any time, a peg can be made to jump over a neighbor peg (horizontally, vertically, or diagonally adjacent) and onto the next grid point in line, provided that point was unoccupied. The jumped peg is then removed.

Can you get a peg arbitrarily far above the X-axis?

Unbreakable Domino Cover

A 6×5 rectangle can be covered with 2×1 dominoes, as in the figure below, in such a way that no line between dominoes cuts all the way across the board. Can you cover a 6×6 square that way?

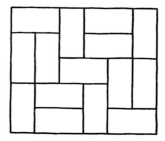

Turning the Die

In the game of Turn-Die, a die is rolled and the number that appears is recorded. The first player then turns the die 90 degrees (giving her four options) and the new value is added to the old one. The second player does the same and the two players alternate until a sum of 21 or higher is reached. If 21 was hit exactly, the player who reached it wins; otherwise, the player who exceeded it loses.

Do you want to be first or second?

Protecting the Statue

Michelangelo's *David*, in Florence, is protected (on the plane) by laser beams in such a way that no-one can approach the statue or any of the laser-beam sources without crossing a beam.

What's the minimum number of lasers needed to accomplish this? (Beams reach 100 meters, say.)

Wild Guess

David and Carolyn are mathematicians who are unafraid of the infinite and cheerfully invoke the Axiom of Choice when needed. They elect to play the following two-move game. For her move, Carolyn chooses an infinite sequence of real numbers, and puts each number in an opaque box. David gets to open as many boxes as he wants—even infinitely many—but must leave one box unopened. To win, he must guess exactly the real number in that box.

On whom will you bet in this game, Carolyn or David?

Laser Gun

You find yourself standing in a large rectangular room with mirrored walls. At another point in the room is your enemy, brandishing a laser gun. You and she are fixed points in the room; your only defense is that you may summon bodyguards (also points) and position them in the room to absorb the laser rays for you. How many bodyguards do you need to block all possible shots by the enemy?

Splitting the Stacks

Your job is to separate a stack of n objects into individual items. At any time you may split a stack into two stacks, and get paid the product of the sizes of the two stacks.

For example, for n=8, you could split into a stack of size 5 and a stack of size 3, getting paid $15; then you can split the 5-stack into 2 and 3 to get $6 more. Then the two threes can each be split ($2 each) and the remaining three twos ($1 each) for a total income of $15 + 6 + (2 \times 2) + (3 \times 1) = 28$ dollars.

What's the most money you can get starting with n objects, and how can you get it?

Chameleons

A colony of chameleons currently contains 20 red, 18 blue, and 16 green individuals. When two chameleons of different colors meet, each of them changes his or her color to the third color. Is it possible that, after a while, all the chameleons have the same color?

Wires under the Hudson

Fifty identical wires run through a tunnel under the Hudson River, but they all look the same, and you need to determine which pairs of wire-ends belong to the same wire. To do this you can tie pairs of wires together at one end of the tunnel and test pairs of wire-ends at the other end to see if they close a circuit; in other words, you can determine whether two wires are tied together at the other end.

How many trips across the Hudson do you need to accomplish your task?

Two Round-robins

The Games Club's 20 members played a round-robin checkers tournament on Monday and a round-robin chess tournament on Tuesday. In each tournament, a player scored one point for each other player he or she beat, and half a point for each tie.

Suppose every player's scores in the two tournaments differed by at least 10. Show that in fact, the differences were all exactly 10.

Factorial Coincidence

Suppose that a, b, c, and d are positive integers, all different, all greater than one. Can it be that $a!^b = c!^d$?

Conversation on a Bus

Ephraim and Fatima, two colleagues in the Mathematics Department at Zorn University, wind up seated together on the bus to campus.

Ephraim begins a conversation with "So, Fatima, how are your kids doing? How old are they now, anyway?"

"It turns out," says Fatima as she is putting the $1 change from the bus driver into her wallet, "that the sum of their ages is the number of this bus, and the product is the number of dollars that happen to be in my purse at the moment."

"Aha!" replies Ephraim. "So, if I remembered how many kids you have, and you told me how much money you are carrying, I could deduce their ages?"

"Actually, no," says Fatima.

"In that case," says Ephraim, "I know how much money is in your purse."

What is the bus number?

Coins on the Table

One hundred quarters lie on a rectangular table, in such a way that no more can be added without overlapping. (We allow a quarter to extend over the edge, as long as its center is on the table.)

Prove that you can start all over again and cover the whole table with 400 quarters! (This time we allow overlap *and* overhang).

Roulette for the Unwary

Elwyn is in Las Vegas celebrating his 21st birthday, and his girlfriend has gifted him with twenty-one $5 bills to gamble with. He saunters over to the roulette table, noting that there are 38 numbers (0, 00, and 1 through 36) on the wheel. If he bets $1 on a single number, he will win with probability 1/38 and collect $36 (in return for his $1 stake, which still goes to the bank). Otherwise, of course, he just loses the dollar.

Elwyn decides to use his $105 to make 105 one-dollar bets on the number 21. What, approximately, is the probability that Elwyn will come out ahead? Is it better than, say, 10%?

Uniformity at the Bakery

A baker's dozen (thirteen) bagels have the property that any twelve of them can be split into two piles of six each, which balance perfectly on the scale. Suppose each bagel weighs an integer number of grams. Must all the bagels have the same weight?

Slabs in 3-Space

A "slab" is the region between two parallel planes in three-dimensional space. Prove that you cannot cover all of 3-space with a set of slabs the sum of whose thicknesses is finite.

Pegs in a Square

Suppose we begin with n^2 pegs on a plane grid, one peg occupying each vertex of an n-vertex by n-vertex square. Pegs jump only horizontally or vertically, by passing over a neighboring peg and into an unoccupied vertex; the jumped peg is then removed. The goal is to reduce the n^2 pegs to only 1.

Prove that if n is a multiple of 3, it can't be done!

Filling a Bucket

Before you are 12 two-gallon buckets and a 1-gallon scoop. At each turn, you may fill the scoop with water and distribute the water any way you like among the buckets.

However, each time you do this your opponent will empty two buckets of her choice.

You win if you can get one of the big buckets to overflow. Can you force a win? If so, how long will it take you?

Polygon on the Grid

A convex polygon is drawn on the coordinate plane with all its vertices on integer points, but no side parallel to the x- or y-axis. Let h be the sum of the lengths of the horizontal line segments at integer height that intersect the (filled-in) polygon, and v the equivalent for the vertical line segments. Prove that $h = v$.

Game of Desperation

On a piece of paper is a row of n empty boxes. Tristan and Isolde take turns, each writing an "S" or an "O" into a previously blank box. The winner is the one who completes an "SOS" in consecutive boxes. For which n does the second player (Isolde) have a winning strategy?

Gluing Pyramids

A solid square-base pyramid, with all edges of unit length, and a solid triangle-base pyramid (tetrahedron), also with all edges of unit length, are glued together by matching two triangular faces.

How many faces does the resulting solid have?

Exponent upon Exponent

Part I: If $x^{x^{x^{\cdot^{\cdot^{\cdot}}}}} = 2$, what is x?

Part II: If $x^{x^{x^{\cdot^{\cdot^{\cdot}}}}} = 4$, what is x?

Part III: How do you explain getting the same answer to Parts I and II?

Random Intervals

The points $1, 2, \ldots, 1000$ on the number line are paired up at random, to form the endpoints of 500 intervals. What is the probability that among these intervals is one which intersects all the others?

North by Northwest

If you've never seen the famous Alfred Hitchcock movie *North by Northwest* (1959), you should. But what direction is that, exactly? Assume North is $0°$, East $90°$, etc.

Missing Digit

The number 2^{29} has 9 digits, all different; which digit is missing?

Coins in a Row

On a table is a row of 50 coins, of various denominations. Alix picks a coin from one of the ends and puts it in her pocket; then Bert chooses a coin from one of the (remaining) ends, and the alternation continues until Bert pockets the last coin.

Prove that Alix can play so as to guarantee at least as much money as Bert.

Bugs on Four Lines

You are given four lines in a plane in general position (no two parallel, no three intersecting in a common point). On each line a ghost bug crawls at some constant velocity (possibly different for each bug). Being ghosts, if two bugs happen to cross paths they just continue crawling through each other uninterrupted.

Suppose that five of the possible six meetings actually happen. Prove that the sixth does as well.

First-grade Division

On the first day of class Miss Feldman divides her first-grade class into k working groups. On the second day, she picks the working groups a different way, this time ending up with $k+1$ of them.

Show that there are at least two kids who are in smaller groups on the second day than they were on the first day.

Deterministic Poker

Unhappy with the vagaries of chance, Alice and Bob elect to play a completely deterministic version of draw poker. A deck of cards is spread out face-up on the table. Alice draws five cards, then Bob draws five cards. Alice discards any number of her cards (the discarded cards will remain out of play) and replaces them with a like number of others; then Bob does the same. All actions are taken with the cards face-up in view of the opponent. The player with the better hand wins; since Alice goes first, Bob is declared to be the winner if the final hands are equally strong. Who wins with best play?

Precarious Picture

Suppose that you wish to hang a picture with a string attached at two points on the frame. If you hang it by looping the string over two nails in the ordinary way, as shown below, and one of the nails comes out, the picture will still hang (albeit lopsidedly) on the other nail.

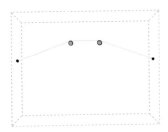

Can you hang it so that the picture falls if *either* nail comes out?

Find the Robot

At time $t = 0$ a robot is placed at some unknown grid point in 3-dimensional space. Every minute, the robot moves a fixed, unknown distance in a fixed, unknown direction, to a new grid point. Each minute, you are allowed to probe any single point in space. Devise an algorithm that is guaranteed to find the robot in finite time.

Early Commuter

A commuter arrives at her home station an hour early and walks toward home until she meets her husband driving to pick her up at the normal time. She ends up home 20 minutes earlier than usual. How long did she walk?

Subtracting Around the Corner

A sequence of n positive numbers is written on a piece of paper. In one "operation," a new sequence is written below the old one according to the following rules: The (absolute) difference between each number and its successor is written below that number; and the absolute difference between the first and last numbers is written below the last number. For example, the sequence 4 13 9 6 is followed by 9 4 3 2.

Try this for $n = 4$ using random numbers between, say, 1 and 100. You'll find that after remarkably few operations, the sequence degenerates to 0 0 0 0 where it of course remains. Why? Is the same true for $n = 5$?

Powers of Roots

What is the first digit after the decimal point in the number $(\sqrt{2}+\sqrt{3})$ to the billionth power?

An Attractive Game

You have an opportunity to bet $1 on a number between 1 and 6. Three dice are then rolled. If your number fails to appear, you lose your $1. If it appears once, you win $1; if twice, $2; if three times, $3.

Is this bet in your favor, fair, or against the odds? Is there a way to determine this without pencil and paper (or a computer)?

Circular Shadows II

Show that if the projections of a solid body onto two planes are perfect disks, then the two projections have the same radius.

Finding the Rectangles

Prove that any tiling of a regular 400-gon by parallelograms must contain at least 100 rectangles.

Figure Eights in the Plane

How many disjoint topological "figure 8s" can be drawn on the plane?

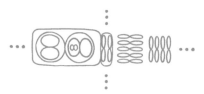

Alternate Connection

100 points lie on a circle. Alice and Bob take turns connecting pairs of points by a line, until every point has at least one connection. The last one to play wins; which player has a winning strategy?

Tiling a Polygon

A "rhombus" is a quadrilateral with four equal sides; we consider two rhombi to be different if you can't translate (move without rotation) one to coincide with the other. Given a regular polygon with 100 sides, you can take any two non-parallel sides, make two copies of each and translate them to form a rhombus. You get $\binom{50}{2}$ different rhombi that way. You can use translated copies of these to tile your 100-gon; show that if you do, you will use each different rhombus exactly once!

Bindweed and Honeysuckle

Two vines, a bindweed and a honeysuckle, climb a tree trunk, starting and ending at the same place. The bindweed circles the trunk three times counter-clockwise while the honeysuckle circles five times clockwise. Not counting the top and bottom, how many times do they cross?

Uniting the Loops

Can you re-arrange the 16 square tiles below into a 4×4 square—no rotation allowed!—in such a way that the curves form a single loop? Lines that exit from the bottom are assumed to continue at the same point on the top, and similarly right to left; in other words, imagine that the square is rolled up into a doughnut.

Splitting a Polygon

A *chord* of a polygon is a straight line segment that touches the polygon's perimeter at the segment's endpoints and nowhere else.

Show that every polygon, convex or not, has a chord such that each of the two regions into which it divides the polygon has area at least 1/3 the area of the polygon.

Two Monks on a Mountain

Remember the monk who climbed Mt. Fuji on Monday and descended on Tuesday? This time, he and a fellow monk climb the mountain on the same day, starting at the same time and altitude, but on different paths. The paths go up and down on the way to the summit (but never dip below the starting altitude); you are asked to prove that they can vary their speeds (sometimes going backwards) so that at *every* moment of the day they are at the same altitude!

Alternative Dice

Can you design two different dice so that their sums behave just like a pair of ordinary dice? That is, there must be two ways to roll a 3, 6 ways to roll a 7, one way to roll a 12, and so forth. Each die must have 6 sides, and each side must be labeled with a positive integer.

Even Split

Prove that from every set of $2n$ integers, you can choose a subset of size n whose sum is divisible by n.

Coconut Classic

Five men and a monkey, marooned on an island, collect a pile of coconuts to be divided equally the next morning. During the night, however, one of the men decides he'd rather take his share now. He tosses one coconut to the monkey and removes exactly 1/5 of the remaining coconuts for himself. A second man does the same thing, then a third, fourth, and fifth.

The following morning the men wake up together, toss one more coconut to the monkey, and divide the rest equally. What's the least original number of coconuts needed to make this whole scenario possible?

Infected Checkerboard

An infection spreads among the squares of an $n \times n$ checkerboard in the following manner: If a square has two or more infected neighbors, then it becomes infected itself. (Neighbors are orthogonal only, so each square has at most four neighbors.)

For example, suppose that we begin with all n squares on the main diagonal infected. Then the infection will spread to neighboring diagonals and eventually to the whole board.

Prove that you *cannot* infect the whole board if you begin with fewer than n infected squares.

Alternating Powers

Since the series $1 - 1 + 1 - 1 + 1 - \cdots$ does not converge, the function $f(x) = x - x^2 + x^4 - x^8 + x^{16} - x^{32} + \cdots$ makes no sense when $x = 1$. However, $f(x)$ does converge for all positive real numbers $x < 1$. If we wanted to give $f(1)$ a value, it might make sense to let it be the limit of $f(x)$ as x approaches 1 from below. Does that limit exist? If so, what is it?

Service Options

You are challenged to a short tennis match, with the winner to be the first player to win four games. You get to serve first. But there are options for determining the sequence in which the two of you serve:

1. Standard: Serve alternates (you, her, you, her, you, her, you).

2. Volleyball Style: The winner of the previous game serves the next one.

3. Reverse Volleyball Style: The winner of the previous game receives in the next one.

Which option should you choose? You may assume it is to your advantage to serve. You may also assume that the outcome of any game is independent of when the game is played and of the outcome of any previous game.

Conway's Immobilizer

Three cards, an ace, a deuce, and a trey, lie face-up on a desk in some or all of three marked positions ("left," "middle," and "right"). If they are all in the same position, you see only the top card of the stack; if they are in two positions, you see only two cards and do not know which of them is concealing the third card.

Your objective is to get the cards stacked on the left with ace on top, then deuce, then trey on bottom. You do this by moving one card at a time, always from the top of one stack to the top of another (possibly empty) stack.

The problem is, you have no short-term memory and must, therefore, devise an algorithm in which each move is based entirely on what you see, and not on what you last saw or did, or on how many moves have transpired. An observer will tell you when you've won. Can you devise an algorithm that will succeed in a bounded number of steps, regardless of the initial configuration?

Same Sum Dice

You roll a set of n red n-sided dice, and a set of n black n-sided dice. Each die is labeled with the numbers from 1 to n. Show that there must always be a nonempty subset of the red dice, and a nonempty subset of the black dice, showing the same sum.

Matching Coins

Sonny and Cher play the following game. In each round, a fair coin is tossed. Before the coin is tossed, Sonny and Cher *simultaneously* declare their guess for the result of the coin toss. They win the round if both guessed correctly. The goal is to maximize the fraction of rounds won, when the game is played for many rounds.

So far, the answer is obviously 50%: Sonny and Cher agree on a sequence of guesses (for example, they decide to always declare "heads"), and they can't do any better than that. However, before the game begins, the players are informed that just prior to the first toss, Cher will be given the results of all the coin tosses in advance! She has a chance now to discuss strategy with Sonny, but once she gets the coinflip information, there will be no further opportunity to conspire. Show how Sonny and Cher can guarantee to get at least six wins in 10 flips.

(If that's too easy for you, show how they can guarantee at least six wins in only nine flips!)

Next Card Bet

Cards are turned face up one at a time from the top of a well-shuffled deck. You begin with a bankroll of $1, and can bet any fraction of your current worth, prior to each revelation, on the color of the next card. You get even odds regardless of the current composition of the deck. Thus, for example, you can decline to bet until the last card, whose color you will of course know, then bet everything and be assured of going home with $2.

Is there any way you can *guarantee* to finish with more than $2? If so, what's the maximum amount you can assure yourself of winning?

Summing Fractions

Gail asks Henry to think of a number n between 10 and 100, but not to tell her what it is. She now tells Henry to find all (unordered) pairs of numbers j, k that are relatively prime and less than n, but add up to more than n. He now adds all the fractions $1/jk$.

Whew! Finally, Gail tells Henry what his sum is. How does she do it?

Box in a Box

Suppose the cost of shipping a rectangular box is given by the sum of its length, width, and height. Might it be possible to save money by fitting your box into a cheaper box?

Option Hats

One hundred prisoners are told that at midnight, in the dark, each will be fitted with a red or black hat according to a fair coinflip. The prisoners will be arranged in a circle and the lights turned on, enabling each prisoner to see every other prisoner's hat color. Once the lights are on, the prisoners will have no opportunity to signal to one another or to communicate in any way.

Each prisoner will then be taken aside and given the option of trying to guess whether his own hat is red or black, *but he may choose to pass.* The prisoners will all be freed if (1) at least one prisoner chooses to guess his hat color, and (2) every prisoner who chooses to guess guesses correctly.

As usual, the prisoners have a chance to devise a strategy before the game begins. Can they achieve a winning probability greater than 50%?

Impressionable Thinkers

The citizens of Floptown meet each week to talk about town politics, and in particular whether or not to support the building of a new shopping mall downtown. During the meetings each citizen talks to his friends—of whom there are always an odd number, for some reason—and the next day, changes (if necessary) his opinion regarding the mall so as to conform to the opinion of the majority of his friends.

Prove that eventually, the opinions held every *other* week will be the same.

One-bulb Room

Each of n prisoners will be sent alone into a certain room, infinitely often, but in some arbitrary order determined by their jailer. The prisoners have a chance to confer in advance, but once the visits begin, their only means of communication will be via a light in the room which they can turn on or off. Help them design a protocol which will ensure that *some* prisoner will eventually be able to deduce that everyone has visited the room.

Sphere and Quadrilateral

A quadrilateral in space has all of its edges tangent to a sphere. Prove that the four points of tangency lie on a plane.

Bluffing with Reals

Consider the following simple bluffing game. Louise and Jeremy ante $1 each and each is given a secret random real number between 0 and 1. Louise may decide to pass in which case the $2 pot goes to the player with the higher number. However, if she chooses, Louise may add another dollar to the pot. Jeremy may then "call" by adding another dollar himself, in which case the pot, now with $4 in it, goes again to the player with the higher number. Or Jeremy may fold, ceding the pot, with his $1 in it, to Louise.

Surely Louise has the advantage in this game, or at least equality, since she can break even by always passing. How much is the game worth to her? What are the players' equilibrium strategies?

Tiling with Crosses

Can you tile the plane with 5-square crosses? Can you tile 3-space with 7-cube crosses?

Y's in the Plane

Prove that only countably many disjoint **Y**'s can be drawn in the plane.

Infected Cubes

An infection spreads among the n^3 unit cubes of an $n \times n \times n$ cube, in the following manner: If a unit cube has 3 or more infected neighbors, then it becomes infected itself. (Neighbors are orthogonal only, so each little cube has at most 6 neighbors.)

Prove that you *can* infect the whole big cube starting with just n^2 sick unit cubes.

Worst Route

A postman has deliveries to make on a long street, to addresses 2, 3, 5, 7, 11, 13, 17, and 19. The distance between any two houses is proportional to the difference of their addresses.

To minimize the distance traveled, the postman would of course make his deliveries in increasing (or decreasing) order of address. But our postman is overweight and would like to *maximize* the distance traveled making these deliveries, so as to get the most exercise he can. But he can't just wander around town; to do his job properly he is obligated to walk directly from each delivery to the next one.

In what order should he make his deliveries?

Factorials and Squares

Consider the product $100! \cdot 99! \cdot 98! \cdot \ \cdots \ \cdot 2! \cdot 1!$. Call each of the 100 factors $k!$ a "term." Can you remove one term and leave a perfect square?

Halfway Points

Let S be a finite set of points in the unit interval $[0,1]$, and suppose that every point $x \in S$ lies either halfway between two other points in S (not necessarily the nearest ones) or halfway between another point in S and an endpoint.

Show that all the points in S are rational.

Who Won the Series?

Two evenly-matched teams meet to play a best-of-seven World Series of baseball games. Each team has the same small advantage when playing at home. As is customary for this event, one team (say, Team A) plays games 1 and 2 at home, and, if necessary, plays games 6 and 7 at home. Team B plays games 3, 4 and, if needed, 5 at home.

You go to a conference in Europe and return to find that the series is over, and six games were played. Which team is more likely to have won the series?

Tiling with L's

Can you tile the positive quadrant of the plane grid with unrotated tri-ominoes each of which is either shaped like a letter L, or a backwards L?

Dishwashing Game

You and your spouse flip a coin to see who washes the dishes each evening. "Heads" she washes, "tails" you wash.

Tonight she tells you she is imposing a different scheme. You flip the coin 13 times, then she flips it 12 times. If you get more heads than she does, she washes; if you get the same number of heads or fewer, you wash.

Should you be happy?

Random Judge

After a wild night of shore leave, you are about to be tried by your US Navy superiors for unseemly behavior. You have a choice between accepting a "summary" court-martial with just one judge, or a "special" court-martial with three judges who decide by majority vote.

Each possible judge has the same (independent) probability—65.43%—of deciding in your favor, except that one officer who would be judging your special court-martial (but not the summary) is notorious for flipping a coin to make his decisions.

Which type of court-martial is more likely to keep you out of the brig?

Angles in Space

Prove that among any set of more than 2^n points in $\mathbb{R}^n$, there are three that determine an obtuse angle.

Wins in a Row

You want to join a certain chess club, but admission requires that you play three games against Ioana, the current club champion, and win two in a row.

Since it is an advantage to play the white pieces (which move first), you alternate playing white and black.

A coin is flipped, and the result is that you will be white in the first and third games, black in the second.

Should you be happy?

Chessboard Guess

Troilus is engaged to marry Cressida but threatened with deportation, and Immigration is questioning the legitimacy of the proposed marriage. To test their connection, Troilus will be brought into a room containing a chessboard, one of whose squares is designated as special. On every square will be a coin, either heads-up or tails-up. Troilus gets to turn over one coin, after which he will be ejected from the room and Cressida brought in.

Cressida, after examining the chessboard, must guess the designated square. If she gets it wrong, Troilus is deported.

Can Troilus and Cressida save their marriage?

Split Games

You are a rabid baseball fan and, miraculously, your team has won the pennant—thus, it gets to play in the World Series. Unfortunately, the opposition is a superior team whose probability of winning any given game against your team is 60%.

Sure enough, your team loses the first game in the best-of-seven series and you are so unhappy that you drink yourself into a stupor. When you regain consciousness, you discover that two more games have been played.

You run out into the street and grab the first passer-by. "What happened in games 2 and 3 of the World Series?"

"They were split," he says. "One game each."

Should you be happy?

Frames on a Chessboard

You have an ordinary 8×8 chessboard with red and black squares. A genie gives you two "magic frames," one 2×2 and one 3×3. When you place one of these frames neatly on the chessboard, the 4 or 9 squares they enclose instantly flip their colors.

Can you reach all 2^{64} possible color configurations?

Angry Baseball

As in *Split Games*, your team is the underdog and wins any given game in the best-of-seven World Series with probability 40%. But, hold on: This time, whenever your team is behind in the series, the players get angry and play better, raising your team's probability of winning that game to 60%.

Before it all begins, what is your team's probability of winning the World Series?

Bugs on a Polyhedron

Associated with each face of a solid convex polyhedron is a bug which crawls along the perimeter of the face, at varying speed, but only in the clockwise direction. Prove that no schedule will permit all the bugs to circumnavigate their faces and return to their initial positions without incurring a collision.

Serious Candidates

Assume that, as is often the case, no one has any idea who the next nominee for President of the United States will be, of the party not currently in power. In particular, at the moment no person has probability as high as 20% of being chosen.

As the politics and primaries proceed, probabilities change continuously and some candidates will exceed the 20% threshold while others will never do so. Eventually one candidate's probability will rise to 100% while everyone else's drops to 0. Let us say that after the

convention, a candidate is entitled to say he or she was a "serious" candidate if at some point his or her probability of being nominated exceeded 20%.

Can anything be said about the expected number of serious candidates?

Sums and Differences

Given 25 different positive numbers, can you always choose two of them such that none of the other numbers equals either their sum or their difference?

Two-Point Conversion

You, coach of the Hoboken Hominids football team, were 14 points behind your rivals (the Gloucester Great Apes) until, with just a minute to go in the game, you scored a touchdown. You have a choice of kicking an extra point (95% success rate) or going for two (45% success rate). Which should you do?

Swedish Lottery

In a proposed mechanism for the Swedish National Lottery, each participant chooses a positive integer. The person who submits the lowest number not chosen by anyone else is the winner. (If no number is chosen by exactly one person, there is no winner.)

If just three people participate, but each employs an optimal, equilibrium, randomized strategy, what is the largest number that has positive probability of being submitted?

Random Chord

What is the probability that a random chord of a circle is longer than a side of an equilateral triangle inscribed in the circle?

Cube Magic

Can you pass a cube through a hole in a smaller cube?

Random Bias

Suppose you choose a real number p between 0 and 1 uniformly at random, then bend a coin so that its probability of coming up heads when you flip it is precisely p. Finally you flip your bent coin 100 times. What is the probability that after all this, you end up flipping exactly 50 heads?

Gladiators, Version 1

Paula and Victor each manage a team of gladiators. Paula's gladiators have strengths $p_1, p_2, \ldots, p_m$ and Victor's, $v_1, v_2, \ldots, v_n$. Gladiators fight one-on-one to the death, and when a gladiator of strength x meets a gladiator of strength y, the former wins with probability $x/(x+y)$, and the latter with probability $y/(x+y)$. Moreover, if the gladiator of strength x wins, he gains in confidence and inherits his opponent's strength, so that his own strength improves to $x+y$; similarly, if the other gladiator wins, his strength improves from y to $x+y$.

After each match, Paula puts forward a gladiator (from those on her team who are still alive), and Victor must choose one of his to face Paula's. The winning team is the one which remains with at least one live player.

What's Victor's best strategy? For example, if Paula begins with her best gladiator, should Victor respond from strength or weakness?

Gladiators, Version II

Again Paula and Victor must face off in the Colosseum, but this time, confidence is not a factor, and when a gladiator wins, he keeps the same strength he had before.

As before, prior to each match, Paula chooses her entry first. What is Victor's best strategy? Whom should he play if Paula opens with her best man?

Rolling a Six

How may rolls of a die does it take, on average, to get a 6—given that you didn't roll any odd numbers *en route*?

Traveling Salesmen

Suppose that between every pair of major cities in Russia, there's a fixed one-way air fare for going from either city to the other. Traveling salesman Alexei Frugal begins in St. Petersburg and tours the cities, always choosing the cheapest flight to a city not yet visited (he does not need to return to St. Petersburg). Salesman Boris Lavish also needs to visit every city, but he starts in Kaliningrad, and his policy is to choose the most *expensive* flight to an unvisited city at each step.

It looks obvious that Lavish's tour costs at least as much as Frugal's, but can you prove it?

Napkins in a Random Setting

At a conference banquet for a meeting of the Association for Women in Mathematics, the participants find themselves assigned to a big circular table. On the table, between each pair of adjacent settings, is a coffee cup containing a cloth napkin. As each mathematician sits down, she takes a napkin from her left or right; if both napkins are present, she chooses randomly. If the seats are occupied in random

order and the number of mathematicians is large, what fraction of them (asymptotically) will end up without a napkin?

Lame Rook

A *lame rook* moves like an ordinary rook in chess—straight up, down, left, or right—but only one square at a time. Suppose that the lame rook begins at some square and tours the 8×8 chessboard, visiting each square once and returning to the starting square on the 64th move. Show that the number of horizontal moves of the tour, and the number of vertical moves of the tour, are *not* equal!

Coin Testing

The Unfair Advantage Magic Company has supplied you, a magician, with a special penny and a special nickel. One of these is supposed to flip "Heads" with probability 1/3, the other 1/4, but UAMCO has not bothered to tell you which is which.

Having limited patience, you plan to try to identify the biases by flipping the nickel and penny one by one until one of them comes up Heads, at which point that one will be declared to be the 1/3-Heads coin.

In what order should you flip the coins to maximize the probability that you get the correct answer, and at the same time to be fair, that is, to give the penny and the nickel the same chance to be designated the 1/3-Heads coin?

Curve and Three Shadows

Is there a simple closed curve in 3-space, all three of whose projections onto the coordinate planes are trees? This means that the shadows of the curve, from the three coordinate directions, may not contain any loops.

Majority Hats

One hundred prisoners are told that at midnight, in the dark, each will be fitted with a red or black hat according to a fair coin-flip. The prisoners will be arranged in a circle and the lights turned on, enabling each prisoner to see every other prisoner's hat color. Once the lights are on, the prisoners will have no opportunity to signal to one another or to communicate in any way.

Each prisoner is then taken aside and *must* try to guess his own hat color. The prisoners will all be freed if a majority (here, at least 51) get it right.

As before, the prisoners have a chance to devise a strategy before the game begins. Can they achieve a winning probability greater than 50%? Would you believe 90%? How about 95%?

Bugs on a Line

Each positive integer on the number line is equipped with a green, yellow, or red light. A bug is dropped on "1" and obeys the following rules at all times: If it sees a green light, it turns the light yellow and moves one step to the right; if it sees a yellow light, it turns the light red and moves one step to the right; if it sees a red light, it turns the light green and moves one step to the *left*.

Eventually, the bug will fall off the line to the left, or run out to infinity on the right. A second bug is then dropped on "1," again following the traffic lights starting from the state the last bug left them in; then, a third bug makes the trip.

Prove that if the second bug falls off to the left, the third will march off to infinity on the right.

Prisoner and Dog

A woman is imprisoned in a large field surrounded by a circular fence. Outside the fence is a vicious guard dog that can run four times as fast as the woman, but is trained to stay near the fence. If the woman can contrive to get to an unguarded point on the fence, she can quickly scale the fence and escape. But can she get to a point on the fence ahead of the dog?

Pegs on the Corners

Four pegs begin on the plane at the corners of a square. At any time, you may cause one peg to jump over a second, placing the first on the opposite side of the second, but at the same distance as before. The jumped peg remains in place. Can you maneuver the pegs to the corners of a larger square?

Circles in Space

Can you partition all of 3-dimensional space into circles?

Fifteen Bits and a Spy

A spy's only chance to communicate with her control lies in the daily broadcast of 15 zeros and ones by a local radio station. She does not know how the bits are chosen, but each day she has the opportunity to *alter* any one bit, changing it from a 0 to a 1 or vice-versa.

How much daily information can she communicate?

Flipping the Pentagon

The vertices of a pentagon are labeled with integers, the sum of which is positive. At any time, you may change the sign of a negative label, but then the new value is subtracted from both neighbors' values so as to maintain the same sum.

Prove that, inevitably, no matter which negative labels are flipped, the process will terminate after finitely many flips, with all values non-negative.

Love in Kleptopia

Jan and Maria have fallen in love (via the internet) and Jan wishes to mail her a ring. Unfortunately, they live in the country of Kleptopia where anything sent through the mail will be stolen unless it is sent in a padlocked box. Jan and Maria each have plenty of padlocks, but none to which the other has a key. How can Jan get the ring safely into Maria's hands?

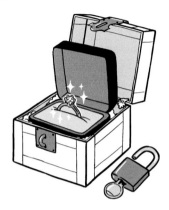

Touring an Island

Aloysius is lost while driving his Porsche on an island in which every intersection is a meeting of three (two-way) streets. He decides to adopt the following algorithm: Starting in an arbitrary direction from his current intersection, he turns right at the next intersection, then left at the next, then right, then left, and so forth.

Prove that Aloysius must return eventually to the intersection at which he began this procedure.

Badly Designed Clock

The hour and minute hands of a certain clock are indistinguishable. How many moments are there in a day when it is not possible to tell from this clock what time it is?

Fibonacci Multiples

Show that every positive integer has a multiple that's a Fibonacci number.

Worms and Water

Lori is having trouble with worms crawling into her bed. To stop them, she places the legs of the bed into pails of water; since the worms can't swim, they can't reach the bed via the floor. But they instead crawl up the walls and across the ceiling, dropping onto her bed from above. Yuck!

How can Lori stop the worms from getting to her bed?

Light Bulbs in a Circle

In a circle are light bulbs numbered clockwise 1 through n, $n > 1$, all initially on. At time t, you examine bulb number t (modulo n), and if it's on, you change the state of bulb $t+1$ (modulo n); i.e., you turn off the clockwise-next bulb if it's on, and on if it's off. If bulb t is off, you do nothing.

Prove that if you continue around and around the ring in this manner, eventually all the bulbs will again be on.

Generating the Rationals

You are given a set S of numbers that contains 0 and 1, and contains the mean of every finite nonempty subset of S. Prove that S contains all the rational numbers between 0 and 1.

Emptying a Bucket

You are presented with three large buckets, each containing an integral number of ounces of some non-evaporating fluid. At any time, you may double the contents of one bucket by pouring into it from a fuller one; in other words, you may pour from a bucket containing x ounces into one containing $y \leq x$ ounces until the latter contains $2y$ ounces (and the former, $x-y$).

Prove that no matter what the initial contents, you can, eventually, empty one of the buckets.

Funny Dice

You have a date with your friend Katrina to play a game with three dice, as follows. She chooses a die, then you choose one of the other two dice. She rolls her die while you roll yours, and whoever rolls the higher number wins. If you roll the same number, Katrina wins.

Wait, it's not as bad as you think; you get to design the dice! Each will be a regular cube, but you can put any number of pips from 1 to 6 on any face, and the three dice don't have to be the same.

Can you make these dice in such a way that you have the advantage in your game?

Picking the Athletic Committee

The Athletic Committee is a popular service option among the faculty of Quincunx University, because while you are on it, you get free tickets to the university's sports events. In an effort to keep the committee from becoming cliquish, the university specifies that no one with three or more friends on the committee may serve on the committee—but, in compensation, if you're not on the committee but have three or more friends on it, you can get free tickets to any athletic event of your choice.

To keep everyone happy, it is therefore desirable to construct the committee in such a way that even though no one on it has three or more friends on it, everyone *not* on the committee *does* have three or more friends on it.

Can this always be arranged?

Ice Cream Cake

On the table before you is a cylindrical ice-cream cake with chocolate icing on top. From it you cut successive wedges of the same angle θ. Each time a wedge is cut, it is turned upside-down and reinserted into the cake. Prove that, regardless of the value of θ, after a finite number of such operations all the icing is back on top of the cake!

Sharing a Pizza

Alice and Bob are preparing to share a circular pizza, divided by radial cuts into some arbitrary number of slices of various sizes. They will be using the "polite pizza protocol": Alice picks any slice to start; thereafter, starting with Bob, they alternate taking slices but always from one side or the other of the gap. Thus after the first slice, there are just two choices at each turn until the last slice is taken (by Bob if the number of slices is even, otherwise by Alice).

Is it possible for the pizza to have been cut in such a way that Bob has the advantage—in other words, so that with best play, Bob gets more than half the pizza?

Moth's Tour

A moth alights on the "12" of a clock face, and begins randomly walking around the dial. Each time it hits a number, it proceeds to the next clockwise number, or the next counterclockwise number, with equal probability. It continues until it has been at every number.

What is the probability that the moth finishes at the number "6"?

Names in Boxes

The names of 100 prisoners are placed in 100 wooden boxes, one name to a box, and the boxes are lined up on a table in a room. One by one, the prisoners are led into the room; each may look in at most 50 boxes, but must then leave the room exactly as he found it and is permitted no further communication with the others.

The prisoners have a chance to plot their strategy in advance, and they are going to need it, because unless *every single prisoner finds his own name* all will subsequently be executed.

Find a strategy that gives the prisoners a decent chance of survival.

Garnering Fruit

Each of 100 baskets contains some number (could be zero) of apples, some number of bananas, and some number of cherries. Show that you can collect 51 of those baskets that together contain at least half the apples, at least half the bananas, and at least half the cherries!

Coin Game

You and a friend each pick a different heads-tails sequence of length 4 and a fair coin is flipped until one sequence or the other appears; the owner of that sequence wins the game.

For example, if you pick HHHH and she picks TTTT, you win if a run of four heads occurs before a run of four tails.

Do you want to pick first or second? If you pick first, what should you pick? If your friend picks first, how should you respond?

Roulette for Parking Money

You're in Las Vegas with only $2 and in desperate need of $5 to feed a parking meter. You run in through the nearest door and find yourself at a roulette table. You can bet any whole dollar amount on any allowable set of numbers. What roulette-betting strategy will maximize your probability of walking out with $5?

Invisible Corners

Can it be that you are standing outside a polyhedron and can't see any of its vertices?

Sleeping Beauty

Sleeping Beauty agrees to the following experiment. On Sunday she is put to sleep and a fair coin is flipped. If it comes up Heads, she is awakened on Monday morning; if Tails, she is awakened on Monday morning and again on Tuesday morning. In all cases, she is not told the day of the week, is put back to sleep shortly after, and will have no memory of any Monday or Tuesday awakenings.

When Sleeping Beauty is awakened on Monday or Tuesday, what—to her—is the probability that the coin came up Heads?

Boardroom Reduction

The Board of Trustees of the National Museum of Mathematics has grown too large—50 members, now—and its members have agreed to the following reduction protocol. The board will vote on whether to (further) reduce its size. A majority of ayes results in the immediate ejection of the newest board member; then another vote is taken, and so on. If at any point half or more of the surviving members vote nay, the session is terminated and the board remains as it currently is.

Suppose that each member's highest priority is to remain on the board, but given that, agrees that the smaller the board, the better.

To what size will this protocol reduce the board?

Buffon's Needle

A needle one inch in length is tossed onto a large mat marked with parallel lines one inch apart. What is the probability that the needle lands across a line?

Seven Cities of Gold

In 1539 Friar Marcos de Niza returned to Mexico from his travels in what is now Arizona, famously reporting his finding of "Seven Cities of Gold." Coronado did not believe the "liar friar" and when subsequent expeditions came up empty, Coronado gave up the quest.

My (unreliable) sources claim that Coronado's reason for disbelieving de Niza is the latter's claim that the cities were laid out in the desert in a manner such that of any three of the cities, at least one pair were exactly 100 furlongs apart.

Coronado's advisors told him no such pattern of points existed. Were they right?

Life-Saving Transposition

There are just two prisoners this time, Alice and Bob. Alice will be shown a deck of 52 cards spread out in some order, face-up on a table. She will be asked to transpose two cards of her choice. Alice is then dismissed, with no further chance to communicate to Bob. Next, the cards are turned down and Bob is brought into room. The warden names a card and to stave off execution for both prisoners, Bob must find the card after turning over, sequentially, at most 26 of the cards.

As usual the prisoners have an opportunity to conspire before-hand. This time, they can guarantee success. How?

More Magnetic Dollars

We return to *Magnetic Dollars*, but we strengthen their attractive power just a bit.

This time, an infinite sequence of coins will be tossed into the two urns. When one urn contains x coins and the other y, the next coin will fall into the first urn with probability $x^{1.01}/(x^{1.01} + y^{1.01})$, otherwise into the second urn.

Prove that after some point, one of the urns will never get another coin!

Covering the Stains

Just as a big event is about to begin, the queen's caterer notices, to his horror, that there are 10 tiny gravy stains on the tablecloth. All he has the time to do is to cover the stains with non-overlapping plates. He has plenty of plates, each a unit disk. Can he do it, no matter how the stains are distributed?

Zero-Sum Vectors

On a piece of paper, you have (for some reason) made an array whose rows consist of all 2^n of the n-dimensional vectors with coordinates in $\{+1, -1\}^n$—that is, all possible strings of $+1$'s and -1's of length n.

Notice that there are lots of nonempty subsets of your rows which sum to the zero vector, for example any vector and its complement; or the whole array, for that matter.

However, your 2-year-old nephew has got hold of the paper and has changed some of the entries in the array to zeroes.

Prove that no matter what your nephew did, you can find a nonempty subset of the rows in the new array that sums to zero.

Colors and Distances

In the town of Hoegaarden, Belgium, exactly half the houses are occupied by Flemish families, the rest by French-speaking Walloons.

Can it be that the town is so well mixed that the sum of the distances between pairs of houses of like ethnicity exceeds the sum of the distances between pairs of houses of different ethnicity?

Two Sheriffs

Two sheriffs in neighboring towns are on the track of a killer, in a case involving eight suspects. By virtue of independent, reliable detective work, each has narrowed his list to only two. Now they are engaged in a telephone call; their object is to compare information, and if their pairs overlap in just one suspect, to identify the killer.

The difficulty is that their telephone line has been tapped by the local lynch mob, who know the original list of suspects but not which pairs the sheriffs have arrived at. If they are able to identify the killer with certainty as a result of the phone call, he will be lynched before he can be arrested.

Can the sheriffs, who have never met, conduct their conversation in such a way that they both end up knowing who the killer is (when possible), yet the lynch mob is still left in the dark?

Painting the Fence

Each of n industrious people chooses a random point on a circular fence, and begins painting toward her farthest neighbor until she encounters a painted section. On average, how much of the fence gets painted? How about if instead, each person paints toward her *nearest* neighbor?

Leading All the Way

Running for local office against Bob, Alice wins with 105 votes to Bob's 95. What is the probability that as the votes were counted (in random order), Alice led the whole way?

Self-Referential Number

The first digit of a certain 8-digit integer N is the number of zeroes in the (ordinary, decimal) representation of N. The second digit is the number of ones; the third, the number of twos; the fourth, the number of threes; the fifth, the number of fours; the sixth, the number of fives; the seventh, the number of sixes; and, finally, the eighth is the total number of *distinct* digits that appear in N. What is N?

Filling the Cup

You go to the grocery store needing one cup of rice. When you push the button on the machine, it dispenses a uniformly random amount of rice between nothing and one cup. On average, how many times do you have to push the button to get your cup?

Two Balls and a Wall

On a line are two identical-looking balls and a vertical wall. The balls are perfectly elastic and friction-free; the wall is perfectly rigid; the ground is perfectly level. If the balls are the same mass and the one farther from the wall is rolled toward the closer one, it will knock the close ball toward the wall; that ball will bounce back and hit the first ball, which will then roll (forever) away from the wall. Three bounces altogether.

Now assume the farther ball has mass a million times greater than the closer one. How many bounces now? (You may ignore the effects of angular momentum, relativity, and gravitational attraction.)

Bulgarian Solitaire

Fifty-five chips are organized into some number of stacks, of arbitrary heights, on a table. At each tick of a clock, one chip is removed from each stack and those collected chips are used to create a new stack.

What eventually happens?

The Hints

Below are hints and/or comments for each puzzle, together with the number of the chapter where you can find a solution. (But to get the most out of these puzzles, don't go there until you've tried everything!)

Adding, Multiplying, and Grouping: Multiply sums of two numbers and of five numbers. (Chapter 12)

Air Routes in Aerostan: What can be said about the two-airline networks? (Chapter 4)

All Right or All Wrong: Use both choice and parity. (Chapter 23)

Alternate Connection: If you number the points in the order they get connected, which one do you not want to connect? (Chapter 1)

Alternating Powers: Note that $f(x) = x - f(x^2)$. (Chapter 9)

Alternative Dice: Represent a die by a polynomial in which the coefficient of x^k is the number of faces with k pips. (Chapter 5)

An Attractive Game: Look at the game from the bank's point of view. (Chapter 14)

Angles in Space: Translate the convex closure of the points. (Chapter 16)

Angry Baseball: What happens after the last tie? (Chapter 10)

Ants on the Circle: Conservation of momentum applies! (Chapter 20)

Area–Perimeter Match: Factoring a polynomial might help here. (Chapter 5)

Ascending and Descending: Consider, for each number in a good list, the length of the longest decreasing subsequence that ends there. (Chapter 12)

Attention Paraskevidekatriaphobes: What is the cycle for our Gregorian calendar? (Chapter 1)

Attic Lamp: You'll need more than one bit of information. (Chapter 13)

Baby Frog: Cover the field with a $2^n \times 2^n$ grid. (Chapter 15)

Bacteria on the Plane: Give each child half the potential of its parent. (Chapter 18)

Bacterial Reproduction: Experiment, and keep track of the numbers of females. (Chapter 18)

Badly Designed Clock: Note that you *can* tell what time it is when the hands coincide. (Chapter 19)

Bags of Marbles: Nothing to prevent you from putting other things in bags... (Chapter 1)

Bat and Ball: Do the math! (Chapter 5)

Belt Around the Earth: Use a letter for the earth's circumference (Chapter 5)

Bias Test: In which case can the second flip change your mind? (Chapter 13)

Biased Betting: Fix a win–loss sequence that would result in going broke. (Chapter 10)

Bidding in the Dark: Compute your expectation after a given bid. (Chapter 14)

Big Pairs in a Matrix: Suppose not, and look separately at the biggest and next-biggest entries in each row. (Chapter 9)

Bindweed and Honeysuckle: Untwist one while you twist the other. (Chapter 1)

Birthday Match: Be careful; you want a match for yourself, not just any pair of people. (Chapter 10)

Bluffing with Reals: Jeremy has one threshold, Louise two. (Chapter 21)

Boarding the Manhole: Archimedes may be able to help you here. (Chapter 16)

Boardroom Reduction: Start by thinking about what would happen if the board came down to just its three most senior members. (Chapter 21)

Box in a Box: Assume you can, and compute the volume of the set of all points within distance ε of the boxes. (Chapter 16)

Boy Born on Tuesday: Count the cases carefully. (Chapter 10)

Bracing the Grid: Make a graph whose vertices are the rows and columns of the grid, with edges where the intersecting square is braced. (Chapter 4)

Breaking a Chocolate Bar: Try this, and see the light. (Chapter 18)

Broken ATM: What's the most George can hope to get, considering that the machine's only options are multiples of $6? (Chapter 6)

Buffon's Needle: What is expected number of lines crossed, per unit length, by a randomly placed unit-diameter circle? (Chapter 14)

Bugging a Disk: Assume the ceiling is flat, and inscribe a hexagon in it. (Chapter 9)

Bugs on Four Lines: Add a time axis. (Chapter 16)

Bugs on a Line: First, check that the bug cannot wander forever without going off to infinity. (Chapter 18)

Bugs on a Polyhedron: Draw an arrow from each face through its bug to the next face. (Chapter 18)

Bugs on a Pyramid: Number the bugs and ask: When one looks at the triangle made by the others, does she see them in clockwise or counterclockwise order? (Chapter 3)

Bulgarian Solitaire: Rotate the configuration 45° and think about gravity. (Chapter 18)

Candles on a Cake: What does the puzzle tell you about the distance from a fixed point on the circumference to the ith candle? (Chapter 7)

Cards from Their Sum: Big numbers work better. (Chapter 6)

Chameleons: Look not just at numbers of chameleons of a given color, but differences between numbers. (Chapter 2)

Chessboard Guess: Assign a 6-digit nimber to each square of the chessboard. (Chapter 17)

Chinese Nim: Starting from (1,2), find a pattern linking the positions you don't want to be in. (Chapter 21)

Chomp: Someone must have a winning strategy; what if it's Bob? (Chapter 9)

Circles in Space: Note that you can tile a sphere, minus any two points, with circles. (Chapter 22)

Circular Shadows I: Can you alter a ball slightly without affecting its coordinate projections? (Chapter 22)

Circular Shadows II: Close in on the body with parallel planes. (Chapter 16)

Coconut Classic: Start by allowing yourself negative numbers of coconuts! (Chapter 7)

Coin Game: Who would you want to be if one of you picks HHHH and the other THHH? (Chapter 10)

Coin Testing: You need to keep the number of flips of each coin as equal as possible—but there's more. (Chapter 10)

Coins in a Row: You are not asked for an optimal strategy, only one that gets Alix half the money. (Chapter 7)

Coins on the Table: The shape of the table plays a part. (Chapter 7)

Colors and Distances: Notice that the distance between two points in the plane is proportional to the probability that a random line, suitably constrained, passes between them. (Chapter 14)

Comparing Numbers, Version I: Have Victor pick a threshold. (Chapter 10)

Comparing Numbers, Version II: What would Paula do if Victor used 1/2 as his threshold? (Chapter 10)

Competing for Programmers: You might have good use for a "friendship graph," mentioned earlier in the same chapter. (Chapter 4)

Conversation on a Bus: One part of the conversation keeps the bus number from being too small, while another part keeps it from being too big. (Chapter 13)

Conway's Immobilizer: A good start is to make sure your algorithm does the right thing when it's about to win. (Chapter 11)

Covering the Stains: How would you try to cover the stains if you had left your glasses at home, and couldn't see where the stains were? (Chapter 14)

Crossing the River: In general: Get the men across first. (Chapter 19)

Cube Magic: Orient the cube to get a large projection. (Chapter 22)

Curve and Three Shadows: Try curves made from grid lines. (Chapter 16)

Curve on a Sphere: Pick two points on the curve halfway around from each other, and imagine that the midpoint on the great circle connecting them is the North Pole. (Chapter 9)

Curves on Potatoes: Replace the potatoes by holograms. (Chapter 16)

Cutting the Necklace: Imagine the necklace laid out in a circle, with a line through the center cutting it twice. (Chapter 3)

Deterministic Poker: Alice must, at the very least, stop Bob from making an ace-high straight flush. (Chapter 21)

Dishwashing Game: Begin with 12 flips each. (Chapter 10)

Divisibility Game: Bob gets the most out of small numbers. (Chapter 6)

Domino Task: Notice that a vertical domino can be turned horizontal if one of its rows has a hole. (Chapter 7)

Dots and Boxes Variation: Think about your choice from your opponent's point of view. (Chapter 9)

Dot-Town Exodus: Reason carefully with a population of 3. (Chapter 13)

Double Cover by Lines: What happens if you try to build your double-cover one line at a time? (Chapter 23)

Drawing Socks: Order all the socks randomly and condition on the color of the last one. (Chapter 14)

Early Commuter: How much time did the husband save? (Chapter 1)

Easy Cake Division: Straight, vertical cuts can do the job. (Chapter 16)

Efficient Pizza-cutting: How can you make the next cut produce the maximum number of new pieces? (Chapter 1)

Emptying a Bucket: One way to do it, paradoxically, is to show that the contents of one bucket can be increased until another is empty. (Chapter 21)

Even Split: Prove it first when n is prime. (Chapter 6)

Even-sum Billiards: What would your answer be if there were no 9-ball? (Chapter 2)

Expecting the Worst: Bend the segment into a circle. (Chapter 24)

Exponent upon Exponent: What happens to the expression if you cut off the bottom x? (Chapter 23)

Factorial Coincidence: Consider the power of a big prime that factors each side of the expression. (Chapter 6)

Factorials and Squares: Use the fact that the product of perfect squares is itself a perfect square. (Chapter 6)

Fair Play: Flip more than once, and look for equiprobable events. (Chapter 8)

Falling Ants: Note that ants are interchangeable, for the purposes of this puzzle. (Chapter 20)

Faulty Combination Lock: Try dividing the numbers from 1 to 8 into two groups, and using the fact that any combination must contain at least two from one of them. (Chapter 1)

Fewest Slopes: Try regular polygons. (Chapter 11)

Fibonacci Multiples: Keep track of the remainders, modulo n, of two successive Fibonacci numbers. (Chapter 21)

Fifteen Bits and a Spy: Let the "nimsum" of the broadcast be the message. (Chapter 17)

Figure Eights in the Plane: Note that there are only countably many rational points in the plane—or pairs of them, for that matter. (Chapter 23)

Filling a Bucket: How far can you get trying to keep all the buckets level? (Chapter 19)

Filling the Cup: Note that the fractional parts of the amount of rice you have got so far also constitutes a sequence of independent uniform random variables. (Chapter 14)

Find the Robot: How many ways are there to start and point the robot? (Chapter 23)

Finding a Jack: The jacks divide the deck into five parts. (Chapter 8)

Finding the Counterfeit: Each weighing can have three possible outcomes. (Chapter 13)

Finding the Missing Number: How much information must be kept? (Chapter 19)

Finding the Rectangles: What do you encounter as you walk across the tiled 400-gon? (Chapter 22)

First Odd Number: List the alphabetically early words that may arise in a number. (Chapter 11)

First-grade Division: Imagine that each project requires the same total effort. (Chapter 18)

Flipping the Pentagon: A reasonable thing to try as a potential might be some measure of how close the numbers are to one another. (Chapter 18)

Flying Saucers: It's important that the person beamed back down might not be the first one beamed up by the next saucer. (Chapter 7)

Fourth Corner: Each grid point can have one of four possible parities. (Chapter 2)

Frames on a Chessboard: Look for a set of cells the parity of whose number of red cells doesn't change. (Chapter 18)

Funny Dice: How can you give one die an advantage over another, even though they both have the same average roll? (Chapter 19)

Game of Desperation: What position can you put your opponent in that will force her to let you win on your next move? (Chapter 21)

Gaming the Quilt: If you've already bought k tickets, when is one more worth buying? (Chapter 5)

Garnering Fruit: Try it for just apples and bananas; adding cherries brings a ham sandwich(!) into the picture. (Chapter 3)

Gasoline Crisis: Start by imagining a trip that starts with plenty of extra gas aboard. (Chapter 7)

Generating the Rationals: Note that you easily get all fractions whose denominators are powers of 2. (Chapter 19)

Gladiators, Version I: If you think of strength as money, each bout becomes a fair game. (Chapter 24)

Gladiators, Version II: Think of each gladiator as a light bulb trying to outlast his opponent. (Chapter 24)

Gluing Pyramids: How can faces disappear other than by hiding them? (Chapter 22)

Guarding the Gallery: Try putting guard-posts at corners. (Chapter 15)

Half Grown: Think about your own young relatives. (Chapter 1)

Half-right Hats: Prisoners don't all need to have the same instructions. (Chapter 2)

Halfway Points: If S has n points, they satisfy n linear equations with rational coefficients. (Chapter 9)

Handshakes at a Party: What numbers did Alexandra hear during her survey? (Chapter 4)

Hats and Infinity: Have the prisoners choose a special hat sequence from each class of similar sequences. (Chapter 23)

Hazards of Electronic Coinflipping: Flip several times, and note that some numbers of heads will allow you to decide between Alice and Bob. Can you give the rest to Charlie? (Chapter 3)

Home-field Advantage: What if all seven games were always played? (Chapter 10)

Hopping and Skipping: What's the probability that the frog never regresses from its current pad? (Chapter 5)

Ice Cream Cake: It's easier to keep track if you cut the cake in the same place each time, but rotate the cake between cuts. (Chapter 21)

Identifying the Majority: Use the counter for reinforcement of your current name. (Chapter 19)

Impressionable Thinkers: Consider the number of current instances where influence fails, that is, a voter thinks one way and her acquaintance disagrees the next day. (Chapter 18)

Increasing Routes: Note that in a graph the average degree is twice the number of edges divided by the number of vertices. (Chapter 8)

Infected Checkerboard: Try the process with various starting positions. What prevents the infected area from getting too "complicated"? (Chapter 18)

Infected Cubes: You need some way to generalize the diagonals that work in dimension 2. (Chapter 24)

Integral Rectangles: There are many ways to do this one; one of them begins by nesting the big rectangle into the positive quadrant of the plane grid, and making a graph whose vertices are grid points that lie on corners of rectangles. (Chapter 24)

Invisible Corners: You can start by putting yourself in a room made from six non-touching planks. (Chapter 22)

King's Salary: Try to reduce the number of salaried citizens. (Chapter 11)

Lame Rook: What happens if you try this on an $n \times n$ board for various (even) n? (Chapter 15)

Laser Gun: Cover the plane with reflected copies of the room. (Chapter 24)

Lattice Points and Line Segments: When does the line segment between two lattice points contain another lattice point? (Chapter 12)

Leading All the Way: Try putting the ballots in random *circular* order, instead of random linear order. (Chapter 10)

Lemming on a Chessboard: If the lemming stays on the board, it must eventually cycle. (Chapter 9)

Life Is a Bowl of Cherries: Consider the general two-bowl game first. (Chapter 17)

Life Isn't a Bowl of Cherries?: Start with the nim strategy itself, rather than its reverse. (Chapter 17)

Life-saving Transposition: If you can do *Names in Boxes*, you can do this one. (Chapter 19)

Light Bulbs in a Circle: Don't forget to keep track of which bulb you're looking at, as well as the state of all bulbs. (Chapter 21)

Line Up by Height: Keep track of the lengths of the descending and ascending subsequences each player is at the end of, when lined up alphabetically. (Chapter 12)

Lines Through a Grid: Nineteen parallel diagonals will do it. Can you do better? (Chapter 12)

Locker Doors: What numbers have an odd number of divisors (including 1 and themselves)? (Chapter 6)

Losing at Dice: You'll need to consider two patterns to compute the number of ways to "roll a 4" by this strange method. (Chapter 1)

Lost Boarding Pass: Try it with three passengers. What is the probability that third passenger ends up in the second passenger's seat? (Chapter 7)

Love in Kleptopia: Can Jan somehow get one of his locks on the box? (Chapter 19)

Magnetic Dollars: Try it with, say, six coins instead of a million. (Chapter 24)

Majority Hats: Can you beat the "preponderance" strategy, where each prisoner guesses the color he sees most of? (Chapter 17)

Matching Coins: Cher needs to use her coinflips to communicate information about coming flips. (Chapter 13)

Mathematical Bookworm: Visualize! (Chapter 22)

Measuring with Fuses: Try lighting both fuses at once. (Chapter 11)

Meet the Williams Sisters: How many pairs of players play each other? (Chapter 1)

Meeting the Ferry: Draw a picture! Note that all times are GMT so you don't have to worry about timezones. (Chapter 22)

Men with Sisters: What's the answer among you and your siblings? (Chapter 8)

Missing Card: Make use of the ordering of the four remaining cards. (Chapter 13)

Missing Digit: Do you remember "casting out nines"? (Chapter 2)

Monk on a Mountain: Superimpose the days. (Chapter 3)

More Magnetic Dollars: Try to set this up so that each urn acquires coins at random moments. (Chapter 24)

Moth's Tour: Start from when the moth first gets to 5 o'clock or 7 o'clock. (Chapter 21)

Names in Boxes: Nothing prevents a prisoner from using what he finds in a box to decide what box to open next. (Chapter 19)

Napkins in a Random Setting: It's helpful to think of each diner having flipped a coin to decide which napkin she would prefer to take. (Chapter 14)

Natives in a Circle: What would happen to their answers if we changed all the truth-tellers to liars and vice-versa? (Chapter 11)

Next Card Bet: The best guarantee, for given expectation, is when the outcome is no longer random. (Chapter 14)

Next Card Red: Try it with a smaller deck. (Chapter 24)

No Twins Today: It's not a trick. (Chapter 11)

Non-repeating String: Let the number of letters in the alphabet be a variable, and use induction. (Chapter 15)

North by Northwest: An easy enough computation, if you know what it means. (Chapter 1)

Not Burning Brownies: Try rectangles, then some sort of triangle. (Chapter 22)

Numbers on Foreheads: Consider the sum of the numbers, modulo 10. (Chapter 6)

Odd Light Flips: Note that the order in which switches are flipped is irrelevant. (Chapter 15)

Odd Run of Heads: Use a variable to stand for the answer. (Chapter 5)

One-bulb Room: Consider first the situation in which the room is known to be dark at the start. (Chapter 19)

Option Hats: Since each guess has probability only 1/2 of being correct, to improve the odds the prisoners need to arrange things so that either very few guess and are right, or many guess and are wrong. (Chapter 17)

Other Side of the Coin: Consider labeling the sides of the coins. (Chapter 10)

Packing Slashes: In showing your solution is optimal, it may pay to consider the 12 outer vertices of the inside 3×3 grid. (Chapter 11)

Painting the Cubes: Paint space and carve out the big cube! (Chapter 16)

Painting the Fence: How does the question of whether an interval between painters gets painted depend on its length, relative to the lengths of its neighbor intervals? (Chapter 14)

Painting the Polyhedron: Assume you can inscribe a sphere, and triangulate the sphere using the tangent points as vertices. (Chapter 16)

Pairs at Maximum Distance: Note that line segments connecting two max-distance pairs must cross. (Chapter 9)

Pancake Stacks: Making the stacks close in size limits one's opponent. (Chapter 21)

Path Through the Cells: Recolor greedily with high-numbered colors, and start your path with a cell that still has color 1. (Chapter 15)

Peek Advantage: Could the second peek change your mind? (Chapter 13)

Pegs in a Square: Look for a useful way to two-color the holes. (Chapter 18)

Pegs on the Corners: Notice that if the pegs begin at grid points, they stay on grid points. (Chapter 21)

Pegs on the Half-Plane: Weigh the holes so that the entire lower half-plane has finite weight. (Chapter 18)

Phone Call: Tackle this one hour at a time. (Chapter 19)

Picking the Athletic Committee: What happens if you just pick a committee at random and try to fix it? (Chapter 18)

Pie in the Sky: The moon is about one degree in diameter. (Chapter 20)

Ping-Pong Progression: How many points does Bob win, on average, while Alice wins 21? (Chapter 14)

Players and Winners: Think about communication in both directions. (Chapter 13)

Points on a Circle: Pick random diameters first, then endpoints. (Chapter 10)

Poker Quickie: Assume no wild cards, one opponent. (Chapter 11)

Polygon Midpoints: What happens if you pick some arbitrary point on the plane and assume it is a vertex? (Chapter 22)

Polygon on the Grid: What is the relationship of h or v to the polygon's area? (Chapter 19)

Polyhedron Faces: Think about the face with the most sides. (Chapter 12)

Poorly Placed Dominoes: You might need two tries. (Chapter 19)

Portrait: Who is "my father's son"? (Chapter 21)

Powers of Roots: Try the 10th power, and see if you can guess what's going on. (Chapter 7)

Powers of Two: Start from the middle of the phrase. (Chapter 1)

Precarious Picture: Somehow you must arrange the string so that it ultimately passes over each nail, when the other is ignored. (Chapter 22)

Prime Test: If you don't find any small prime factors, what else can you try? (Chapter 6)

Prisoner and Dog: When is the woman close enough to the fence to make a straight run for it, if the dog is at the opposite point? (Chapter 19)

Prisoners and Gloves: For this you might want to consider parity of permutations. (Chapter 2)

Profit and Loss: Observe first that it can't be as long as 40 months. (Chapter 15)

Protecting the Statue: Observe that the protected area cannot be convex. (Chapter 22)

Raising Art Value: Compare the picture's value to the averages for the galleries. (Chapter 8)

Random Bias: Can you somehow "do the flips" before the coin is chosen? (Chapter 10)

Random Chord: Does it matter how the random chord is chosen? (Chapter 10)

Random Intersection: What is the probability that a particular point in one ball will end up inside the other as well? (Chapter 14)

Random Intervals: Try small numbers, and look for an explanation for what you find. (Chapter 24)

Random Judge: Imagine that there's one judge that would take part either way. (Chapter 10)

Rating the Horses: You'll certainly need to see all the horses. (Chapter 1)

Recovering the Polynomial: Think of the coefficients as the expansion of a number in base x. (Chapter 5)

Rectangles Tall and Wide: Start by showing that if the tile's sides have irrational ratio, then any tiling must be all horizontal or all vertical. (Chapter 6)

Red and Black Sides: What's the total area of the rectangles? (Chapter 5)

Red and Blue Hats in a Line: How can the first one to guess help the others? (Chapter 2)

Red Points and Blue: What improves if you uncross two segments? (Chapter 18)

Returning Pool Shot: How many angles can the ball take after it is shot? (Chapter 20)

Righting the Pancakes: Represent the state of the pancake stack as a binary number. (Chapter 18)

Rolling All the Numbers: Think of the experiment in stages. (Chapter 14)

Rolling Pencil: Needs more than one second of thought, but two will do. (Chapter 22)

Rolling a Six: Watch out—this is not the same as ignoring odd rolls. (Chapter 14)

Rotating Coin: Try it! (Chapter 20)

Roulette for Parking Money: Every bet loses money, so you want speed as well as accuracy. (Chapter 14)

Roulette for the Unwary: How many wins does Elwyn need to come out ahead? (Chapter 14)

Rows and Columns: You have to try this to see what's happening. (Chapter 1)

Salaries and Raises: How much does each person's *annual* salary go up, each year? (Chapter 1)

Same Sum Dice: It's easier to prove a stronger statement about ordered dice. (Chapter 12)

Same Sum Subsets: What if you found two overlapping sets with the same sum? (Chapter 12)

Second Ace: Compute the conditional probabilities. (Chapter 10)

Self-referential Number: If you try a number and it doesn't work, what's the simplest way to try to fix it? (Chapter 19)

Sequencing the Digits: There's more than one way to go about choosing such a sequence. (Chapter 1)

Serious Candidates: Imagine betting on each candidate as he or she becomes serious. (Chapter 14)

Service Options: It may be convenient to assume that lots of games are played, regardless of outcomes. (Chapter 10)

Seven Cities of Gold: Equilateral triangles are your best friends for this puzzle. (Chapter 11)

Sharing a Pizza: The even case is easy; try the odd case with pieces of size 0(!) and 1. (Chapter 19)

Shoelaces at the Airport: Think about time spent walking. (Chapter 1)

Shoes, Socks, and Gloves: What are the worst choices you can make? (Chapter 12)

Signs in an Array: Is there anything good that must happen when you flip a line that had negative sum? (Chapter 18)

Sinking 15: You may assume the players can sink whatever balls they want, so this is a deterministic game. Does it seem familiar? (Chapter 24)

Skipping a Number: Missy's free-throw percentage jumps up when she hits (and down when she misses), so it should be easy to set up a situation where it jumps over 80%, right? (Chapter 3)

Slabs in 3-Space: Infinity is tricky—try covering a big finite piece instead. (Chapter 16)

Sleeping Beauty: Imagine repeating the whole experiment 100 times. (Chapter 10)

Slicing the Cube: How would you actually do it? Can you do better? (Chapter 24)

Snake Game: Tile the board with dominoes! (Chapter 4)

Soldiers in a Field: Think about the two closest soldiers. (Chapter 9)

Spaghetti Loops: What's the probability that your ith tying operation will make a loop? (Chapter 14)

Sphere and Quadrilateral: Weigh the vertices so as to balance the edges at the tangent points. (Chapter 20)

Spiders on a Cube: How many spiders would be needed if the game were played on a tree? (Chapter 4)

Spinning Switches: Try the two-switch version first. (Chapter 7)

Split Games: Imagine that for some reason the split games *might* not be counted. (Chapter 10)

Splitting a Hexagon: Start with the triangles. (Chapter 22)

Splitting a Polygon: Non-convex polygons are a problem, but you can start by passing a line through that's not of the same slope as any side. (Chapter 3)

Splitting the Stacks: A little experimentation might help you see the light. (Chapter 1)

Sprinklers in a Field: What points are closer to a given grid vertex than other grid vertices? (Chapter 19)

Squaring the Mountain State: What is meant by inscribing West Virginia in a square is drawing a square around the shape of the state, on the plane, in such a way that the state is inside the square but touches all four sides. (Chapter 3)

Stopping After the Boy: Assume twins are rare. (Chapter 10)

Strength of Schedule: Consider the strength-of-schedule scores for the teams with the most wins, and for the teams with the fewest wins. (Chapter 5)

Subsets with Constraints: Group the numbers so that each group can supply only one number to your set. (Chapter 6)

Subtracting Around the Corner: Notice that odd numbers eventually disappear. (Chapter 2)

Summing Fractions: Try small values for n, then prove your conclusion by comparing gains with losses going from n to $n+1$. (Chapter 15)

Sums and Differences: Suppose otherwise and start by considering the highest number. (Chapter 19)

Sums of Two Squares: Think of pairs of integers as lattice points in a circle. (Chapter 8)

Swapping Executives: Start by getting an executive who belongs at one of the ends into her correct place. (Chapter 15)

Swedish Lottery: Think about betting the highest allowed number, versus "cheating" and betting higher. (Chapter 21)

Testing Ostrich Eggs: Think about Oskar's situation after the first egg breaks. (Chapter 21)

Three Natives at the Crossroads: You will get no information from the random native. (Chapter 13)

Three Negatives: Consider the lowest two numbers in the set. (Chapter 5)

Three Sticks: Can you pick the lengths so that an operation leaves their ratios unchanged? (Chapter 3)

Three-way Duel: Compare Alice's survival probabilities when she hits Bob, hits Carol, or misses both. (Chapter 21)

Three-way Election: Consider the majority second choice among each candidate's fans. (Chapter 1)

Tiling a Polygon: How do you find a tile with sides parallel to two given sides of the 100-gon? (Chapter 22)

Tiling with Crosses: Do the plane with diagonals, and space with plane slabs that have holes and pegs. (Chapter 22)

Tiling with L's: Try it one row at a time, left to right. (Chapter 15)

Tipping the Scales: How much of the time does a given weight spend on each side of the scale? (Chapter 8)

Touring an Island: What do you need to know about Aloysius' position at a particular moment? (Chapter 22)

Trapped in Thickland: Try regular tetrahedra. (Chapter 22)

Traveling Salesmen: Compare Lavish's kth most costly flight to Frugal's. (Chapter 15)

Truly Even Split: Observe that 16 is a power of 2. (Chapter 15)

Turning the Die: Your position depends on the current sum, and also which numbers are available on the die. (Chapter 21)

Two Balls and a Wall: Plot the velocities of the balls, making use of conservation of energy and momentum. (Chapter 20)

Two Different Distances: You need to be very systematic to get them all! (Chapter 11)

Two Monks on a Mountain: You can assume each path consists of a finite number of alternately ascending and descending line segments. (Chapter 4)

Two-Point Conversion: You may as well assume the Hominids will score again, and that if the game goes to overtime, either team is equally likely to win. (Chapter 10)

Two Round-Robins: Divide the players into the good chess players and the good checkers players. (Chapter 5)

Two Runners: Relate sums and difference of distances covered to meeting time. (Chapter 5)

Two Sheriffs: Have Lew and Ralph agree on a list of partitions of the suspect set into pairs, such that each pair of suspects appears once on the list. (Chapter 13)

Unanimous Hats: What would the prisoners do if they knew the number of red hats was even? (Chapter 2)

Unbreakable Domino Cover: Can you cross a line with just one domino? (Chapter 19)

Curves: Start with the busiest tiles. (Chapter 11)

Unbroken Lines: Do the horizontal and vertical lines first. (Chapter 11)

Uniform Unit Distances: How might you extend a configuration that works for n, so that it could work for $n+1$? (Chapter 15)

Uniformity at the Bakery: Start with integer weights, and make one odd. (Chapter 15)

Uniting the Loops: Look at the more general problem of getting a single loop from tilings involving *some* of the given tiles, with some repetitions. (Chapter 2)

Waiting for Heads: There's a bit more to this than the probability of flipping 5 heads. (Chapter 8)

Watches on the Table: Draw a picture and use geometry. (Chapter 8)

Watermelons: Try some numbers. (Chapter 1)

Whim-Nim: Notice that prior to any declarations, handing your opponent either a Nim-0 position or a bunch of singletons is fatal. (Chapter 17)

Who Won the Series?: What would the answer be if instead you heard that *at most* six games were played? (Chapter 10)

Whose Bullet?: Check your intuition with math. (Chapter 10)

Wild Guess: You'll need representative real sequences for this one. (Chapter 23)

Winning at Wimbledon: How many chances to win the match can you arrange to get? (Chapter 10)

Wins in a Row: Imagine playing four games instead of three. (Chapter 10)

Wires under the Hudson: Link wires in pairs at each end. (Chapter 4)

Worms and Water: You'll need something over Lori that the worms can't crawl underneath. (Chapter 19)

Worst Route: Try it for just four houses; use various addresses, and experiment. (Chapter 4)

Y's in the Plane: Put small rational circles around each Y's endpoints. (Chapter 23)

Zero-sum Vectors: Try to build a sequence of new rows whose partial sums are tightly controlled. (Chapter 12)

Zeroes and Ones: What happens when you subtract from a number another number with the same value modulo n? (Chapter 12)

1. Out for the Count

There's nothing more basic in mathematics than counting—but at the same time, counting mathematical objects can be dauntingly difficult. We'll start with some simple puzzles and gradually introduce some of the special tools that we sometimes need for counting.

Half Grown

At what age is the average child half the height that he or she will be as an adult?

Solution: Most people guess too high, maybe thinking that if full height is reached around age 16, then half-height should be around 8.

There are two problems with that reasoning: (1) the rate of growth for a human being is not constant, and (2) babies have a substantial head start of 20 inches or so (when you prop them up).

The right answer: two years old! (For a girl, it's actually about $2\frac{1}{4}$ years, for a boy $2\frac{1}{2}$.)

Powers of Two

How many people are "two pairs of twins twice"?

Solution: There are four words in the phrase suggesting the number "2" and most people rightly divine that these 2's should be multiplied, not added. So the answer is 16, right?

Not so fast—a twin is only one person, thus a pair of twins, only two. So the correct answer is 8.

Watermelons

Yesterday a thousand pounds of watermelons lay in the watermelon patch. They were 99% water, but overnight they lost moisture to evaporation and now they are only 98% water. How much do they weigh now?

Solution: 500 lb. The watermelons contain 10 pounds of solid matter that now comprises 2% of their final weight; divide 10 pounds by 0.02, and there you are. Apparently, quite a bit of evaporation took place!

Bags of Marbles

You have 15 bags. How many marbles do you need so that you can have a different number of marbles in each bag?

Solution: Counting 0 as a number, you might reasonably deduce that you should put no marbles in the first bag, 1 in the second, 2 in the third, etc., and finally 14 in the last. How many marbles is that?

The quick way to answer that is to observe that the *average* number of marbles in one of your 15 bags is 7. Thus the total number of marbles is $15 \times 7 = 105$.

But there's a trick: You can put bags inside bags! If you put the empty first bag inside the second along with one marble, then the second inside the third along with another marble, etc., you end with the last bag containing all the marbles. So you only need 14 marbles in all.

Salaries and Raises

Wendy, Monica, and Yancey were hired at the beginning of calendar 2020.

Wendy is paid weekly, $500 per week, but gets a $5 raise each week; Monica is paid monthly, $2500 per month, and gets a $50 raise each month; Yancey is paid yearly, $50,000 per year, and gets a $1500 raise each year.

Who will make the most money in 2030?

Solution: Yancey gains $1500 per year, thus he makes $50,000 + 10 × $1500 = $65,000, 10 years later.

Monica gains $50 × 78(!) = 3900 per year, making $30,000 + 10 × $39,000 = $69,000$ at the end of the 10th year (which is when 2030 is about to start). So Monica out-earns Yancey in 2030, without even considering the additional raises she is due that year.

Wait, why 78? Because Monica's monthly raises add up to $50 × (1 + 2 + 3 + \cdots + 12) = \$50 \times 13 \times 12/2 = \$50 \times 78$.

By similar reasoning, figuring 52 weeks in a year, Wendy gains $5 × 53 × 52/2 = 6890 per year, making $26,000 + 10 × $6890 = $94,000$ at end of her 10th year. Yay Wendy!

Here's yet another chance to add up consecutive numbers.

Efficient Pizza-cutting

What's the maximum number of pieces you can get by cutting a (round) pizza with 10 straight cuts?

Solution: There are a number of ways to tackle this one, but perhaps the easiest is to note that the nth cut can at best cross each of the $n-1$ previous cuts, and between each pair of crossings, split a previous piece in two. Since you also get a new piece before the first crossing and after the last one, the cut ends up adding n new pieces.

It follows that with n cuts you can create at most $1 + 2 + \cdots + n = n(n+1)/2$ new pieces, but remember you started with one piece (the whole pizza), so the answer is $1 + n(n+1)/2$ pieces. For 10 slices, that works out to 56 pieces.

Wait, we haven't actually shown that you can *achieve* that many pieces—that would require having every two cuts cross, with never any more than two cuts crossing at the same place. But if we just mark $2n$ random points along the edge of the pizza, number them clockwise (say) and cut from the first to the $n+1$st, then the second to the $n+2$nd, etc., we'll get what we want with probability 1.

You don't like random cuts? I'll leave it to you to come up with a deterministic way to make your 10 (or n) cuts subject to the above constraints.

Attention Paraskevidekatriaphobes

Is the 13th of the month more likely to be a Friday than any other day of the week, or does it just *seem* that way?

Solution: Amazingly, it is true—in the sense that in the 400-year cycle of our Gregorian calendar, more months have their 13th falling on a Friday than on any other day of the week.

It turns out that in 688 out of 4800 months in the 400-year cycle, the 13th falls on a Friday. Sunday and Wednesday claim 687 each, Monday and Tuesday 685 each, and Thursday and Saturday only 684 each. To check this you need to remember that years which are multiples of 100 are not leap years unless (like 2000) they are divisible by 400.

Incidentally, with some practice you can quickly determine the day of the week of any date in history—even accounting for past calendar glitches. For lazier or more present-time-oriented mortals, a useful fact to remember is that in any year 4/4, 6/6, 8/8, 10/10, 9/5, 5/9, 7/11, 11/7, and the last day of February all fall on the same day of the week. (This is even easier to remember if you happen to play craps daily from 9 to 5.) Note that is doesn't matter whether you like to put the month-number before the day-number, or vice-versa.

The late John H. Conway (whose name you will see often in this volume) liked to call that day of the week "doomsday" for that year. Doomsday advances one day each year, two before a leap year. For 2020 doomsday was Saturday, thus Sunday for 2021, Monday for 2022 Monday, Tuesday for 2023 but Thursday for 2024, a leap year.

For the next puzzle, we introduce some basic combinatorics—techniques for the art of counting.

Most readers will know that if you construct something in two steps, and you have n_1 choices for the first step and n_2 for the second, then you have $n_1 \times n_2$ ways to do it. This extends to multiple stages; I call this the "multiplication rule." For example, the number of different license plates you can make that display three letters followed by two digits is $26 \times 26 \times 26 \times 10 \times 10 = 26^3 \times 100$. The number of subsets of an n-element set is 2^n, because you can construct the subset by listing elements of the big set and one by one, deciding whether or not to put that element into the subset.

If you want to count the subsets of a particular size k, you'll need the *division* rule, which says that if you "accidentally" count each object m times, then you have to divide by m at the end. To count the subsets of size k of an n-element set, we first count *ordered* subsets— subsets with a designated first element, second, etc. There are n ways to pick the first element of your ordered subset, $n-1$ to pick the second, down to $n-k+1$ for the last, so $n(n-1)(n-2)\cdots(n-k+1)$ in all. But any subset of size k can be ordered $k(k-1)(k-2)\cdots 1 = k!$ ways, so we've counted each subset that many times. It follows that the number of subsets of size k is really

$$\frac{n(n-1)(n-2)\cdots n-k+1}{k(k-1)(k-2)\cdots 1} = \frac{n!}{k!(n-k)!} .$$

The latter expression is written $\binom{n}{k}$ and usually pronounced "n choose k."

Here's an example where k is just 2.

Meet the Williams Sisters

Some tennis fans get excited when Venus and Serena Williams meet in a tournament. The likelihood of that happening normally depends on seeding and talent, so let's instead construct an idealized elimination tournament of 64 players, each as likely to win as to lose any given match, with bracketing chosen uniformly at random. What is the probability that the Williams sisters wind up playing each other?

Solution: This looks complicated and in fact, in Frederick Mosteller's delightful book *Fifty Challenging Problems in Probability*, the solution (of an equivalent problem) is obtained by working it out for 2^k players when k is small, guessing the general answer, and then proving the guess to be correct by induction on k.

But there's an easy way to think of it: There are $\binom{64}{2} = (63 \times 64)/2$ pairs of players, out of which 63 will meet (since 63 players need to be eliminated to produce the winner), so the probability that a particular pair meets must be $63/((63 \times 64)/2) = 1/32$.

Rating the Horses

You have 25 horses and can race them in groups of five, but having no stopwatch, you can only observe the order of finish. How many heats of five horses do you need to determine the fastest three of your 25?

Solution: Hmm. It's clear that all the horses need to be run, since an untested horse could easily be one of the fastest three. You could do that in five heats but that would not clinch the deal, so you'll need at least six heats for sure.

Well, suppose you do partition the horses into five heats; you'd then need to test the winners against one another. What will you know then?

Suppose (without loss of generality) that the winner of your sixth heat came from Heat 1, the runner-up from Heat 2, and the placer from Heat 3. Of course the fastest horse overall is the winner of Heat 6, but the second-fastest could either be the runner-up of Heat 6 or the runner-up of Heat 1. The third-fastest horse could be the placer in Heat 6, the runner-up or placer in Heat 1, or the runner-up in Heat 2.

So the only horses "in the running" (sorry about that) for second and third place are the runner-up and placer from Heat 1, the winner and runner-up from Heat 2, and the winner from Heat 3. That's five horses; run then in Heat 7, and you're done. It's not hard to see you can't beat this elegant solution.

Shoelaces at the Airport

You are walking to your gate at O'Hare and need to tie your shoelace. Up ahead is a moving walkway that you plan to utilize. To minimize your time to get to the gate, should you stop and tie your shoelace now, or wait to tie it on the walkway?

Solution: You should tie your shoelace on the walkway. Either way, you spend the same amount of time walking on solid ground, and the same amount of time tying your shoelace. But if you tie your shoelaces on solid ground, you spend more time *walking* on the walkway.

Rows and Columns

Prove that if you sort each row of a matrix, then each column, the rows are still sorted!

Solution: To prove this rather surprising assertion, let us imagine that the matrix has m rows and n columns, and that after each row is sorted (smallest values to the left, say), the entry on the ith row and jth column is a_{ij}; after the columns are sorted, it is b_{ij}. We need to show that if $j < k$ then $b_{ij} \leq b_{ik}$.

This is one of those things which alternates between being mysterious and obvious, each time you think about it. One way to argue the point is to note that b_{ik} is the ith smallest entry in the old column $\{a_{1k}, a_{2k}, \ldots, a_{mk}\}$. For every entry $a_{i'k}$ which ends up below a_{ik}, the entry $a_{i'j}$ that was on the same row but in the jth column is no larger; thus, counting a_{ij} as well, there are *at least* i entries from the old jth column that are no bigger than b_{ik}. Thus, the ith smallest entry in the old jth column (namely, b_{ij}) is itself no bigger than b_{ik} and we are done.

Is this more convincing than just trying an example? You be the judge.

Three-way Election

Alison, Bonnie, and Clyde run for class president and finish in a three-way tie. To break it, they solicit their fellow students' second choices, but again there is a three-way tie. The election committee is stymied until Alison steps forward and points out that, since the number of voters is odd, they can make two-way decisions. She therefore proposes that the students choose between Bonnie and Clyde, and then the winner would face Alison in a runoff.

Bonnie complains that this is unfair because it gives Alison a better chance to win than either of the other two candidates. Is Bonnie right?

Solution: Bonnie is correct—in fact, she has understated the case; assuming no voters change their minds, Alison will win for sure! To see this, suppose Alison's supporters mostly prefer Bonnie to Clyde (so that Bonnie would beat Clyde in the proposed two-candidate race). Then Bonnie's supporters must prefer Clyde to Alison, otherwise Clyde would have garnered fewer than 1/3 of the second-place votes; similarly Clyde's supporters prefer Alison to Bonnie. Thus, in this case, Alison will beat Bonnie in the runoff.

If Alison's supporters prefer Clyde to Bonnie, a symmetric argument shows that Alison will beat Clyde in the runoff. ♡

This puzzle serves as a warning: There may be more to some tiebreakers than meets the eye!

Let us return to the basic multiplication rule, but with some additional cleverness required.

Sequencing the Digits

In how many ways can you write the numbers 0 through 9 in a row, such that each digit other than the leftmost is within one of some digit to the left of it?

Solution: To apply the multiplication rule when you build something in stages, you need the number of choices in a stage not to depend on choices made in previous stages. That seems not to be the case here; for example, if you start your sequence with 8 you have two choices for the next digit (7 or 9) but if you start with 9, you have only one choice for the next digit (8) and in fact now the whole sequence is determined.

You can get around this with some fanciness involving binomial coefficients (those $\binom{n}{k}$ guys mentioned above), but there's a nicer way: Create the sequence in reverse! The last digit must be 0 or 9; if it's 9, the next-to-last must be 0 or 8, while if it's 0 the next-to-last must be 1 or 9. No matter what you do there are two choices at every stage until you reach the leftmost digit, so the number of sequences is $2^9 = 512$.

Faulty Combination Lock

A combination lock with three dials, each numbered 1 through 8, is defective in that you only need to get two of the numbers right to open the lock.

What is the minimum number of (three-number) combinations you need to try in order to be sure of opening the lock?

Solution: Often the easiest way to tackle a problem of this sort is geometrically. The space of all possible combinations is an $8 \times 8 \times 8$ combinatorial cube, and each time you try a combination, you cover all combinations on the three orthogonal lines which intersect at that point.

Once you see the problem this way, you are likely to discover that the best way to cover all the points in the cube is to concentrate all your test points in just two of the eight $4 \times 4 \times 4$ octants. You will then arrive at a solution equivalent to the one described below.

Try all combinations with numbers from $\{1, 2, 3, 4\}$ whose sum is a multiple of 4; there are 16 of those, since if you pick the numbers on the first two (or any two) of the dials, the number on the third is determined. Now try all the combinations you get by adding $(4,4,4)$, that is, by adding 4 to each of the three numbers; there are 16 more of those, and we claim that together the 32 choices cover all possibilities.

It is easy to see that this works. The correct combination must have either two (or more) values in the set $\{1, 2, 3, 4\}$, or two or more values in the set $\{5, 6, 7, 8\}$. If the former is the case, there is a unique value for the third dial (the one whose number may not be among $\{1, 2, 3, 4\}$) such that the three were among the first 16 tested combinations. The other case is similar.

To see that we can't cover with 31 or fewer test-combinations, suppose that S is a cover and $|S| = 31$. Let $S_i = \{(x, y, z) \in S : z = i\}$ be the ith level of S.

Let A be the set $\{1, 2, 3\}$, $B = \{4, 5, 6, 7, 8\}$, and $C = \{2, 3, 4, 5, 6, 7, 8\}$. At least one level of S must contain 3 or fewer points; we might as well assume S_1 is this level and $|S_1| = 3$. (If $|S_1| \leq 2$, there's an easy contradiction.) The points of S_1 must lie in a $3 \times 3 \times 1$ subcube; we may assume that they lie in $A \times A \times \{1\}$.

The 25 points in $B \times B \times \{1\}$ must be covered by points not in S_1. No two of them can be covered by the same point in S, thus, $S \setminus S_1$ (that is, the elements of S that are not in S_1) has a subset T

of size 25 that lies in the subcube $B \times B \times C$. Now consider the set $P = \{(x, y, z) : z \in C, (x, y, 1) \notin S_0, (x, y) \notin B \times B\}$. A quick count shows that $|P| = (64 - 3 - 25) \times 7 = 252$. The points in P are not covered by S_1, and each point in T can cover at most $3 + 3 = 6$ points in P. Therefore there are at least $252 - (6 \times 25) = 102$ points in P that must be covered by points in $S \setminus S_1 \setminus T$.

However, $|S \setminus S_1 \setminus T| = 31 - 3 - 25 = 3$, and each point in the cube covers exactly 22 points. Since $22 \times 3 = 66 < 102$, we have our contradiction. ♡

Losing at Dice

Visiting Las Vegas, you are offered the following game. Six dice are to be rolled, and the number of *different* numbers that appear will be counted. That could be any number from one to six, of course, but they are not equally likely.

If you get the number "four" this way you win $1, otherwise you lose $1. You decide you like this game, and plan to play it repeatedly until the $100 you came with is gone.

At one minute per game, how long, on the average, will it take before you are wiped out?

Solution: This is a trick, of course; otherwise it would be in Chapter 10, *In All Probability*, not this one. On the average, it'll take forever for you to be wiped out—the game is in your favor, and thus you may never even dip below your $10 stake.

There are $6^6 = 46{,}656$ ways to roll the dice. For four different numbers to appear, you need either the pattern AABBCD or AAABCD. There are

$$\binom{6}{2} \cdot \binom{4}{2} / 2 = 45$$

versions of the former pattern, keeping the equinumerous labels in alphabetical order; e.g., AABBCD, ABABCD, ACDABB, but not BBAACD or AABBDC.

For the latter pattern, there are $\binom{6}{3} = 20$ versions.

In either case, there are $6 \times 5 \times 4 \times 3 = 360$ ways to assign numbers to the letters, for a total of $360 \times 65 = 23{,}400$ rolls. Thus, the probability of winning is $23400/46656 = 50.154321\%$. ♡

If you win some bets with this game, don't forget to send 5% of your profits to me c/o Taylor & Francis.

Splitting the Stacks

Your job is to separate a stack of n objects into individual items. At any time you may split a stack into two stacks, and get paid the product of the sizes of the two stacks.

For example, for $n = 8$, you could split into a stack of size 5 and a stack of size 3, getting paid $15; then you can split the 5-stack into 2 and 3 to get $6 more. Then the two threes can each be split ($2 each) and the remaining three twos ($1 each) for a total income of $15 + 6 + (2 \times 2) + (3 \times 1) = 28$ dollars.

What's the most money you can get starting with n objects, and how can you get it?

Solution: Try a few other ways to deal with eight objects, and you will find that they all earn the same $28. Does it really not matter how you split the stacks?

On the web, this problem has been used to demonstrate something called "strong induction." But there's an easy way to see that the outcome is $\binom{n}{2}$ dollars no matter what you do: Each (unordered) pair of objects earns $1!

Imagine that there is a string connecting every pair of objects, and when you split a stack, you cut every string between pairs of objects that straddle the split. If there are i objects on one side of the split and j on the other, you cut ij strings and get paid $\$ij$. Eventually you cut every string and get paid $1 for each one, so you earn $\binom{n}{2}$ dollars regardless of strategy.

North by Northwest

If you've never seen the famous Alfred Hitchcock movie *North by Northwest* (1959), you should. But what direction is that, exactly? Assume North is $0°$, East $90°$, etc.

Solution: Sounds like it should be halfway between north ($0°$) and northwest ($315°$), thus $337.5°$, but that's actually "north–northwest," not "north *by* northwest." The preposition "by" moves the needle just $11.25°$ in the direction of its object; thus "north by northwest" puts you at $(360 - 11.25)° = 348.75°$. Northwest by north, on the other hand, is $(315 + 11.25)° = 326.25°$. To make things even more confusing, "north by northwest" is commonly designated more simply as "north by west." So the movie title was not actually a phrase in use; in fact the movie's original title was "In a Northwesterly Direction."

Doesn't roll off the tongue quite as well, though, does it?

The next puzzle is, really, just a bit of arithmetic.

Early Commuter

A commuter arrives at her home station an hour early and walks toward home until she meets her husband driving to pick her up at the normal time. She ends up home 20 minutes earlier than usual. How long did she walk?

Solution: This kind of timing problem can drive you nuts, until you think about it the "right way." Here, you notice that the commuter's husband apparently saved 10 minutes each way, relative to normal, so must have picked up his wife 10 minutes earlier than usual. So she walked for 50 minutes.

To solve the next puzzle, we return to those binomial coefficients.

Alternate Connection

One hundred points lie on a circle. Alice and Bob take turns connecting pairs of points by a line, until every point has at least one connection. The last one to play wins; which player has a winning strategy?

Solution: The player who connects the 98th point is doomed, since there will be either one or two unconnected points left after her move, which her opponent can then deal with. There are $\binom{97}{2}$ moves that can be made before the 98th needs to be connected, and since this number is even, the second player can always win.

Bindweed and Honeysuckle

Two vines, a bindweed and a honeysuckle, climb a tree trunk, starting and ending at the same place. The bindweed circles the trunk three times counter-clockwise while the honeysuckle circles five times clockwise. Not counting the top and bottom, how many times do they cross?

Solution: By mentally twisting the tree trunk with the vines still on it, we see that the answer is the same as it would be if the bindweed went straight up while the honeysuckle went around $5 + 3 = 8$ times.

Therefore there are 9 meetings in total, and the answer, with the top and bottom not counted, is 7.

Do you remember phone trees? Suppose you belonged to an organization that needed a method of spreading information by phone. It was once common in such a case to set up a *phone tree*. The idea is this: As a member, you will have a special list of certain other members (at least one), and everyone on your list will have your name on their list. If you gain possession of outside information of interest to the other members, you call the members on your list. If, instead, you get information from another member, you pass it on to all the *other* members on your list.

For the phone tree to work, it must be that there is a path between any two members; in other words, if any member gets information, then eventually the whole organization must hear it. If we think of each member as a point, with a line between two points when the corresponding members are on each others' phone lists, then the collection of all points and lines must form a *connected graph*. (There'll be more about graphs in Chapter 4, *Graphography*.)

Why is it called a phone tree and not a phone graph? Because if it is efficiently constructed, there are only $n-1$ lines needed in the graph—equivalently, members will never get their own information relayed back to them.

Here is an example of a phone tree for a 10-member organization whose names happen to be the numbers from 0 to 9.

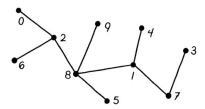

The question we address is: How many ways are there to set up a phone tree for an organization with n members? This question was answered by German mathematician Carl Wilhelm Borchardt in 1860, the same year Johann Philipp Reis built (arguably) the first telephone prototype.

Borchardt's contribution has been mostly forgotten and the name of an English mathematician, Arthur Cayley, is associated with the formula to come below.

Before we state and prove "Cayley's formula," roughly how many phone trees might you *guess* would be possible if for a 10-member organization? Hundreds? Thousands?

The answer is a hundred million. Exactly!

Theorem. *The number of trees on a set of n labeled points is n^{n-2}.*

Proof. We will prove the formula by showing that there is a one-to-one correspondence between the set of trees on points $\{0, 1, 2, \ldots, n-1\}$, and the set of sequences of length $n-2$ of the numbers $\{0, 1, 2, \ldots, n-1\}$.

Since there are n ways to choose each entry of the sequence, we know from the multiplication rule above that there are n^{n-2} such sequences. Thus, if we can establish the aforementioned one-to-one correspondence, we are done.

A *leaf* of a tree is a point (better, a "vertex") of the tree that's attached to only one other vertex. To go from tree to sequence, we find the lowest-numbered leaf and start the sequence not with the label of that leaf, but with the label of its one neighbor.

Then that leaf is erased and the procedure is repeated to determine the next entry in the sequence, and this continues until there are only two vertices left in the tree. That's it!

Let's try it with the tree shown above. We get the following chain of ever-longer sequences and ever-smaller trees; in each tree, all leaves are circled and the vertex attached to the current lowest-numbers leaf is boxed.

To show that this is a one-to-one correspondence, we need to get from any sequence to the unique tree that would yield that sequence. It turns out, that's quite easy and logical; let's do it for the sequence we just got, and check that we get the same tree back.

Note that the leaves of a tree will not themselves appear in that tree's sequence. Thus, given a sequence, we look for the lowest number that doesn't appear, and start building our tree by attaching that vertex to the vertex corresponding to the leftmost number in the sequence.

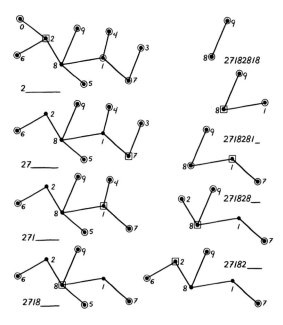

Then we erase that leftmost number and repeat, now looking for the smallest number missing from the new, shorter sequence that has not already been used as a leaf. This continues until the sequence is gone, and then we connect the two vertices that were never used as leaves to complete the tree.

We will not formally prove here that the two algorithms described above are inverses of each other, but hopefully the example is convincing. ♡

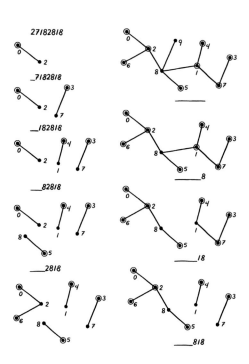

27182818

_7182818

_182818

_82818

_2818

_____

_____8

_____18

_____818

2. Achieving Parity

To a mathematician, "parity" often means just odd versus even. Typically, the concept of parity is not mentioned in a puzzle; but when you start to play with the puzzle, the concepts of odd and even emerge. And when they do, you are often half the way to solving it.

Here's a simple example.

Bacterial Reproduction

When two pixo-bacteria mate, a new bacterium results; if the parents are of different sexes the child is female, otherwise it is male. When food is scarce, matings are random and the parents die when the child is born.

It follows that if food remains scarce a colony of pixo-bacteria will eventually reduce to a single bacterium. If the colony originally had 10 males and 15 females, what is the probability that the ultimate pixo-bacterium will be female?

Solution: Here a bit of experimentation might lead to the observation that the parity of the number of females cannot change—in other words, the number of females must remain odd.

Thus the probability that the last pixo-bacterium will be female is 1.

For the next puzzle, you have to come up with the parity idea on your own.

Fourth Corner

Pegs occupy three corners of a square. At any time a peg can jump over another peg, landing an equal distance on the other side. Jumped pegs are not removed. Can you get a peg onto the fourth corner of the square?

Solution: The first thing to do here is think of the initial square as a cell of the plane grid, for example, the points (0,0), (0,1), (1,0), and (1,1) on the XY-plane. Then the pegs will always be on grid points.

Grid points, however, have four possible parities: Each coordinate can be even or odd. When a peg jumps, its parity is preserved; its X-coordinate goes up by even number (2, 0, or -2), and likewise its Y-coordinate.

The points of the unit cell above have all four parities among them, so the corner that starts without a peg can never be occupied.

Unanimous Hats

100 prisoners are offered the chance to be freed if they can win the following game. In the dark, each will be fitted with a red or black hat according to a fair coinflip. When the lights are turned on, each will see the others' hat colors but not his own; no communication between prisoners will be permitted.

Each prisoner will be asked to write down his guess at his own hat color, and the prisoners will be freed if they all get it right.

The prisoners have a chance to conspire beforehand. Can you come up with a strategy for them that will maximize their probability of winning?

Solution: The probability that Bob (a specified one of the prisoners) guesses correctly is exactly $\frac{1}{2}$, no matter what the general strategy is, so the probability that they *all* get it right cannot exceed $\frac{1}{2}$. But, amazingly, they can achieve $\frac{1}{2}$.

The key is: Suppose Bob knows there are an even number of red hats. Then he is home free, because if he sees an even number of red hats on his fellow prisoners' heads, he knows his hat is black; if he sees an odd number of red hats, he knows his hat is red.

So the prisoners simply agree to all *guess* that the number of red hats will be even. Either they're all right, or they're all wrong; and the probability that the number of red hats is even is exactly $\frac{1}{2}$. (That's true even if only *one* hat is determined by coinflip, since no matter what the other hat colors are, that hat color determines the parity.)

Once you've got that puzzle, the next one is easy.

Half-Right Hats

One hundred prisoners are offered the chance to be freed if they can win the following game. In the dark, each will be fitted with a red or black hat according to a fair coinflip. When the lights are turned on, each will see the others' hat colors but not his own; no communication between prisoners will be permitted.

Each prisoner will be asked to write down his guess at his own hat color, and the prisoners will be freed if at least half get it right.

The prisoners have a chance to conspire beforehand. Can you come up with a strategy for them that will maximize their probability of winning?

Solution: Yes, they just arrange for 50 prisoners to guess their color by assuming the total number of red hats is even, while the other 50 assume the number of red hats is odd. Then one group or the other is bound to be right, assuming no poor fellow screws up!

Red and Blue Hats in a Line

There are n prisoners this time, and again each will be fitted with a red or black hat according to flips of a fair coin. The prisoners are to be arranged in a line, so that each prisoner can see only the colors of the hats in front of him. Each prisoner must guess the color of his own hat, and is executed if he is wrong; however, the guesses are made sequentially, from the back of the line toward the front. Thus, for example, the ith prisoner in line sees the hat colors of prisoners $1, 2, \ldots, i-1$ and hears the guesses of prisoners $n, n-1, \ldots, i+1$ (but he isn't told which of those guesses were correct—the executions take place later).

The prisoners have a chance to collaborate beforehand on a strategy, with the object of guaranteeing as many survivors as possible. How many prisoners can be saved in the worst case?

Solution: It's clear that no more than $n-1$ prisoners can be sure of surviving, because the first one to guess—namely, the last in line—has no clue. But he can pass a clue to the prisoner in front of him, by letting him know whether the number of red hats among prisoners 1 through $n-1$ is even or odd. How does he communicate? By his guess, of course. The prisoners can agree, for example, that the last prisoner in line guesses "red" if he sees an odd number of red hats, "black" otherwise.

So that saves the life of prisoner $n-1$; what about the rest? They're saved too! Everyone has heard the last prisoner, and knows that after him, all guesses will be correct. For example, if the ith prisoner to guess hears that there are an even number of reds among prisoners 1 through $n-1$, and after the first guess he hears 5 "red" guesses, he concludes that the number of red hats from himself forward is odd. If he actually sees an even number of red hats in front of him, he knows that his own hat must be red and guesses accordingly.

The next one is a bit trickier. Luckily, the prisoners are getting pretty good with these games.

Prisoners and Gloves

Each of 100 prisoners gets a different real number written on his forehead, a black glove, and a white glove. Seeing the numbers on the other prisoners' foreheads, he must put one glove on each hand in such a way that when the prisoners are later lined up in real-number order and hold hands, the gloves match. How can the prisoners, who are permitted to conspire beforehand, ensure success?

Solution: This seems impossible: Each prisoner sees 99 different real numbers and has no idea where the number on his own forehead fits in, so how does he know what to do? What information has he got?

The parity here is of a more subtle sort: It's the parity of a *permutation*. Permutations are classified as "even" or "odd" according to the parity of the number of swaps needed to sort it. If we write a permutation in "one-line form," so that $(p_1, p_2, \ldots, p_n)$ represents the permutation that sends i to p_i, then (1,2,3,4,5) is even since it is already the identity (requiring 0 swaps), but (4,2,3,1,5)—which can be sorted by swapping "1" with "4"—is odd.

Assume the prisoners themselves are numbered from 1 to 100 (e.g., alphabetically). Then the numbers on the prisoners' foreheads induce a permutation of the numbers from 1 to 100: If the ith prisoner's forehead shows the jth smallest real number, i is mapped to j by the permutation.

For example, suppose the prisoners are Able, Baker, Charlie, and Dog, numbered 1 to 4 in that order. Say their numbers, respectively, are 2.4, 1.3, 6.89, and π. Then the permutation is (2,1,4,3) where the "2" at the front represents the fact that the first prisoner got the second-lowest number, etc.

This permutation happens to have even parity, since interchanging the 2 and the 1, then the 4 and the 3, restores it to the identity (1,2,3,4).

Of course, Charlie cannot see the 6.89 on his own forehead so he doesn't know the parity of the permutation. But suppose he *assumes* that his own number is the lowest of all; then he can compute the parity. He'll be right if his number is the lowest, or the third lowest, or the fifth lowest, and so forth; wrong if his number happens to be the ith lowest, i being an even number.

Let's have Charlie put the white glove on his left hand (and black on right) if his assumption leads him to an even permutation, and the reverse if he gets an odd permutation. (In the example, Charlie's assumption leads him to the permutation (3,2,1,4) which is odd, so he puts the black glove on his left hand.)

That works! If the permutation actually *is* even, then the prisoner with lowest forehead number would have computed "even" and would have a white glove on his left hand. The next-lowest-forehead-numbered prisoner would have computed "odd" and his black-gloved left hand would end up holding the first guy's black-gloved right hand, and similarly down the line. If the actual permutation is odd, the glove colors will be the complement of these, and will still work.

Even-sum Billiards

You pick 10 times, with replacement (necessarily!), from an urn containing nine billiard balls numbered 1 through 9. What is the probability that the sum of the numbers of the balls you picked is even?

Solution: If there were no ball number 9, an even sum would have probability exactly $\frac{1}{2}$ since no matter what happens in your first 9 picks, the parity of the tenth ball will determine whether the sum is even or odd; and that ball is equally likely to be either.

Now put ball 9 back in. No matter what order you choose the balls in, you can ask your assistant to reveal the numbers to you with all the 9's first. Then the above argument still applies: The parity of the last ball shown you, whose number is between 1 and 8, will determine the parity of the sum; so again, the sum is equally likely to be even or odd.

Oops, wait, there's just one possibility we haven't considered: that you got the 9 ball every time! That has probability $(1/9)^{10}$ and results in an even sum. So altogether, the probability of an even sum is

$$\frac{1}{2}\left(1-\frac{1}{9}\right)^{10} + \left(\frac{1}{9}\right)^{10} = \frac{1}{2} + \frac{1}{2}\left(\frac{1}{9}\right)^{10}$$

which is about 0.50000000014. Probably not enough bias to be worth betting on!

Of course, "even" means that the remainder after dividing by 2 is 0; "odd," that the remainer is 1. For the next puzzle, you'll need to stretch your definition of parity to the remainder after dividing by 3; and for the puzzle after that, after dividing by 9.

Chameleons

A colony of chameleons currently contains 20 red, 18 blue, and 16 green individuals. When two chameleons of different colors meet, each of them changes his or her color to the third color. Is it possible that, after a while, all the chameleons have the same color?

Solution: The key thing is to observe that, after each encounter of two chameleons, the difference between the number of individuals of any two colors remains the same modulo 3. In symbols, letting N_R stand for the number of red chameleons and N_B and N_G similarly, we claim that, for example, $N_R - N_B$ has the same remainder after division by 3, after any two chameleons meet, that it did before. This is easy to verify by checking cases. Thus these differences remain the same modulo 3 forever, and since in the given colony none of those differences is equal to zero modulo 3, we can never get two of the color populations to be zero.

If, on the other hand, the difference of two color populations (say, $N_R - N_B$) is a positive multiple of 3, we can lower that difference by having a red chameleon meet with a green one (or if there are no green ones, by first having a red one meet a blue one). We repeat until $N_R = N_B$ and then have reds meet blues until only green chameleons remain.

Putting all this together, and noting that if two differences are multiples of 3 then the third must be as well, we see that:

- If all three differences are multiples of 3, then any color can take over the colony;

- If just one of the differences is a multiple of 3, then the *remaining* color is the only one which can take over the colony; and finally,

- If none of the differences are multiples of 3, as in the given problem, the colony can never become monochromatic and will remain fluid until other circumstances (e.g., birth, death) intervene.

Now for a quickie.

Missing Digit

The number 2^{29} has 9 digits, all different; which digit is missing?

Solution: What to do? You can type "2^29" into your computer or calculator and look at the number for yourself, but is there a way to do it in your head without getting a headache?

Hmm ... maybe you remember a technique from grade school called "casting out nines," where you keep adding up digits and always end up with a number's value modulo 9 (that is, its remainder after dividing by 9). It uses the fact that $10 \equiv 1 \mod 9$, thus $10^n \equiv 1^n \equiv 1 \mod 9$ for every n. If we denote by x^* the sum of the digits of the number x, then we get $(xy)^* \equiv x^* y^* \mod 9$ for every x and y.

It follows in particular that $(2^n)^* \equiv 2^n \mod 9$. The powers of 2 mod 9 begin 2, 4, 8, 7, 5, 1, and then repeat; since $29 \equiv 5 \mod 6$, $2^n \mod 9 \equiv$ the fifth number in this series, which happens to be 5.

Now the sum of *all* the digits is $10 \times 4.5 = 45 \equiv 0 \mod 9$, so the missing digit must be 4. Indeed, $2^{29} = 536{,}870{,}912$.

Subtracting Around the Corner

A sequence of n positive numbers is written on a piece of paper. In one "operation," a new sequence is written below the old one according to the following rules: The (absolute) difference between each number and its successor is written below that number; and the absolute difference between the first and last numbers is written below the last number. For example, the sequence 4 13 9 6 is followed by 9 4 3 2.

Try this for $n = 4$ using random numbers between, say, 1 and 100. You'll find that after remarkably few operations, the sequence degenerates to 0 0 0 0 where it of course remains. Why? Is the same true for $n = 5$?

Solution: Considering values modulo 2 solves both problems. In the $n = 4$ case, up to rotation and reflection, 1 0 0 0 and 1 1 1 0 become 1 1 0 0, then 1 0 1 0, then 1 1 1 1, then 0 0 0 0. Since this covers all cases, we see that when working with ordinary integers, at most four steps are required to make all the numbers even; at that point, we may as well divide out by the largest common power of two before proceeding. Since the value M of the largest number in the sequence can never increase, and drops by a factor of 2 or more at least once every four steps, the sequence must hit 0 0 0 0 after at most $4(1 + \lceil \log_2 M \rceil)$ steps. (The symbol $\lceil x \rceil$ stands for the "ceiling" of x, that is, the least integer greater than or equal to x.)

On the other hand, for $n = 5$, the sequence 1 1 0 0 0 (considered either as binary or ordinary numbers) cycles via 1 0 1 0 0, 1 1 1 1 0, 1 1 0 0 0. ♡

A little analysis, using polynomials over the integers modulo 2, shows that the salient issue is whether n is a power of 2.

For our next puzzle, we return to even versus odd, but now you'll need some keen powers of observation to see how to apply this idea.

Uniting the Loops

Can you re-arrange the 16 square tiles below into a 4×4 square—no rotation allowed!—in such a way that the curves form a single loop? Lines that exit from the bottom are assumed to continue at the same point on the top, and similarly right to left; in other words, imagine that the square is rolled up into a doughnut.

Solution: This puzzle (along with two others in the book) was inspired by the work of artist Sol LeWitt, who loved to construct paintings (and sculptures) from pieces that consisted of every possible way to do some combinatorial task.

In the case of the present puzzle, examination of the tiles shows that each side of a tile has two entry/exit points. The four points belonging to two adjacent sides are paired up in one of two ways: with the two points closest to the common corner paired, and the other two paired (each by a quarter of a circle); or by pairing each point close to the corner to the one farther from the corner on the other side (using two crossing quarter-ellipses).

Thus, for each corner there's a decision to be made: to cross, or not to cross. These four binary decisions result exactly in the 16 different tiles you see.

Since when we wrap around the doughnut, none of the curved lines terminate, the curves are forced to form loops. The question is whether we can arrange the tiles to form just one loop.

As with many puzzles, our initial strategy is just to try various arrangements and count the loops. You will find that no matter how you do it, there always seem to be an even number of loops. If that holds in general, then, of course, we can never make a single loop. But why should the number of loops always be even?

One natural way to try to prove it might be to show that if we switch two adjacent tiles, the number of loops always changes by an even number. Alas, there are a lot of curved lines that enter a pair of adjacent tiles (12 in all) and it's awfully hard to deal with all the different things that can happen when two adjacent tiles are switched.

We need to consider some much smaller change. How about crossing or uncrossing one pair of curves on a tile? Of course, if we do that, it would effectively destroy one tile and duplicate another. We would then be in a different universe where we have as many as we like of each tile, and can choose any 16 tiles to arrange into a square.

A little experimentation shows that if we take any configuration involving the 16 original tiles—for example, the one above—and cross or uncross curved lines on one side of one tile, the number of loops always goes up or down by one. If that's universally true, we arrive at a conjecture: If the total number of crossings in the picture is even, the number of loops will be even; if it is odd, the number of loops will be odd. More experimentation seems to confirm this; can we prove it?

The figure below shows the three things that can happen when you make or undo a crossing. The first two create or destroy one loop, thus they add or subtract one from the loop total. The third is impossible; if a loop is oriented (say, clockwise), and the tiles are colored like a checkerboard, then a loop alternately enters gray tiles vertically and white tiles horizontally, or vice-versa. The loops shown violate this condition.

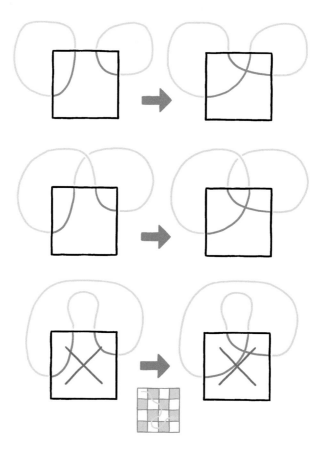

Now, if you start with the first figure, which has 16 crossings and four loops, you can create or delete crossings to get any desired configuration. But every time you create or delete a crossing, the number of crossings and the number of loops remain the same parity—both odd, or both even. Eventually when you get to any arrangement of the 16 original tiles, you are back to an even number of loops. ♡

Our theorem was an open problem for some time, until parity was unleashed on it. It concerns partitions of an abstract combinatorial construction.

Fix a positive number n. A *box* is a Cartesian product of n finite sets; if the sets are $A_1, \ldots, A_n$, then the box consists of all sequences $(a_1, \ldots, a_n)$ such that $a_i \in A_i$ for each i.

A box $B = B_1 \times \cdots \times B_n$ is a *proper sub-box* of $A = A_1 \times \cdots \times A_n$ if B_i is a proper nonempty subset of A_i for each i.

If each factor A_i has at least two elements, so that each factor A_i can be partitioned into two non-empty sets, then by taking every product of one set from each A_i we could partition the box A into 2^n proper sub-boxes. However,

Theorem. *A box can never be partitioned into fewer than 2^n proper sub-boxes.*

Proof. Call a sub-box *odd* if it has an odd number of elements, and let $\mathbb{O}$ be the set of all odd sub-boxes of the big box A, and if B is a sub-box of A, let $\mathbb{O}_B$ consist of the elements of $\mathbb{O}$ that intersect B in an odd set.

If $B = B_1 \times \cdots \times B_n$ then in order for a set C to intersect B in an odd number of elements, it must intersect every factor B_i in an odd number of elements (because $|C \cap B| = |C \cap B_1| \times |C \cap B_2| \times \cdots \times |C \cap B_n|$).

But the probability that a *random* odd subset C_i of A_i intersects B_i in an odd number of elements is exactly $\frac{1}{2}$. Why? Because we can flip a coin to determine whether each $a \in A_i$ is in C_i or not, ending with some element a' that's *not* in A_i (remember, B_i is supposed to be a *proper* subset of A_i). This last element a' is put into C_i or not so as to assure C_i is odd. The elements of B_i that we flipped for constitute a random subset of B_I, and since exactly half the subsets of any nonempty finite set are odd, the probability that $B_i \cap C_i$ is odd is $\frac{1}{2}$ as claimed.

Each $C \in \mathbb{O}$ is partitioned by the B's, and since C is odd, at least one of the parts of the partition must be odd. If there are m parts of the partition, it follows that the probability that $C \in \mathbb{O}_B$ is at least $1/m$. But we just showed that probability is $1/2^n$, so $m \geq 2^n$! $\heartsuit$

3. Intermediate Math

The *Intermediate Value Theorem* (IVT) states that if a real number changes continuously as it moves from a to b, then it must pass through every value between a and b. This simple and intuitive fact—arguably, more of a property of the real numbers than a theorem—is a powerful tool in mathematics, and shows up often in puzzle solutions.

A useful variation of IVT addresses the case when one real number is moving continuously from a to b while another goes from b to a; they must cross. For example:

Squaring the Mountain State

Can West Virginia be inscribed in a square?

Solution: We are referring here to the *shape* of West Virginia as projected (any way you like) onto a plane. We begin by drawing a vertical line west of the state and moving it east until it just touches the state boundary. Then we draw another vertical line east of the state, moving it *west* until it too touches the state boundary. Similar operations with two horizontal lines inscribe the state in a rectangle whose width w is somewhat greater than its height h.

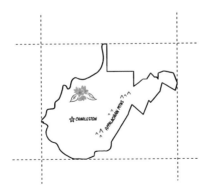

To get a square, we rotate the rectangle continuously, maintaining its 90° angles while keeping all four sides tangent to the state. After 90° of rotation, the figure is the same as it was at the start, but now w has become h and h has become w. It follows that at some angle, h and w were equal; and at that point, the state was inscribed in a square. ♡

Of course, this works for any state, and indeed for any bounded shape in the plane.

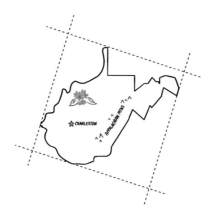

The same technique has a practical application: You can always rotate your square four-legged picnic table so as to have all four legs in contact with the ground, assuming the spot you chose doesn't have any actual surface discontinuities. Just put the table down with three legs touching and rotate until the fourth touches; this is guaranteed to happen before you get to 90°, since when you *do* get to 90°, a ground-contacting leg will be where your fourth leg began.

You might find the following classic puzzle even simpler.

Monk on a Mountain

A monk begins an ascent of Mt. Fuji on Monday morning, reaching the summit by nightfall. He spends the night at the summit and starts down the mountain on the same path the following morning, reaching the bottom by dusk on Tuesday.

Prove that at some precise time of day, the monk was at exactly the same spot on the path on Tuesday as he was on Monday.

Solution: If we plot the monk's position as a function of time of day, we get two curves connecting opposing corners, and the conclusion follows from IVT. But here's another way to think of it, underscoring the intuitiveness of IVT: Superimpose Monday and Tuesday, so now the Monk climbs at the same time that his *doppelgänger* descends. They must meet!

Cutting the Necklace

Two thieves steal a necklace consisting of 10 red rubies and 14 pink diamonds, fixed in some arbitrary order on a loop of golden string. Show that they can cut the necklace in two places so that when each thief takes one of the resulting pieces, he gets half the rubies and half the diamonds.

Solution: We could check all possible necklaces meeting the above description, but there are more than 40,000 of them, and, anyway, we'd like to solve the problem for any (even) number of rubies and of diamonds.

To apply the IVT requires a bit of subtlety; we first have to set up the problem geometrically, then make a couple of key observations.

The idea is to represent the necklace as a circle cut into 24 equal arcs, each colored red or pink according to the pattern of rubies and diamonds in the necklace. Then we draw a line through the center of the circle, cutting the circle twice.

Our line will certainly cut the full set of gems into equal portions, since any diameter cuts the circle into equal-length arcs (i.e., semicircles). Thus, if a diameter happens to cut the rubies in half, it will cut the diamonds in half as well.

The line pictured in the figure above does not work; it's got more red on one side than the other. But as we rotate the line, the fraction of

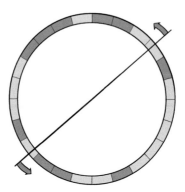

red on the redder side changes continuously, and after 180° of rotation it has switched from redder to pinker. The IVT tells us that there is a point when there was the same amount of red on both sides of the line.

The remaining subtlety: What if that position involves cutting through a gem? Well, it can't cut a ruby on one end while it cuts a diamond at the other end, because then the number of rubies on one side of the line would not be an integer—and half of 10 is an integer.

So either both ends of the line cut through a ruby or both through a diamond, and if we move the line a bit until it no longer cuts through a gem, the number of rubies on one side will not change.

Another way to look at it: You can apply the IVT to integers, if you replace the requirement of continuity by the requirement that the integer value changes only by one. If the rotating line always points to the boundary between two gems (at both ends), and moves one gem at a time, the number of rubies on one side can never change by more than one. We can thus conclude that as the line rotates 180°, it must at some point have half the rubies on each side.

In some puzzles the need for IVT is not at first obvious.

Three Sticks

You have three sticks that can't make a triangle; that is, one is longer than the sum of the lengths of the other two. You shorten the long one by an amount equal to the sum of the lengths of the other two, so you again have three sticks. If they also fail to make a triangle, you again shorten the longest stick by an amount equal to the sum of the lengths of the other two.

You repeat this operation until the sticks do make a triangle, or the long stick disappears entirely.

Can this process go on forever?

Solution: You can certainly make the process go on indefinitely if you design it so that the ratios of the lengths of the sticks are unchanged after one operation. For that to happen with lengths $a < b < c$ you'll need $b/a = c/b = a/(c-a-b)$; let that ratio be $r > 1$, so that the lengths are proportional to 1, r, r^2 and satisfy $1/(r^2 - r - 1) = r$, $r^3 - r^2 - r - 1 = 0$.

Writing $f(r) = r^3 - r^2 - r - 1$, we need an $r > 1$ for which $f(r) = 0$. Does such an r exist? Yes, because $f(1) = -2 < 0$ while $f(2) = 1 > 0$; since all polynomials are continuous functions, we can apply the IVT to deduce that there is a root of f between 1 and 2.

It remains only to check that our root gives us sticks that don't make a triangle, but that's easy, because $r^2 = r^3/r = (r^2 + r + 1)/r = r + 1 + (1/r) > r + 1$. ♡

Here's another case where IVT saves you the trouble of finding a root of a polynomial.

Hazards of Electronic Coinflipping

You have been hired as an arbitrator and called upon to award a certain indivisible widget randomly to either Alice, Bob, or Charlie, each with probability $\frac{1}{3}$. Fortunately, you have an electronic coinflipping device with an analog dial that enables you to enter any desired probability p. Then you push a button and the device shows "Heads" with probability p, else "Tails."

Alas, your device is showing its "low battery" light and warning you that you may set the probability p only once, and then push the coinflip button at most 10 times. Can you still do your job?

Solution: If you could set p twice, two flips would be enough: Set $p = \frac{1}{3}$ and on Heads give the widget to Alice, else reset to $\frac{1}{2}$ and use the result to decide between Bob and Charlie. Since you can only set p once, it might make sense to try to design a scheme first, then pick p to make the scheme work.

For example, you could flip your coin three times and if the flips are not all the same, you could use the "different" flip to decide to whom to award the widget (e.g., "HTT" means Alice gets it). But if you get all heads or all tails, you'll have to do it again, and maybe yet again—so, not in any bounded number of flips.

But now suppose we flip the coin *four* times. There are four ways to get one head, six to get two, four to get three: Even numbers all, so anytime you get non-uniform results you can use the flips to decide between Alice and Bob. If the coin is fair, the remaining probability—that is, the probability $q = p^4 + (1 - p)^4$ of "HHHH or TTTT"—is only $\frac{1}{8}$, too small for Charlie.

As you reduce p, however, q changes continuously and approaches 1. Thus, by IVT, there is some p for which the probability of "HHHH or TTTT" is exactly the $\frac{1}{3}$ you need. ♡

Bugs on a Pyramid

Four bugs live on the four vertices of a triangular pyramid (tetrahedron). Each bug decides to go for a little walk on the surface of the tetrahedron. When they are done, two of them are back home, but the other two find that they have switched vertices.

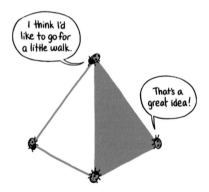

Prove that there was an instant when all four bugs lay on the same plane.

Solution: Let the bugs be A, B, C, and D. Initially when D looks down on the ABC plane she sees the ABC triangle labeled clockwise, but in final position she sees it counterclockwise (or vice-versa). Since the bugs move continuously, there must either be some position where she is on the plane ABC, or some position where ABC do not determine a plane. But in the latter case they are collinear and thus they are coplanar with any other point, thus with D's position in particular.

In the next puzzle, IVT's continuity hypothesis is missing.

Skipping a Number

At the start of the 2019 season WNBA star Missy Overshoot's lifetime free-throw percentage was below 80%, but by the end of the season it was above 80%. Must there have been a moment in the season when Missy's free-throw percentage was exactly 80%?

Solution: Let $X(t)$ be the fraction of Missy's free-throws that have gone in, up to time t. This value jumps up or down each time Missy attempts a free throw, and is always a rational number. The IVT clearly does not apply, so your immediate reaction is probably "no"—there's no reason why $X(t)$ should have to be exactly 4/5 at some point.

But it seems annoyingly difficult to actually construct a "history" for Missy that misses 80% (try it). We could certainly get her past 70%; perhaps she's 2 for 3 at the start of the season, then sinks her next free throw, jumping from $66\frac{2}{3}$% directly to 75%. But 80% seems impossible to skip. What's going on here?

As is often the case, it pays to try simpler numbers. Can we get Missy past 50% without hitting it? Ah—the answer is no, because if $H(t)$ is Missy's number of hits up to time t and $M(t)$ her number of misses, then the *difference* $H(t) - M(t)$ starts off negative and ends positive. Since this difference is an integer which changes by only 1 when a free throw is attempted, it must (by applying integer-IVT) hit 0; at that moment, $H(t) = M(t)$ and therefore Missy's success rate is exactly 50%.

Returning to 80%, we observe that at the moment when $X(t) = .8$, if there is one, we have $H(t) = 4M(t)$. Now, let t_0 be the first time at which Missy's percentage hits *or exceeds* 80%. That moment was, obviously, marked by a successful free throw; in other words, at time t_0, $H(t)$ went up by one while $M(t)$ stayed fixed. But $H(t)$ can't have skipped from below $4M(T)$ to above $4M(t)$; again by integer-IVT, it must hit $4M(t)$. So 80% can't be skipped!

Observe: (1) Our argument required that Missy went from below 80% to above it. In fact, she can easily skip 80% *going down*. (2) The argument uses the fact that at 80%, the number of hits will be an integer multiple of the number of misses. Thus, it works for any percentage that represents a fraction of the form $(k-1)/k$, and you can readily verify that any other fraction between 0 and 1 is skippable going up. Going down, it is the fractions of the form $1/k$ that can't be skipped.

Some processes are *piecewise* continuous, that is, continuous except at a discrete set of points. To apply IVT in such a case, you might have to "limit the damage" at jumping points.

Splitting a Polygon

A *chord* of a polygon is a straight line segment that touches the polygon's perimeter at the segment's endpoints and nowhere else.

Show that every polygon, convex or not, has a chord such that each of the two regions into which it divides the polygon has area at least $\frac{1}{3}$ the area of the polygon.

Solution: Call the polygon P and scale it so that it has area 1. If P is convex we can always find a chord that cuts P into two pieces of area $\frac{1}{2}$, as you, dear reader, now an expert with IVT, can readily prove. In fact, you can choose arbitrarily the chord's angle to the horizontal. Just move a line at that angle across P; since P is convex the intersection of its interior with the line consists of a single chord. Behind that chord is a piece of P whose area changes continuously from 0 to 1.

Try that when P is nonconvex, and things get messy: The line's intersection with P's interior may consist of several chords. Indeed, some polygons simply can't be cut into two equal-area parts by a chord—see the figure below for an example.

When P is nonconvex, we simplify things by first choosing an angle for our moving line L that doesn't encounter any more trouble than it needs to. We do that by ensuring that L is not parallel to any line through two vertices of P. Then the worst that can happen as we move our chord is that it encounters some concave vertex; it can't hit two vertices (or, therefore, any whole edge of P) at the same time.

Suppose we've moved L across P but so far the area of the smallest piece into which P has been cut has not reached $\frac{1}{3}$. If we hit a concave vertex v head-on (as pictured on the left side of the figure), we may be forced to split the chord. At that point we have cut P into three pieces, with, say A (of area a) behind the chord and B and C (areas b and c, respectively) ahead. At least one of b and c (say b, as in the figure below) exceeds $\frac{1}{3}$; we keep that part of the chord. If in fact $\frac{1}{3} \leq b \leq \frac{2}{3}$, we stop there; else we keep moving L forward.

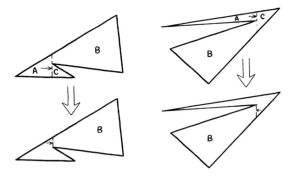

There is one other possibility: L hits v "from behind," for example, as pictured on the right, forcing the chord to suddenly jump in length. If the region ahead (C) is of area at least $\frac{1}{3}$, we can stop here or plow ahead as before. The fun case is when, as pictured, a and c are each less than $\frac{1}{3}$: Then we must flip our chord over v and send it back the other way into B!

In this fashion we continue pushing our chord through P, always with more than area $\frac{2}{3}$ ahead of us, until we have the desired split.

The last (and toughest) puzzle in this chapter doesn't look at all like a case for IVT, and even less like IVT's three-dimensional analogue. But let's see how it plays out.

Garnering Fruit

Each of 100 baskets contains some number (could be zero) of apples, some number of bananas, and some number of cherries. Show that you can collect 51 of those baskets that together contain at least half the apples, at least half the bananas, and at least half the cherries!

Solution: If we had just apples, then we could clearly get half the fruit in only 50 baskets by just taking the 50 baskets with the most apples. Or we could divide the baskets into two groups of 50 any way we want, and just take the part with more apples. Or—more geometrically—we could line up the baskets, divide the line (possibly through the middle of a basket, maybe cutting apples in the basket) in such a way that exactly half the apples are on either side of the cut, and note that there at most 49 whole baskets on one side of the cut or the other. Take those 49 plus all of the cut basket to get your winning 50 baskets. (If no basket is cut, just take a side with at most 50 baskets.)

With two fruits, we can't guarantee half of each with 50 baskets out of 100; for example, one basket with just one apple in it, and the remaining 99 with one lonely banana in each, would require 51. But we can get 50 out of 99 in several ways.

For example, we could arrange the baskets by decreasing number of apples, then take basket 1, and from each pair (2,3), (4,5), ..., (98,99) the basket with more bananas. This certainly gets at least half the bananas since it gets at least half from baskets 2 through 99, plus all in basket 1. And it gets at least half the apples since at worst it gets all the apples from baskets 1, 3, 5, 7, etc. and basket k has as many apples as basket $k+1$.

(Note that this algorithm uses only order information about the numbers of apples or bananas in baskets, not the actual amount.)

Another way to do it is to space the baskets equally around a circle, in any order. Now consider the 99 different "arc sets" you get by taking 50 baskets in a contiguous arc from the circle. We claim that most (i.e., at least 50) of the arc sets contain at least half the apples. Why? if they didn't, then more than half of the *complements* of the arc sets would contain more than half of the apples. But the complements are arc sets of size 49, and each of those can be thought of as a subset of a different arc set of size 50 (say, the one you get by adding the next basket clockwise). Thus, if most of the 49-ers contained more than half the apples, the same must be true of the 50's, a contradiction.

But if at least 50 of the size-50 arc sets contain at least half the apples, and (by a similar argument) at least 50 of them contain at least half the bananas, then at least one arc set contains at least half of both fruits.

It's worth noting that this proof shows that even if the baskets are arbitrarily numbered, there is a list of just 99 of the 5×10^{28} subsets of size 50, one of which is guaranteed to have the desired property. It's even more instructive to wonder if the circular arrangement of the baskets on the plane allows a straight-line cut, possibly right through the middle of two baskets and through some apples and bananas, that cuts the fruits *exactly* in two. If so, there are at most 48 baskets on one side or the other of the cut to which we can add the cut baskets to get 50 with at least half of each fruit. And this is the kind of thing we need to solve the problem when there are three (or more) types of fruit.

The combinatorial methods don't seem to generalize to the three-fruit case, so let's try the geometric ones. Suppose we place the baskets in 3-space, and spread them out enough so that no plane cuts more than three baskets. With three degrees of freedom, we can find a plane that cuts the apples, bananas, and cherries each exactly in half. With 100 baskets, there must be at most 48 whole baskets on one side or the other of the line. Put them together with the cut baskets, and we have 51 that contain at least half the apples, half the bananas, and half the cherries. (If fewer than three baskets were cut, it only gets easier.)

We've been a bit glib with this "degrees of freedom" business, but it really is true (and known usually as the *Stone–Tukey Theorem*) that in n-dimensional space, any n sets can be simultaneously bisected by an $n-1$-dimensional hyperplane. The three-dimensional version is known as the *ham-sandwich theorem* and says that any sandwich consisting of a slice of ham and two slices of bread, however sloppily piled, can be cut with a planar cut that exactly halves the ham and each slice of bread. Such facts are usually proved using fancy mathematics (algebraic topology).

Applying Stone–Tukey allows us to generalize to b baskets and n kinds of fruit; the result is there are always the greatest integer in $b/2 + n/2$ baskets with at least half of each fruit type. This is best possible, because if $b = m_1 + \cdots + m_n + x$ where all m_i are odd and $x = 0$ or 1, then we can have m_i boxes with nothing but one sample of fruit i, and possibly one empty box. To get half of each fruit type with this arrangement would costs us at least $n/2$ half-baskets beyond $b/2$.

Of the many theorems that employ IVT in their proofs, the most famous is the Mean Value Theorem of calculus, which says essentially that if you drive 60 miles in an hour then there must have been a moment when your speed was exactly 60 miles per hour. But we're avoiding calculus in this book and we don't need it here: A simple theorem about polynomials (simple compared to some of the above puzzles, for sure!) will do.

For us a "polynomial in the variable x" is an expression $p(x)$ of the form $a_0 + a_1 x + a_2 x^2 + \cdots + a_d x^d$ where d is a non-negative integer, the coefficients a_i are real numbers, and a_d is not zero. The "degree" of $p(x)$ is d, the exponent of the highest power of x having nonzero coefficient.

Theorem. *If a polynomial $p(x)$ has odd degree, then it has a real root, that is, there is a real number r such that $p(r) = 0$.*

The oddness condition for the degree of $p(x)$ is certainly necessary; for example, no real number x satisfies $1 + x^d = 0$ when d is an even non-negative integer.

The proof is a snap if you've gotten this far in the chapter. The idea is to first assume that $a_d = 1$ (dividing $p(x)$ by a_d does not affect its roots). Then, for large enough values of x, $p(x)$ will be positive; for highly *negative* values of x, $p(x)$ will be negative. Thus, since powers and thus polynomials are always continuous, IVT tells us there will be some x_0 with $p(x_0) = 0$.

How big does x have to be to guarantee that $p(x) > 0$? If we let $a = |a_0| + |a_1| + \cdots + |a_{d-1}|$ be the sum of the absolute values of the other coefficients, then $x > \max(a, 1)$ is good enough. The reason is that then the powers of x are increasing, thus

$$p(x) \geq x^d - ax^{d-1} = x^{d-1}(x - a) > 0.$$

Similarly,

$$p(-x) \leq (-x)^d + ax^{d-1} = x^{d-1}(a - x) < 0.$$

This tells us a little more than the theorem, namely, that $p(x)$ has a root in the union of the intervals $[-a, a]$ and $[-1, 1]$, where $a = (|a_0| + |a_1| + \cdots + |a_{d-1}|)/|a_d|$.

4. Graphography

Among the simplest of mathematical abstractions is the one which models objects with points (called "nodes" or "vertices") and indicates some relationship between a pair of nodes by the presence of a line (called an "edge") from one to the other. Such a structure is called a *graph*, not be confused with the graph of a function like $y = x^2 - 4$.

A graph is said to be *connected* if you can get from any node to any other by following a path of edges. It's not hard to see that you need at least $n-1$ edges to connect n nodes; three ways to do that with five nodes are shown in Fig below:

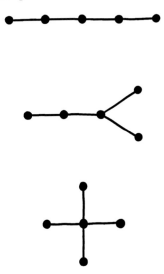

A graph with the minimum number of edges to be connected is called a *tree* (we counted those at the end of Chapter 1). Note that a tree cannot have a *cycle* (a closed path with three or more nodes) because removal of an edge from a cycle cannot disconnect a graph. Thus a tree can also be described as a graph with the *maximal* number of edges to not have a cycle.

Consider the following connectivity problem.

Air Routes in Aerostan

The country of Aerostan has three airlines which operate routes between various pairs of Aerostan's 15 cities. What's the smallest possible total number of routes in Aerostan if bankruptcy of any one of the three airlines still leaves a network that connects the cities?

Solution: Of course our graph will have 15 nodes, one for each city, and we can add edges of three types (say, solid black, dashed black, and solid pink) to represent the routes of the three airlines. From the above remarks, it follows that any two of Aerostan's airlines must operate at least 14 routes, and thus together the airlines must operate at least 21 routes. (Why? Add up the inequalities $a+b \geq 14$, $a+c \geq 14$, and $b+c \geq 14$).

Can 21 be achieved? If so, each airline will need exactly seven routes, and it must be the case that every cycle of routes uses routes from all three airlines (otherwise bankrupting the missing airline would leave a cycle, wasting one of the 14 remaining edges). There are lots of ways to do this; perhaps the simplest is to pick a hub city, let Airline A connect the hub to a set of seven other cities, and let Airline B connect the same hub to the remaining seven cities. Airline C avoids the hub but connects A's seven spoke cities pairwise to B's. The figure below shows this graph and another.

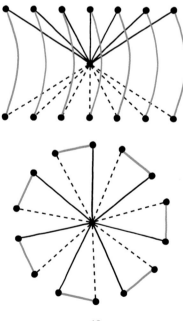

The second of these is known as a "windmill graph" for obvious reasons. If the nodes represent people and the edges friendships, the graph has the property that every two people have a unique friend in common. (There is only one connected graph with this property and a given odd number of nodes. If you want this property with an even number of nodes, you're out of luck!)

Spiders on a Cube

Three spiders are trying to catch an ant. All are constrained to the edges of a cube. Each spider can move one third as fast as the ant can. Can the spiders catch the ant?

Solution: If our protagonists were on a *tree*, then one spider—even a super-slow old codger—could catch the ant by himself, just by moving steadily toward the ant. With no cycles to dash around, the ant is doomed.

We take advantage of this idea by using two of our three spiders to "patrol" one edge each, thereby reducing the edge-graph of the cube to a tree. To patrol an edge PQ, a spider chases the ant off the edge if necessary, then patrols the edge making sure he is at all times at most $\frac{1}{3}$ the distance from P (respectively, Q) that the ant is. This is possible since the distance from P to Q along the edges of the cube, if you are not allowed to use the edge PQ, is three times the length of the edge.

If we choose for our two controlled edges two *opposite* edges (some other choices are equally good), we find that with these edges—including their endpoints—removed, the rest of the edge-network (in black in the figure below) contains no cycles (so it's just a collection of trees). It follows that the third spider can simply chase the ant to the end of a patrolled edge, where the ant will meet her sad fate.

In a graph, the number of edges ending at a particular node is called the *degree* of the node. Since we're not allowing edges from a node to itself, or multiple edges between the same pair of nodes, the degree of a node can never be more than the number of nodes in the graph, minus one.

Handshakes at a Party

Nicholas and Alexandra went to a reception with ten other couples; each person there shook hands with everyone he or she didn't know.

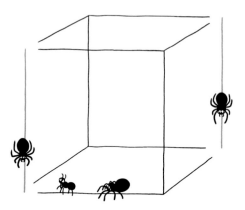

Later, Alexandra asked each of the other 21 partygoers how many people they shook hands with, and got a different answer every time.

How many people did Nicholas shake hands with?

Solution: It seems implausible that Alexandra's survey somehow tells us about Nicholas, but that's what makes this a cute puzzle.

Here our graph has 22 nodes and we put an edge between two if they shook hands during the party. The maximum possible degree of a node is only 20, since each person knows his or her partner, and of course the minimum is 0, so the numbers of handshakes recorded by Alexandra's survey must exactly be the 21 numbers $0, 1, \ldots, 20$. The guest who shook all 20 hands of the other 10 couples must be the partner of the one who shook 0 hands (otherwise, the two extremists would have had to both shake hands and not shake hands). Similarly, the one who shook 19 hands (all but the 0-guest's) must be the partner of the one who shook only 1 hand (only the 20-guest's), and so on. This leaves Nicholas as the one who shook 10 hands, since he's not partnered with anyone who was surveyed.

As sometimes happens with puzzles, there's a cheap way to get this answer, if you trust that the puzzle has a unique solution. You could have set up the graph so that an edge represents two non-coupled guests that did *not* shake hands. Whatever argument might have enabled you to conclude that Nicholas shook hands with n people in the original formulation would also allow you to conclude that he *failed* to shake hands with n people, other than Alexandra, in the second formulation. This is possible only if $n = 10$.

Snake Game

Joan begins by marking any square of an $n \times n$ chessboard; Judy then marks an orthogonally adjacent square. Thereafter, Joan and Judy continue alternating, each marking a square adjacent to the last one marked, creating a snake on the board. The first player unable to play loses.

For which values of n does Joan have a winning strategy, and when she does, what square does she begin at?

Solution: If n is even, Judy has a simple winning strategy no matter where Joan starts. She merely imagines a covering of the board by dominoes, each domino covering two adjacent squares of the board. Judy then plays in the other half of each domino started by Joan.

When n is odd, and Joan begins in a corner, *Joan* wins by imagining a domino tiling that covers every square except the corner in which she starts.

However, Joan loses in the odd-n case if she picks her starting square badly, for example, if she begins at a square adjacent to a corner square. Suppose the corner squares are black in a checkerboard coloring, so that Joan's starting square is white. There is a domino tiling of the whole board minus one black square; Judy wins by completing these dominoes. Joan can never mark the one uncovered black square because all the squares she marks are white! ♡

There'll be more about this game at the end of the chapter.

Bracing the Grid

Suppose that you are given an $n \times n$ grid of unit-length rods, jointed at their ends. You may brace some subset S of the small squares with diagonal segments (of length $\sqrt{2}$).

Which choices of S suffice to make the grid rigid in the plane?

Solution: It helps to turn this into a graph-theoretical problem, but not in the most obvious way (with vertices as joints and edges as rods). Instead, suppose you have put in your braces; now envision a graph G whose vertices correspond to the rows of squares and the columns of squares. Each edge of G corresponds to a row and column whose intersecting square is braced, so that the number of edges of G is the same as the number of braces you used.

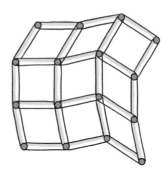

A partly-braced grid, with its associated graph, is pictured below.

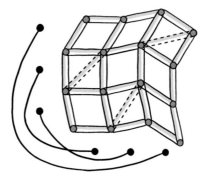

If a row and a column are adjacent in G, the vertical rods in the row are all forced to be perpendicular to the horizontal rods in the column. If G is a *connected* graph, meaning that there is a path from any vertex to any other, then *all* the horizontal rods must be perpendicular to all the vertical ones. Thus all the horizontal rods are parallel to one another, and similarly for the vertical ones, and now it is clear that the grid is rigid.

On the other hand, suppose the graph is disconnected and let C be a "component," that is, a connected piece of G which has no edges to the rest of G. Then there is nothing to prevent any vertical rod in a C-row or any horizontal rod in a C-column from flexing relative to the other rods in the grid.

Thus, the criterion for rigidity is exactly that the graph G be connected. Since G has $2n$ vertices, it must have at least $2n-1$ edges in order to be connected (you can prove this easily by induction if you haven't seen it before), hence you must have at least $2n-1$ braces to make the grid rigid. Notice, however, that they cannot be "just any-where."

The next figure shows an efficiently braced 3×3 grid and its cor-responding graph. For practice, you might want to compute the total number of ways to rigidify the 3×3 grid with the minimum number (five) of braces. A theorem of graph theory (to the effect that every connected graph has a connected subgraph, called a "spanning tree," with the minimum number of edges) tells us that if more than $2n-1$ squares are braced *and the grid is rigid*, then there are ways to remove all but $2n-1$ of the braces and preserve rigidity.

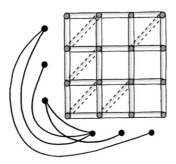

Competing for Programmers

Two Silicon Valley startups are competing to hire programmers. Hir-ing on alternate days, each can begin by hiring anyone, but every sub-sequent hire by a company must be a friend of some previous hire—unless no such person exists, in which case that company can again hire anyone.

In the candidate pool are 10 geniuses; naturally each company would like to get as many of those as possible. Can it be that the com-pany that hires first cannot prevent its rival from hiring nine of the geniuses?

Solution: Yes. Imagine a friendship graph that contains a (non-genius) hub Hubert with 10 friends each of whom knows a differ-ent genius, plus one lonely friend (Lonnie) who knows only Hubert. That's it; the geniuses (black vertices on the periphery) each have only one friend.

Suppose Company A goes first and hires Hubert. Then B hires Lonnie and after that, every time A hires a friend of Hubert, B hires that person's genius friend.

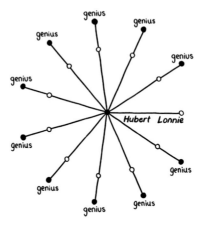

If A hires Lonnie, B hires a genius. A must hire Hubert and then, after A hires its genius' friend, the game proceeds as before.

If A hires a genius, B hires another genius; three days later A hires Hubert and then B hires Lonnie, after which the remaining geniuses go to B.

Finally, if A hires a genius' friend, B hires that genius. After A takes Hubert and B takes Lonnie, A winds up with no genius at all.

Wires under the Hudson

Fifty identical wires run through a tunnel under the Hudson River, but they all look the same, and you need to determine which pairs of wire-ends belong to the same wire. To do this you can tie pairs of wires together at one end of the tunnel and test pairs of wire-ends at the other end to see if they close a circuit; in other words, you can determine whether two wires are tied together at the other end.

How many trips across the Hudson do you need to accomplish your task?

Solution: Suppose the wires are labeled $w_1, w_2, \ldots, w_{50}$ at the west end of the tunnel and $e_1, \ldots, e_{50}$ at the east end. Supposing that you start on the west side of the river, you tie w_1 and w_2 together, w_3 and w_4, w_5 and w_6, etc. until all except w_{49} and w_{50} are paired.

You then test pairs of wire-ends at the east end of the tunnel until you have identified all of the tied pairs. For example, you might find that (say) e_4 and e_{29} are tied, e_2 and e_{15} are tied, e_8 and e_{31} are tied, and so forth, and finally that e_{12} and e_{40} are bachelors.

Next you return to the west end, untie all pairs, and instead tie w_2 to w_3, w_4 to w_5, etc. until all but w_1 and w_{50} are paired.

Finally you test pairs at the east end until, as before, you have identified all the paired wire-ends. To continue the example, the new pairs might include e_{12} and e_{15}, e_{29} and e_2, and e_4 and e_{31}, with e_{40} and e_8 as bachelors.

This simple procedure suffices to identify all the wires.

Observe that the one east-wire-end which was paired the first time but not the second (in our example, this is e_8) must belong to w_1. The east-wire-end with which e_8 was paired the first time (here, e_{31}) must therefore belong to w_2. But then, w_3 must belong to the east-wire-end to which e_{31} was paired the *second* time, namely e_4. Proceeding in this fashion, you find that w_4 belongs to e_{29} (e_4's mate on the first round), w_5 belongs to e_2 (e_{29}'s mate on the second round), and so forth. Eventually the sequence will end with w_{50} belonging to e_{40}.

If the number of wires (say, n) had been odd, you'd have left only w_n out of the pairings the first time, and w_1 the second time; the rest works pretty much the same way.

Two Monks on a Mountain

Remember the monk who climbed Mt. Fuji on Monday and descended on Tuesday? This time, he and a fellow monk climb the mountain on the same day, starting at the same time and altitude, but on different paths. The paths go up and down on the way to the summit (but never dip below the starting altitude); you are asked to prove that they can vary their speeds (sometimes going backwards) so that at *every* moment of the day they are at the same altitude!

Solution: It is convenient to divide each path into a finite sequence of monotone "segments" within which the path is always ascending, or always descending. (Level segments cause no problem as we can have one monk pause while the other traverses such a segment). Then we can assume that each such segment is just a straight ascent or descent, since we can have the monks modulate their rates so that their rate of altitude change is constant on any segment.

(There are mathematical curves of finite length that cannot be divided into a finite number of monotone segments, but in our problem we needn't worry about segments that are shorter than a monk's step size.)

Label the X-axis on the plane by positions along the first monk's path, and the Y-axis by positions along the second monk's path. Plot all points where the two positions happen to be at the same altitude; this will include the origin (where both paths begin) and the summit (where they end, say at $(1,1)$). Our objective is to find a path along plotted points from $(0,0)$ to $(1,1)$; the monks can then follow this path, slowly enough to make sure that no monk is anywhere asked to move faster than he can.

Any two monotone segments—one from each path—which have some common altitude show up on the plot as a (closed) line segment, possibly of zero length. If we regard as a vertex any point on the plot which maps back to a segment endpoint (for either or both monks), the plot becomes a graph (in the combinatorial sense); and an easy checking of cases shows that except for the vertices at $(0,0)$ and $(1,1)$, all the vertices are incident to either 0, 2, or 4 edges.

Once we begin a walk on the graph at $(0,0)$, there is no place to get stuck or be forced to retrace but at $(1,1)$. Hence we can get to $(1,1)$, and any such route defines a successful strategy for the monks. ♡

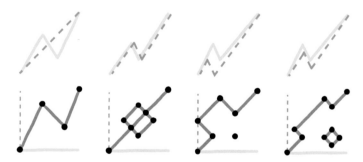

The figure shows four possible landscapes, with one monk's path shown as a solid line, the other as dashed. Below each landscape is the corresponding graph. Note that, as in the last case, there may be detached portions of the graph which the monks cannot access (without breaking the common altitude rule).

Worst Route

A postman has deliveries to make on a long street, to addresses 2, 3, 5, 7, 11, 13, 17, and 19. The distance between any two houses is proportional to the difference of their addresses.

To minimize the distance traveled, the postman would of course make his deliveries in increasing (or decreasing) order of address. But our postman is overweight and would like to *maximize* the distance traveled making these deliveries, so as to get the most exercise he can. But he can't just wander around town; to do his job properly he is obligated to walk directly from each delivery to the next one.

In what order should he make his deliveries?

Solution: The "greedy algorithm" for maximizing the postman's distance would have him start at #2, go all the way to #19, then back to #3, then to #17 and so forth, ending at #11. That would result in total distance $17 + 16 + 14 + 12 + 8 + 6 + 4 = 77$ units. Is this the best he can do?

Here's a useful way to look at the problem. Call the addresses A, B, etc. up to H, in numerical order, and represent the trek from house X to house Y by XY. No matter what the postman does, his trip can be represented as a list of treks if the form AB, BC, CD, DE, EF, FG, and GH.

How many times can the postman perform the trek AB? At most twice, once on his way to A and once back. A similar argument works for GH. How about BC? That one he might be able to do four times: to A and back (but not from B), and to B and back (not from A). Continuing this reasoning, we see that the postman cannot possibly do better than $2AB + 4BC + 6CD + 8DE + 6EF + 4FG + 2GH$, which adds up to 86 units in our puzzle.

Wait—the postman can't actually do the EF trek eight times because he only makes seven trips between consecutive deliveries. So in fact he can do no better than $2AB + 4BC + 6CD + 7DE + 6EF + 4FG + 2GH$, which is 82 units in our puzzle. Can he achieve that number?

Yes, he can, but only if he begins at D or E and ends at the other one, after zigzagging between low-numbered houses (A, B, C, D) and high (E, F, G, H). Example: D, H, C, G, B, F, A, E. There are $8 \times 4 \times 3 \times 3 \times 2 \times 2 \times 1 \times 1 = 1{,}152$ ways to do this, and each one gives the maximum possible length, 82 units.

The given addresses don't matter a bit, nor even the number of houses (provided that remains even). The reasoning above is quite general, the conclusion being that the postman should always (a) begin at one of the middle two houses and end at the other, and (b) zigzag between the lower-half addresses and the upper-half addresses.

For our theorem, let's take another look at the *Snake Game* puzzle. We can play that game on any graph G: The rules are that the players alternate marking vertices, and after Player 1 (Joan) marks her first vertex, every subsequent vertex must be unmarked and adjacent to the vertex just marked by the opponent. The first player unable to move loses.

The proof above that the second player wins on an 8×8 chessboard actually shows that Player 2 wins on any graph with a *perfect matching*, that is, a set of disjoint edges that cover all the vertices (as the dominoes do on the chessboard).

In fact, when G has a perfect matching, Player 2 wins even when Player 1 is allowed to flout the main rule of the game, marking *any* (unmarked) vertex instead of one adjacent to Player 1's last move! The strategy is the same: Whenever Player 1 marks a vertex v, Player 2 simply marks the vertex on the other end of the matching edge that covers v.

So you might think that having a perfect matching is a far stronger condition than is needed to assure a win for Player 2 in the original game. But no.

Theorem. *With best play, Player 2 wins on G if and only if G has a perfect matching.*

Proof. We've already proved the "if" part of the theorem, so let us suppose that G does not have a perfect matching, with the intent of providing a winning strategy for Player 1.

The key is to pick some *maximum-size* matching M, that is, some collection of disjoint edges which is as large as possible. Since there is no perfect matching, there will be some vertex u_1 of G that's not in M, and we'll let Player 1 start there. After that, Player 1 intends to follow the "matching strategy" from above, playing the other end of whatever matching edge Player 2 marks a vertex of. If Player 1 can always do that, she of course will win, but how do we know she can always do that? What if Player 2 marks a vertex that's not in an edge of M?

Well, Player 2 can't *immediately* do that, because her first play (say, v_1) must be adjacent to v_0. If v_1 were not in the matching, the edge (u_1, v_1) could have been added to M to make a bigger matching—and M was supposed to be as big as possible.

And she can't do it later, either. Suppose Player 1 has been able to execute her matching strategy through her kth move, and now must reply to Player 2's kth move. The moves so far have been $u_1, v_1, u_2, v_2, \ldots, u_k, v_k$; by assumption, all these vertices are distinct, all the pairs (u_i, v_i) are edges of G, and all the pairs (v_i, u_{i+1}) are not only edges of G but also members of the matching M.

But now we claim that Player 2's latest move, v_k, belongs to an edge of M. Why? Because if it didn't, we could make M bigger by replacing the $k-1$ edges $(v_1, u_2), (v_2, u_3), \ldots, (v_{k-1}, u_k)$ by the k edges $(u_1, v_1), (u_2, v_2), \ldots, (u_k, v_k)$—again contradicting the maximality of M.

So Player 1 is never stuck for a move, and the theorem is proved!
♡

5. Algebra Too

The single most useful technique for solving puzzles that ask for a quantity is simply to assign a letter (x and n are popular choices) to the quantity, write down what the puzzle tells you, and then solve using algebra.

Our first example is simple, but people get it wrong all the time.

Bat and Ball

A bat costs $1 more than a ball; together they cost $1.10. How much does the bat cost?

Solution: Did you say $1? Oops! Let x be the price of the bat, so the price of the ball is $x - \$1$. Then we have the equation $x + (x - \$1) = \$1.10$, $2x = \$2.10$, $x = \$1.05$.

Note that we could have assigned another variable (y, perhaps) to the price of the ball, then solved two equations in two unknowns. If you can avoid that by expressing new unknowns in terms of old ones, you can save yourself a lot of trouble.

Two Runners

Two runners start together on a circular track, running at different constant speeds. If they head in opposite directions they meet after a minute; if they head in the same direction, they meet after an hour. What is the ratio of their speeds?

Solution: This is simultaneously confusing and easy. Let L be the length of the track. Then, when going in opposite direcions, the runners meet when, together, they have run a distance L. When going the same direction, they meet when the *difference* of their achieved distances is L.

Thus, $(s + t)L/(s - t)L = 60$, and solving gives $s/t = 61/59$.

A classic:

Belt Around the Earth

Suppose the earth is a perfect sphere and a belt is tightened around the equator. Then the belt is loosened by adding 1 meter to its length, and it now sits the same amount off the ground all the way around.

Is there enough room to slip a credit card underneath the belt?

Solution: The circumference of the earth is about 40 million meters, but who cares? Call it C, and call the earth's radius R, so that $C = 2\pi R$, thus $R = C/(2\pi)$. When we lengthen the belt its circumference is $C+1$ so the new radius is $(C+1)/(2\pi) = C/(2\pi)+1/(2\pi) = R+16$ cm, approximately. There's room for a stack of about 200 credit cards to pass under the belt.

There are dozens of published puzzles along the lines of "John is twice as old as Mary will be when . . . " that can be solved straightforwardly as above. Let's try something a little different.

Odd Run of Heads

On average, how many flips of a coin does it take to get a run of an odd number of heads, preceded and followed by tails?

Solution: First, we need to get a tail, since starting with odd run of heads won't help us. If that takes time x on average, then, since flipping H doesn't help, we have

$$x = 1 + \frac{1}{2} \cdot x$$

giving $x = 2$. Now we need an odd run of heads; say that takes time y on average. If we flip another T (probability $\frac{1}{2}$) or HH (probability $\frac{1}{4}$), we've made no further progress; if we flip HT (probability $\frac{1}{4}$) we're done. Thus,

$$y = \frac{1}{2} \cdot (1 + y) + \frac{1}{4} \cdot (2 + y) + \frac{1}{4} \cdot 2 ,$$

giving $y = 6$. So in all it takes an average of $x + y = 2 + 6 = 8$ flips to get that run.

Randomness enters into the next puzzle as well. This is a toughie but the general technique is the same.

Hopping and Skipping

A frog hops down a long line of lily pads; at each pad, he flips a coin to decide whether to hop two pads forward or one pad back. What fraction of the pads does he hit?

Solution: Let us label the lily pads with integers in order. One way to begin solving the problem is to compute the probability p that a frog starting at 1 regresses, at some point, to 0. To never regress, he must jump ahead (probability $\frac{1}{2}$) to lily pad #3 and not subsequently regress three times from there (probability $(1 - p^3)$). Thus $1 - p = \frac{1}{2}(1 - p^3)$; divide by $(1 - p)/2$ to get $2 = 1 + p + p^2$, giving $p = (\sqrt{5} - 1)/2 \sim$ .618034, the familiar golden ratio.

It seems awkward to try to compute the probability that the frog skips over a particular position (say, 1). You might want to start your calculations from the first time the frog hits 0, if it does, but it might already have hit 1. A better idea is to try to compute the probability that *during a particular jump* the frog leaps over a pad that he never has, and never will, hit.

For that to happen, he must (a) be jumping forward at the moment, (b) never drop back from where he lands, and (c) not have reached the pad he is jumping over in the past.

We may as well assume the lily pad the frog is currently passing over is #0. The key is to note that event (c) is an independent copy of event (a) if you reverse both space and time. Looking at the frog backwards in time, and regarding lower-numbered lily pads as progress, the frog appears to be behaving as before: jumping two forward or one back with equal probability. Event (c) says that when he reaches -1, he must not later "drop back" to 0.

The probability of these three events all occurring is thus $\frac{1}{2} \cdot (1-p) \cdot (1-p) = (1-p)^2/2$. We must not forget, however, that we have not calculated the probability that 0 is skipped; only the probability that a particular jump carries the frog over a skipped pad.

Since the frog travels, on the average, at rate $\frac{1}{2}$, he is creating skipped lily pads at rate $(1 - p)^2$ relative to his spatial progress. It follows that the fraction of pads he hits is $1 - (1 - p)^2 = (3\sqrt{5} - 5)/2 \sim 0.854102$. Whew! ♡

The next puzzle is an *almost* straightforward application of the algebraic technique, but a little extra cleverness is needed—as is often the case when the resulting equation or equations are *Diophantine*, that is, they require integer solutions.

Area–perimeter Match

Find all integer-sided rectangles with equal area and perimeter.

Solution: There are two.

Let x and y be the sides, with $x \geq y$. we need $xy = 2x + 2y$, equivalently $xy - 2x - 2y = 0$, but the left-hand side of the latter reminds us of the product $(x-2)(y-2)$. Rewriting, $(x-2)(y-2) = 4$, so either $x-2 = y-2 = 2$ or $x-2 = 4$ and $y-2 = 1$. This gives $x = y = 4$, or $x = 6$ and $y = 3$.

The remaining puzzles involve algebra in other ways than solving equations.

Three Negatives

A set of 1000 integers has the property that every member of the set exceeds the sum of the rest. Show that the set includes at least three negative numbers.

Solution: Let a and b be the only two negative numbers in the set, and let the sum of all the numbers in the set be S. We are told that $a > S - a$, and $b > S - b$; adding the two inequalities gives $a + b > 2S - a - b$, thus $a + b > S$ which is impossible unless there are other negative numbers in the set.

Red and Black Sides

Bored in math class, George draws a square on a piece of paper and divides it into rectangles (with sides parallel to a side of the square), using a red pen after his black one runs out of ink. When he's done, he notices that every rectangular tile has at least one totally red side.

Prove that the total length of George's red lines is at least the length of a side of the big square.

Solution: Why should this be true? Maybe each rectangle is long and thin and its only red side is a short one. One thing we do know, though, is that the total area of the rectangles is s^2, where s is the side of the big square. If the rectangles are $r_i \times b_i$, where each r_i represents a red side, then (using the convention that $\sum_i x_i$ means the sum of x_i for all salient i):

$$\sum_i r_i b_i = s^2.$$

But of course each $b_i \leq s$ so

$$\sum_i r_i b_i \leq \sum_i r_i s = s \sum_i r_i$$

and it follows that $\sum_i r_i \geq s$. In fact, you get equality only if all the rectangles span the square east–west (or, all north–south) with one end of each rectangle colored red.

Recovering the Polynomial

The Oracle at Delphi has in mind a certain polynomial p (in the variable x, say) with non-negative integer coefficients. You may query the Oracle with any integer x, and the Oracle will tell you the value of $p(x)$.

How many queries do you have to make to determine p?

Solution: It only takes two queries, and the first is needed only to get a bound on the size of the polynomial's coefficients. The Oracle's answer to $x = 1$ (say, n) tells you that no coefficient can exceed n. Then you can send in $x = n+1$, and when you expand the Oracle's answer base $n+1$, you have the polynomial!

For example, suppose (rather conveniently) that $p(1) < 10$, so you know every coefficient is an integer between 0 and 9. You then send in the number 10 and if the oracle says $p(10) = 3{,}867{,}709{,}884$, you know that the polynomial is

$$3x^9 + 8x^8 + 6x^7 + 7x^6 + 7x^5 + 9x^3 + 8x^2 + 8x + 4.$$

Here's a puzzle whose solution is, potentially, a useful strategy.

Gaming the Quilt

You come along just as a church lottery is closing; its prize is a quilt worth $100 to you, but they've only sold 25 tickets. At $1 a ticket, how many tickets should you buy?

Solution: Clearly you should buy at least 1 ticket, assuming you are trying to maximize your expected gain, because then for $1 you have a 1/26 chance to win $100; you will gain on average almost $3 from the invesment.

On the other hand, it would be a losing proposition (except to the church!) to buy 100 or more tickets, because then you could break even at best.

So it looks like some intermediate number of tickets is right, but how many? You can try to maximize your expectation $(x/(25+x)) \cdot \$100 - \$x$ from buying x tickets using calculus, if you happen to know your calculus, but what do you do in the (likely) event that the optimum value of x is not an integer?

Often an effective way to tackle this kind of problem is to figure out whether you gain or lose when you increment your variable by 1. If you increase your purchase from m tickets to $m+1$ tickets, your expected return goes up by $\$100((m+1)/(m+26) - m/(m+25)) - 1$. This value goes steadily down and eventually becomes negative (meaning that you shouldn't buy another ticket). Thus, if you can identify the least m for which this expression is negative, you'll know that the right number of tickets to buy is m.

That's just algebra: You solve the (quadratic) equation $\$100((m+1)/(m+26) - m/(m+25)) - 1 = 0$ to get $m = 24.5025$. Thus 25 is the least value of m with the property that you lose money by buying another ticket, and it follows that 25 is the best number of tickets to buy. Is it a coincidence that 25 was also the number of tickets sold before you got there? Actually, it is.

And if you don't care for solving quadratic equations? You can still get the answer pretty quickly by plugging your guesses for m into the expression $\$100((m+1)/(m+26) - m/(m+25)) - 1$, until you zero in on the least m that makes the expression negative.

Strength of Schedule

The 10 teams comprising the "Big 12" college football conference are all scheduled to play one another in the upcoming season, after which one will be declared conference champion. There are no ties, and each team scores a point for every team it beats.

Suppose that, worried about breaking ties, a member of the governing board of the conference suggests that each team receive in addition "strength of schedule points," calculated as the sum of the scores achieved by the teams it beat.

A second member asks: "What if all the teams end up with the same number of strength-of-schedule points?"

What, indeed? Can that even happen?

Solution: No.

First, we observe that if all the teams end up with the same number of strength-of-schedule points, they must also all have the same score. Why? If not, let b be the biggest score and s the smallest. Each team that won b games racked up at least bs strength-of-schedule points (SSP's, for short), because, at worst, all their wins came against teams that won only s games. Similarly, teams that won only s games got at most sb SSP's. Whoops, that means both the b teams and the s teams got exactly sb SSP's, and therefore that all of the b teams' wins came against s teams and vice-versa. But that's impossible unless there is only one b team and one s team, because otherwise there would be a game between two b teams or between two s teams.

It follows that the b team won only once. But then, any third team would have to have beaten both the b team and the s-team, outscoring the b-team and contradicting the definition of b.

We conclude (since there are more than two teams in the Big 12) that if every team has the same number of SSP's, then every team has the same score. That's not possible in a conference with an even number of teams, because then each team would have to have won half its games, and the number of games it plays is odd.

Round-robin tournaments are excellent puzzle fodder; here's another.

Two Round-robins

The Games Club's 20 members played a round-robin checkers tournament on Monday and a round-robin chess tournament on Tuesday. In each tournament, a player scored 1 point for each other player he or she beat, and half a point for each tie.

Suppose every player's scores in the two tournaments differed by at least 10. Show that in fact, the differences were all exactly 10.

Solution: Let's divide the club members into the good checkers players (who scored higher in the checkers tournament) and the good chess players. One of these groups must have at least 10 members; suppose there are $k \geq 10$ good chess players, and that they scored a total of $t \geq 10k$ points higher in the chess tournament than in the checkers tournament.

This difference must have come entirely at the expense of the good checkers players, because the number of points scored by the good chess players *among themselves* had to be the same in both tournaments—namely, $\binom{k}{2}$, A.K.A. $k(k-1)/2$. Since there are $k(20-k)$ intergroup matches in each tournament, the most they can contribute to t is $k(20-k)$. So we have $10k \leq k(20-k)$, $k \leq 10$.

But k was supposed to be at least 10, so we conclude that $k = 10$ and that all of the above inequalities must be equalities.

In particular, since $t = 10k$ every chess player scored exactly 10 points higher in the chess tournament than in the checkers tournament. Moreover, since $t = k(20-k)$, every intergroup match in the chess tournament must have been won by the good chess player, and in the checkers tournament, by the good checkers player.

Alternative Dice

Can you design two different dice so that their sums behave just like a pair of ordinary dice? That is, there must be two ways to roll a 3, 6 ways to roll a 7, one way to roll a 12, and so forth. Each die must have six sides, and each side must be labeled with a positive integer.

Solution: The unique answer is that the labelings of the two dice are $\{1, 3, 4, 5, 6, 8\}$ and $\{1, 2, 2, 3, 3, 4\}$.

Perhaps you came up with this answer by trial and error, which is a perfectly fine way of solving the problem. However, there is another way in this case, involving a simple example of a powerful mathematical tool called *generating functions*.

The idea is to represent a die by a polynomial in the variable x, in which the coefficient of the term x^k represents the number of appearances of the number k on the die. Thus, for example, an ordinary die would be represented by the polynomial $f(x) = x + x^2 + x^3 + x^4 + x^5 + x^6$.

The key observation is that the result of rolling two (or more) dice is represented by the *product* of their polynomials. For instance, if we roll two ordinary dice, the coefficient of x^{10} in the product (namely, $f(x)^2$) is exactly the number of ways of picking two terms from $f(x)$ whose product is x^{10}. These ways are $x^4 \cdot x^6$, $x^5 \cdot x^5$, and $x^6 \cdot x^4$; and these represent the three ways to roll a sum of 10.

It follows that if $g(x)$ and $h(x)$ are the polynomials representing our alternative dice, then $g(x) \cdot h(x) = f(x)^2$. Now, polynomials, like numbers, have unique prime factorizations; the polynomial $f(x)$ factors as $x(x + 1)(x^2 + x + 1)(x^2 - x + 1)$. To get $g(x)$ and $h(x)$ to multiply out to $f(x)^2$, we need to take each of these four factors and assign either one copy to each of $g(x)$ and $h(x)$, or two copies to one and none to the other. But there are some constraints: In particular, we can't have a non-zero constant term in $g(x)$ or $h(x)$ (which would represent some sides labeled "0") or any negative coefficient, and the coefficients must sum to 6 since we have six sides to label.

The only way to do this (other than $g(x) = h(x) = f(x)$) is to have

$$g(x) = x(x+1)(x^2+x+1)(x^2+x-1)^2 = x+x^3+x^4+x^5+x^6+x^8$$

and

$$h(x) = x(x + 1)(x^2 + x + 1) = x + 2x^2 + 2x^3 + x^4,$$

or vice-versa.

This may seem a bit like trial and error after all, but with this technique you can solve problems that are much more complex than this one. To begin with, you can invent alternatives for a pair of 8-sided dice labeled 1 through 8 (there are three new ways to do it), or for rolling *three* ordinary dice (many ways).

It will not be a surprise to readers that *thousands*, literally, of theorems have been proved using algebra. Here's one that uses just the elementary sort of algebra we have seen above.

Have you ever wondered: What is the probability that your family surname will die out? Francis Galton posed this question in 1873 and an answer posted by the Reverend Henry William Watson led to a joint paper. Galton was interested in the longevity of aristocratic names, but he could (in more modern times) have been thinking equivalently of the preservation of a Y-chromosome.

In Galton and Watson's model of propagation, each individual independently has a random number of offspring, given by a fixed probability distribution: i offspring with probability p_i. (In the case of tracking patrilineal surnames, we would count only male offspring.)

We wish to know: Starting with (say) one individual, what is the "extinction" probability x that the family name dies out? Of course this depends on the probability distribution; as you might expect, if the average number of offspring

$$\mu = \sum_{i=0}^{\infty} i p_i$$

is less than 1, then the name will always die out, but if it's greater than 1, the name might live on forever. In the latter case, how do you determine x?

Theorem. *Let T be a Galton–Watson tree with offspring probabilities $p_0, p_1, \ldots, p_k$. Then the extinction probability x satisfies $x = \sum p_i x^i$.*

Let's do one example before we proceed to the proof. Suppose each individual has zero offspring with probability 0.1, one with probability 0.6, and two with probability 0.3. Then $\mu = 1.2$ and we expect the tree will sometimes be finite (i.e., die out) and sometimes infinite (continue forever). The equation in the theorem becomes $x = 0.1 + 0.6x + 0.3x^2$, which reduces to $3x^2 - 4x + 1 = 0$; factoring gives $(3x - 1)(x - 1) = 0$, so $x = 1$ or $x = \frac{1}{3}$. Since the probability of extinction is not 1, it must be $\frac{1}{3}$.

Proof. We observe that the tree is finite if either the starting individual has no offspring, or he does have offspring but each child is himself the start of a finite tree. If x is the extinction probability, then the probability that each of i individuals leads only to a finite tree is x^i, thus:

$$x = \sum_{i=0}^{k} p_i x^i$$

as claimed. ♡

The equation in the theorem always has $x = 1$ as one solution, since the probabilities sum to one. If $\mu > 1$, there will be one other solution—the correct one. If $\mu \leq 1$, except in the case where $p_1 = 1$ and $p_i = 0$ for all $i \neq 1$, the only solution will be $x = 1$ and thus extinction is assured.

6. Safety in Numbers

Numbers are everywhere, and endlessly fascinating. You probably know that every positive integer can be expressed uniquely as the product of primes (a prime number being one that cannot be divided evenly by anything except itself and 1). Let's use that and some other elementary observations to solve some puzzles, shall we?

Broken ATM

George owns only $500, all in a bank account. He needs cash badly but his only option is a broken ATM that can process only withdrawals of exactly $300, and deposits of exactly $198. How much cash can George get out of his account?

Solution: Since both the deposit and withdrawal amounts are multiples of $6, George cannot hope to get more than $498 (the largest multiple of 6 under $500). Using "–300" to indicate a withdrawal and "+198" a deposit, we see that George can get $6 out by –300, +198, –300, +198, +198. Doing this 16 times gets $96 out, after which –300, +198, –300 finishes the job.

Subsets with Constraints

What is the maximum number of numbers you can have between 1 and 30, such that no two have a product which is a perfect square? How about if (instead) no number divides another evenly? Or if no two have a factor (other than 1) in common?

Solution: These puzzles can all be tackled the same way. The idea is to try to cover the numbers from 1 to 30 by bunches of numbers with the property that from each bunch, you can only take one number for inclusion in your subset. Then, if you *can* construct a subset that consists of one choice from each bunch, you have yourself a maximum-sized subset.

In the first case, fix any number k which is square-free (in other words, its prime factorization contains at most one copy of each prime). Now look at the set S_k you get by multiplying k by all possible perfect squares.

If you take two numbers, say kx^2 and ky^2, from S_k, then their product is $k^2x^2y^2 = (kxy)^2$ so they can't both be in our subset. On the other hand, two numbers from *different* S_k's cannot have a square product since one of the two k's will have a prime factor not found in the other, and that factor will appear an odd number of times in the product.

Now, *every* number is in just one of these sets S_k—given n, you can recover the k for which $n \in S_k$ by multiplying together one copy of each prime that appears with an odd exponent in the factorization of n. Between 1 and 30, the choices for k (i.e., the square-free numbers) are 1, 2, 3, 5, 7, 11, 13, 17, 19, 23, 29, 2×3, 2×5, 2×7, 2×9, 2×11, 2×13, 3×5, 3×7, and finally $2\times3\times5$: 20 in all. You can choose each k itself as the representative from S_k, so a subset of size 20 is achievable and best possible.

To avoid one number dividing another evenly, note that if you fix an odd number j, then in the bunch $B_j = \{j, 2j, 4j, 8j, \dots\}$—that is, j times the powers of 2—you can only take one. If you take for your subset the top half of the numbers from 1 to 30, namely 16 through 30, you have got one from each B_j, and of course no member of this subset divides another evenly since their ratios are all less than 2. So the 15 numbers you get this way is best possible.

Finally, to get a maximum-sized subset all of whose members are relatively prime, you naturally want to look at bunches consisting of all multiples of a fixed prime p. You can take p itself as the representative of its bunch, so you can do no better than to take as your subset all the primes below 30, plus the number 1 itself, for a total of 11 members. ♡

Cards from Their Sum

The magician riffle-shuffled the cards and dealt five of them face down onto your outstretched hand. She asked you to pick any number of them secretly, then tell you their sum (jacks count as 11, queens 12, kings 13). You did that and the magician then told you exactly what cards you chose, including the suits!

After a while you figured out that her shuffles were defective—they didn't change the top five cards, so the magician had picked them

in advance and knew what they were. But she still had to pick the ranks of those cards carefully (and memorize their suits), so that from the sum she could always recover the ranks.

In conclusion, if you want to do the trick yourself, you need to find five distinct numbers between 1 and 13 (inclusive), with the property that every subset has a different sum. Can you do it?

Solution: If only you had a card worth 16, you'd be in fine shape: The powers of two, in particular 1, 2, 4, 8, and 16, have the desired property. In fact when you write a number in binary, the positions of the ones tell you exactly which powers go into the sum. For example, suppose that your victim reports a sum of 13; in binary that's 1101, thus 8 + 4 + 1, so his cards must consist of an eight, a four, and an ace.

There are $2^5 = 32$ ways to choose a subset of the five cards (if you include the null set, which has sum 0). Can we get them all to be different, without having a number bigger than 13?

The first thing to realize is that big numbers are better: It's easier to tell whether a big-numbered card was chosen or not. So let's try making our set using the greedy algorithm: Start with the king and throw in the biggest cards possible without causing the same sum to come up twice.

If you do that you'll end up with K, Q, J, 9, 6; and that works! (It turns out there's only one other way to do it, the same cards but with the 9 replaced by a 3).

What about suits? I recommend 6 of spades, 9 of diamonds, jack of clubs, queen of hearts and king of spades. But pick anything that looks random and that you can remember. Then practice (including that fake riffle shuffle), and you'll be the life of the party. Read Colm Mulcahy's book (see *Notes & Sources*) for more great mathematical card tricks!

Divisibility Game

Alice chooses a whole number bigger than 100 and writes it down secretly. Bob now guesses a number greater than 1, say k; if k divides Alice's number, Bob wins. Otherwise k is subtracted from Alice's number and Bob tries again, but may not use a number he used before. This continues until either Bob succeeds by finding a number that divides Alice's, in which case Bob wins, or Alice's number becomes 0 or negative, in which case Alice wins.

Does Bob have a winning strategy for this game?

Solution: The key is that there are enough small numbers around so that by judicious use of them, Bob can find a divisor of Alice's number. For example, Bob wins using $2, 3, 4, 6, 16, 12$ (or $6, 4, 3, 2, 5, 12$ or others). To see this, we consider the possibilities for Alice's number modulo 12; in other words, the remainder when Alice's number (let's call it A) is divided by 12.

Using the suggested sequence $2, 3, 4, 6, 16, 12$, Alice is immediately lost if her number is even, that is, if $A \equiv 0, 2, 4, 6, 8,$ or 10 modulo 12. If $A \equiv 5$ or 11 mod 12, it will be 3 or 9 mod 12 after 2 is subtracted, allowing Bob to win on his second turn. If $A \equiv 1$ or 9, Alice will succumb at turn 3; if $A \equiv 3$, at turn 4. The only value mod 12 that Alice can start with that has survived so far is 1, which after 2, 3, 4, and 6 have been subtracted, is now 4 modulo 12. Subtracting 16 knocks that down to 0 modulo 12, and now Bob's 12 spells doom for Alice. Of course, we must check that Bob's guesses add up to at most 100, and they do ($2 + 3 + 4 + 6 + 16 + 12 = 43$).

Prime Test

Does $4^9 + 6^{10} + 3^{20}$ happen to be a prime number?

Solution: Since the number is $(2^9)^2 + 2 \cdot 2^9 \cdot 3^{10} + (3^{10})^2 = (2^9 + 3^{10})^2$, it is a perfect square (so, it's not prime). If you guessed correctly that the number was composite but tried to prove it by finding a small prime factor, you were in for a tough time, because $2^9 + 3^{10}$ itself is prime—so our number has no prime factor less than 59,561.

Yes, it is easy to forget that finding a divisor is not the only way to show a number is composite. Showing that it is a square (or a cube, etc.) is another; other, more sophisticated ways exist as well.

Numbers on Foreheads

Each of 10 prisoners will have a digit between 0 and 9 painted on his forehead (they could be all 2's, for example). At the appointed time each will be exposed to all the others, then taken aside and asked to guess his own digit.

In order to avoid mass execution, at least one prisoner must guess correctly. The prisoners have an opportunity to conspire beforehand; find a scheme by means of which they can ensure success.

Solution: It is useful in many problems to introduce probability, even though none is present in the statement. Here, if we assume that the numbers painted on foreheads are chosen independently and uniformly at random, we see that no matter what he does, each prisoner has probability $1/10$ of guessing correctly.

Let the prisoners be numbered from 0 to 9. Since we want the probability that *some* prisoner guesses correctly to be 1, we need the 10 events "Prisoner k guesses correctly" to be mutually exclusive: In other words, no two can occur. Otherwise, the probability of at least one success would be strictly less than $10(\frac{1}{10}) = 1$.

To do this it would behoove us to separate the set of possible configurations into 10 equally likely scenarios, then have each prisoner base his guess on a different scenario. This reasoning may already have led you to the easiest solution: Let s be the sum of the numbers on *all* the prisoners' foreheads, modulo 10 (in other words, the rightmost digit of the sum). Now let Prisoner k guess that $s = k$, in other words, guess that his own number is k minus the sum of the numbers he sees, modulo 10.

This will ensure that Prisoner s, whoever that may be, will be correct (and all others wrong).

Rectangles Tall and Wide

It is easy to check that a 6×5 rectangle can be tiled with 3×2 rectangles, but only if some tiles are horizontally oriented and others are vertically oriented.

Show that this can't happen if the big rectangle is a square—in other words, if a square is tiled with congruent rectangles, then it can be re-tiled with the same rectangles all oriented the same way.

Solution: First, we claim that if the sides of a tile are not commensurable (i.e., if their ratio is irrational) then a much stronger statement holds: *Any* tiling of *any* rectangle either has all tiles horizontal or all vertical. To see this, assume otherwise and draw a horizontal line across the square at the top, and imagine moving it downwards. Suppose that just below the top border of the square, our line cuts a tiles vertically and b horizontally. Then, except when it coincides with some tile boundary, our line must always cut a tiles vertically and b horizontally; for otherwise, if the tiles are $x \times y$, we would have $ax + by = cx + dy$ for $(a, b) \neq (c, d)$, giving $x/y = (b - d)/(a - c)$, a rational ratio.

If a or b is zero all the tiles are oriented the same way and we are done. If the parts of our line that cross tiles vertically never change, then the height of the big rectangle is both a multiple of x and of y, an impossibility. Thus somewhere our line reaches a point where some vertical crossings switch to horizontal while simultaneously some horizontal crossings switch to vertical, but that again is impossible because the distance from the top of the big rectangle to the place where this happens would then be a multiple of both x and y.

If the ratio of the sides of a tile is rational, we may as well assume the sides have integer lengths a and b, and that the tiled square C is $c \times c$. If there is no horizontal tiling, a and b do not both divide c. We may as well suppose that a does not divide c, and write $c = ma + r$, where $0 < r < a$.

We want to show that in this case there is no tiling at all. For example, there can't be any tiling of a 10×10 square by 4×1 rectangles.

We cover C with a grid of c^2 unit squares $S(i, j)$, numbered as in a matrix by pairs i, j, $1 \leq i \leq c$, $1 \leq j \leq c$. We now color all the squares $S(i, j)$ for which $i - j \equiv 0 \mod a$.

This colors all c unit squares on the main diagonal, plus all the squares on other diagonals a multiple of a diagonals away from the main diagonal; and it colors every ath square along each row or column. As a consequence, suppose R is any rectangle carved out of the grid. If either R's height or R's width is a multiple of a (or both) then R is "balanced," in the sense that the colored squares in R comprise exactly the fraction $1/a$ of all the unit squares in R.

Thus the $c \times ma$ rectangle U containing $S(1, 1)$ is balanced, as are also the $ma \times c$ rectangle V containing $S(1, 1)$ and their intersection, the $ma \times ma$ square W again containing $S(1, 1)$. Let L be the little $r \times r$ square containing $S(c, c)$, that is, diagonally opposite the $S(1, 1)$ corner.

The figure shows the $a = 4$, $c = 10$ case, with the rectangles U and V (and thus also the square W) outlined in heavy black.

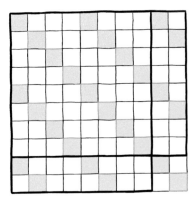

If $|R|$ denotes the area of a rectangle R, then $|C| = |U| + |V| - |W| + |L|$ and a similar statement applies to the numbers of colored squares in each of these rectangles. But L is not balanced: It containes r colored squares on its main diagonal, so has a fraction $r/r^\square = 1/r > 1/a$ of its squares colored. It follows that C is not balanced.

But our tiles are balanced, thus they cannot tile C. ♡

Locker Doors

Lockers numbered 1 to 100 stand in a row in the main hallway of Euclid Junior High School. The first student arrives and opens all the lockers. The second student then goes through and re-closes lockers numbered 2, 4, 6, etc.; the third student changes the state of every locker whose number is a multiple of 3, then the next every multiple of 4, etc., until the last student opens or closes only locker number 100.

After the 100 students have passed through, which lockers are open?

Solution: The state of locker n is changed when the kth student passes through, for every divisor k of n. Here, we make use of the fact that divisors *usually* come in pairs $\{j, k\}$ where $j \cdot k = n$ (including the pair $\{1,n\}$); so the net effect of students j and k on this locker is nil. The exception is when n is a perfect square, in which case there is no other divisor to cancel the effect of the $\sqrt{n}$th student; therefore, the lockers which are open at the end are exactly the perfect squares, 1, 4, 9, 16, 25, 36, 49, 64, 81, and 100. ♡

For the next puzzle you might like to know a certain handy rhyme, attributed to the mathematician Nathan Fine but inspired by a lovely proof by the great, late Paul Erdős:

> Chebyshev said it and I say it again
> There's always a prime between n and $2n$.

This fact is known as Bertrand's Postulate, proved by Chebyshev in 1852, with later (neater) proofs by Ramanujan and Erdős.

Factorial Coincidence

Suppose that a, b, c, and d are positive integers, all different, all greater than one. Can it be that $a!^b = c!^d$?

Solution: Assume that such numbers exist, say with $a < c$, thus $b > d$. Then $c > 2$, and by Bertrand's Postulate, there is a prime p with $c/2 < p < c$. This p appears exactly d times in the prime factorization of $c!^d$ but either b times or not at all in the prime factorization of $a!^b$. The contradiction shows that the desired quartet (a, b, c, d) does not exist.

Even Split

Prove that from every set of $2n$ integers, you can choose a subset of size n whose sum is divisible by n.

Solution: Call a set "flat" if it sums to 0 modulo n. Let us note first that the statement we want to prove implies the following seemingly weaker statement: If S is a flat set of $2n$ numbers, then S can be split into two flat sets of size n. However, that in turn implies that any set of only $2n - 1$ numbers contains a flat subset of size n because we can add a $2n$th number to make the original set flat, then apply the previous statement to split this into *two* flat subsets of size n. One of these (the one without the new number) will do the trick.

So all three of these statements are equivalent. Suppose we can prove the second for $n = a$ and for $n = b$. Then if a set S of size $2n = 2ab$ sums to 0 mod ab, it is, in particular, flat with respect to a, and we can peel off subsets $S_1, \ldots, S_{2b}$ of size a which are also flat with respect to a. Each of these subsets S_i has a sum we can write in the form ab_i. The numbers b_i now constitute a set of size $2b$ which sums to $0 \mod b$, so we can split them into two sets of size b which are flat with respect to b. The unions of the sets S_i in each part are a bipartition of the original S into sets of size ab which are ab-flat, just what we wanted.

It follows that if we can prove the statement for $n = p$ prime, then we have it for all n. Let S be a set of size $2p$, with the idea of creating a p-flat subset of size p.

How can we create such a subset? One natural possibility is to pair up the elements of S and choose one element from each pair. Of course, if we do that, it will behoove us to ensure that the elements in each pair are different mod p, so our choice will not be of Hobson's variety. Can we do that?

Yes, order the elements of S modulo p (say, 0 through $p-1$) and consider the pairs (x_i, x_{i+p}) for $i = 1, 2, \ldots, p$. If x_i were equivalent to x_{i+p} mod p for some i, then $x_i, x_{i+1}, \ldots, x_{i+p}$ would all be equivalent mod p and we could take p of them to make our desired subset.

Now that we have our pairs, we proceed by "dynamic programming." Let A_k be the set of all sums (mod p) obtainable by adding one number from each of the first k pairs. Then $|A_1| = 2$ and we claim $|A_{k+1}| \geq |A_k|$, and moreover, $|A_{k+1}| > |A_k|$ as long as $|A_k| \neq p$. This is because $A_{k+1} = (A_k + x_{k+1}) \cup (A_k + x_{k+1+p})$; thus if $|A_{k+1}| = |A_k|$, these two sets are identical, implying $A_k = A_k + (x_{k+1+p} - x_{k+1})$. This is impossible since p is prime and $x_{k+1+p} - x_{k+1} \not\equiv 0 \mod p$, unless $|A_k| = 0$ or p.

Since there are p pairs, we must eventually have $|A_k| = p$ for some $k \leq p$, hence $|A_p| = p$ and, in particular, $0 \in A_p$. The statement of the puzzle follows. ♡

Factorials and Squares

Consider the product $100! \cdot 99! \cdot 98! \cdot \; \cdots \; \cdot 2! \cdot 1!$. Call each of the 100 factors $k!$ a "term." Can you remove one term and leave a perfect square?

Solution: Perfect squares have some nice properties. For example, the product of any number of perfect squares is itself a perfect square: e.g., $A^2 \cdot B^2 \cdot C^2 = (A \cdot B \cdot C)^2$.

Let's call our big product N and observe that it's not far from already being a product of perfect squares. The product of N's first two terms, for example, is $100 \cdot 99!^2$ (which happens to be a perfect square since $100 = 10^2$).

In fact, we could pair up all the terms and write N in the following form: $100 \cdot 99!^2 \cdot 98 \cdot 97!^2 \cdot 96 \cdot 95!^2 \cdot \ \cdots \ \cdot 4 \cdot 3!^2 \cdot 2 \cdot 1!^2$, which is a perfect square times the number $M = 100 \cdot 98 \cdot 96 \cdot \ \cdots \ \cdot 4 \cdot 2$. But we can rewrite M as $2^{50} \cdot 50 \cdot 49 \cdot 48 \cdot \ \cdots \ \cdot 2 \cdot 1 = (2^{25})^2 \cdot 50!$.

So N is the product of a lot of perfect squares and the number $50!$. It follows that if we remove the $50!$ term from N, we're down to a perfect square. $\heartsuit$

It's time for our theorem. You probably know that $\sqrt{2}$ is irrational, that is, it cannot be written as a fraction a/b for any integers a or b. You might even know a proof, something like this:

Suppose $\sqrt{2}$ is a fraction, and write that fraction in lowest terms as a/b. Then $2 = a^2/b^2$, and therefore $a^2 = 2b^2$, thus a^2 is even and so must be a; write $a = 2k$. Then $b^2 = 2k^2$ so b is also even, but this contradicts the assumption that a/b is in lowest terms.

Theorem. *For any positive integer n, $\sqrt{n}$ is either an integer or irrational.*

Proof. The proof for $n = 2$ extends automatically to the case where n is a prime p: If $\sqrt{p} = a/b$, reduced to lowest terms, then $p = a^2/b^2$, so a is a multiple of p and thus $a^2 = (pk)^2$ is a multiple of p^2, but then b^2 is a multiple of p thus so is b, and we again contradict a/b being reduced to lowest terms.

This doesn't work when n is not prime, though, because we cannot deduce from a^2 being a multiple of n that a is too. But wait, this does work when n is *square-free*, that is, it is a product of distinct primes; then, we can apply the argument for each prime separately.

So we know that $\sqrt{n}$ is either an integer or irrational when n is square-free. But every n is the product of a perfect square and a square-free number (the latter obtained by taking the product of all primes that appear an odd number of times in the factorization of n). So let's write $n = k^2m$, where m is square-free. Then $\sqrt{n} = k\sqrt{m}$ and using the fact that an integer multiple of an integer is an integer, and a non-zero-integer multiple of an irrational number is irrational, we have our proof. $\heartsuit$

With a little work you can extend the above to kth-roots of n, where k and n are integers greater than 1.

7. The Law of Small Numbers

It sounds silly, but many otherwise-very-smart people, given a puzzle with numbers in it, treat those numbers as if underlined in a sacred text. This is math—you're allowed to change the numbers and see what happens! If the numbers in a puzzle are dauntingly large, replace them with small ones. How small? As small as possible, without making the puzzle trivial; if that doesn't give you enough insight, make them gradually bigger.

Domino Task

An 8×8 chessboard is tiled arbitrarily with 32 2×1 dominoes. A new square is added to the right-hand side of the board, making the top row length 9.

At any time you may move a domino from its current position to a new one, provided that after the domino is lifted, there are two adjacent empty squares to receive it.

Can you retile the augmented board so that all the dominoes are horizontal?

Solution: Yes. Let T be the "snake" tiling obtained by placing four vertical tiles on the left column, three on the right column (missing the top and bottom squares), and filling in all but the lower right square with horizontal dominoes.

We proceed to construct the snake tiling one tile at a time, as follows: At each time there's a hole at some square, say S, and adjacent to it is the other square (call it S') covered by the snake tile that covers S; move the domino currently covering S' so that it covers both S and S', thus becoming the snake tile.

In this way the hole zigzags across the board and back, finally reaching the lower right-hand square when the snake tiling is complete.

Now it's easy to shift the snake tiles one at a time, beginning with the rightmost one on the bottom row, to create a horizontal tiling.

How to find this curious two-part algorithm? Try the problem first on a 4 × 4 board!

Spinning Switches

Four identical, unlabeled switches are wired in series to a light bulb. The switches are simple buttons whose state cannot be directly observed, but can be changed by pushing; they are mounted on the corners of a rotatable square. At any point, you may push, simultaneously, any subset of the buttons, but then an adversary spins the square. Is there an algorithm that will enable you to turn on the bulb in at most a fixed number of spins?

Solution: Looking at a simpler version of this puzzle is crucial. Consider the two-switch version, where all you've got are two buttons on diagonally opposite corners of the square. Pushing both buttons will ascertain whether the two switches were both in the same state, since then the bulb will light (if it wasn't already lit). Otherwise, push one button, after which they *will* be in the same state, and at worst one more operation of pushing both buttons will turn on the bulb. So three operations suffice.

Back to the four-switch case. Name the buttons N, E, S, and W after the compass directions, although of course the botton you're calling N now might be the button you will call E, W, or S after a spin. Suppose that at the start, diagonally opposite switches (N and S, E and W) are in the same state—both on or both off. Then you can treat opposite pairs as a single button and use the two-button solution: Push both pairs (i.e., all four switches); then one pair (which may as well be N–S); then both pairs again, and you're done. So begin with those three operations; if the light doesn't go on, then one or both of the opposite pairs must have been mismatched. Try flipping two neighboring switches, say N and E, then going back through your three-move two-button solution. Then you're fine if both pairs were mismatched. If not, push just one button; that'll either make both opposite pairs match, or both mismatch. Run through the two-button solution a third time. If bulb is still off, push N and E again and now you *know* both opposite pairs match, and a fourth application of the two-button solution will get that bulb turned on.

In conclusion, pushing buttons NESW, NS, NESW, NE, NESW, NS, NESW, N, NESW, NS, NESW, NE, NESW, NS, NESW, will at some point turn the light on—15 operations. No sequence of fewer than 15 operations can be guaranteed to work because there are $2^4 = 16$ possible states for the four switches, and they all must be tested; you get to test one state (the starting state) for free.

Seeing the solution for four buttons, you can generalize to the case where the number of buttons is any power of two; if there are 2^k buttons, the solution will take, and require, $2^{2^k} - 1$ steps. (When there are n buttons, they are located at the corners of a spinnable regular n-gon.)

The puzzle is insoluble when the number of buttons, n, is not a power of 2. Let's just prove that for three buttons, no fixed number of operations can guarantee to get that bulb on. (For general n, write n as $m \cdot 2^k$ for some odd number $m > 1$; it is m which plays the role of 3 in what follows.)

You may as well assume the switches are spun before you even make your first move. Suppose that before they are spun, the switches are not all in the same state. Then it is easy to check that *no matter what move you planned*, if you were unlucky with the spin, then after the spin and your move, the switches will still not all be in the same state.

It follows that you can never be sure that you have ever had all the switches in the same state, so no fixed sequence of moves can guarantee to light the bulb. It's curious that you can solve the problem for 32 buttons (albeit in about 136 years, at one second per operation), but not for just three buttons.

Candles on a Cake

It's Joanna's 18th birthday and her cake is cylindrical with 18 candles on its $18''$ circumference. The length of any arc (in inches) between two candles is greater than the number of candles on the arc, excluding the candles at the ends.

Prove that Joanna's cake can be cut into 18 equal wedges with a candle on each piece.

Solution: The conditions give some assurance that the candles are fairly evenly spaced; one way to say that is that as we move around the circumference from some fixed origin 0, the number of candles we encounter is not far from the distance we have traveled. Accordingly, let a_i be the arc-distance from 0 to the ith candle, numbered counterclockwise, and let $d_i = a_i - i$.

We claim that for any i and j, d_i and d_j differ by less than 1. We may assume $i < j$; suppose, for instance, that $d_j - d_i \leq -1$. Then $j - i - 1 \geq a_j - a_i$, but $j - i - 1$ is the number of candles between i and j, contradicting the condition. Similarly, if $d_j - d_i \geq 1$, then $d_i - d_j \leq -1$ and the same argument applies to the other arc, counterclockwise from j to i.

So the "discrepancies" d_i all lie in some interval of length less than 1. Let d_k be the smallest of these, and let ε be some number between 0 and d_k, so that all the d_i's lie strictly between ε and $1 + \varepsilon$. Now cutting the cake at ε, $\varepsilon + 1$, etc. gives the desired result.

How are you supposed to find this proof? By trying two candles, then three, instead of eighteen.

Lost Boarding Pass

One hundred people line up to board a full jetliner, but the first has lost his boarding pass and takes a random seat instead. Each subsequent passenger takes his or her assigned seat if available, otherwise a random unoccupied seat.

What is the probability that the last passenger to board finds his seat unoccupied?

Solution: This is a daunting problem if you insist on working out what happens with 100 passengers; the number of possiblities is astronomical. So let's reduce the number to something manageable. With two passengers it's obvious that the probability that the second (i.e., last) get her own seat is $\frac{1}{2}$. What about three passengers?

It's useful to number the seats according to who was supposed to sit there. If passenger 1 sits in seat 1, his assigned seat, then everyone will be get his or her own seat. If he sits in seat 3, then passenger 2 will get seat 2 and passenger 3 will get seat 1. Finally, if passenger 1 sits in seat 2, then whether passenger 3 gets seat 3 will depend on whether passenger 2 chooses seat 1 or seat 3. Altogether, the probability that passenger 3 gets her own seat is $\frac{1}{3} + \frac{1}{3} \cdot \frac{1}{2} = \frac{1}{2}$.

Interesting! Is it possible that the answer is always $\frac{1}{2}$?

We notice that in the above analysis, the last passenger never ends up in seat 2. In fact, now that we think of it, we see that with n passengers total, the last passenger never ends up in seat i for $1 < i < n$. Why? Because when passenger i came on board, either seat i was already taken, or it is taken now. Thus, seat i will *never* be available to the last passenger. The only seats that passenger n could end up in are seat 1 and seat n.

We can't yet conclude that the probability that passenger n gets seat n is $\frac{1}{2}$—we still need to argue that seat 1 and seat n are equally likely to be available at the end. But that's easy, because every time someone took a random seat, they were equally likely to choose seat 1 or seat n. Putting it another way, seat 1 and seat n were treated identically throughout the process; thus, by symmetry, each has the same likelihood of being open when passenger n finally gets on board.

Flying Saucers

A fleet of saucers from planet Xylofon has been sent to bring back the inhabitants of a certain randomly-selected house, for exhibition in the Xylofon Xoo. The house happens to contain five men and eight women, to be beamed up randomly one at a time.

Owing to the Xylofonians' strict sex separation policy, a saucer cannot bring back earthlings of both sexes. Thus, it beams people up until it gets a member of a second sex, at which point that one is beamed back down and the saucer takes off with whatever it has left. Another saucer then starts beaming people up, following the same rule, and so forth.

What is the probability that the last person beamed up is a woman?

Solution: Let's try some smaller numbers and see what happens. Obviously if the house is all men or all women, the sex of the last person beamed up will be determined. If there are equal numbers of men and women, then by symmetry, the probability that the last person beamed up is a woman would be $\frac{1}{2}$. So the simplest interesting case is, say, one man and two women.

In that case, if the man is beamed up first (probability: $\frac{1}{3}$), the last one will be a woman. Suppose a woman is beamed up first; if she is followed by a man (who is then beamed back down), we are down to the symmetric case where the probability of ending with a woman is $\frac{1}{2}$. Finally, if a second woman follows the first (probability $\frac{2}{3} \cdot \frac{1}{2} = \frac{1}{3}$), the man will be last to be beamed up. Putting the cases together, we get probably $\frac{1}{2}$ that the last person beamed up is a woman. Is it possible that $\frac{1}{2}$ is the answer no matter how many men and women are present, as long as there's at least one of each?

Looking more closely at the above analysis, it seems that the sex of the last person beamed up is determined by the *next-to-last* saucer—the one that reduces the house to one sex. To see why this is so, it is useful to imagine that the Xylofonian acquisition process operates the following way: Each time a flying saucer arrives, the current inhabitants of the house arrange themselves in a uniformly random permutation, from which they are beamed up left to right.

For example, if the inhabitants at one saucer's arrival consist of males m_1 and m_3 and females f_2, f_3, and f_5, and they arrange themselves "f_3, f_5, m_2, f_2, m_3," then the saucer will beam up f_3, f_5, and m_2, then will beam m_3 back down again and take off with just the females f_3 and f_5. The remaining folks, m_3, m_2, and f_2 will now re-permute themselves in anticipation of the next saucer's arrival.

We see that a saucer will be the next to last just when the permutation it encounters consists of all men followed by all women, or all women followed by all men. But no matter how many of each sex are in the house at this point, these two events are equally likely! Why? Because if we simply reverse the order of a such a permutation, we go from all-men-then-all-women to all-women-then-all-men, and vice-versa.

There's just one more observation to make: If both men and women are present initially, then one saucer will never do, thus there always will be a next-to-last saucer. When that comes—even though we do not know in advance which saucer it will be—it is equally likely to depart with the rest of the men, or the rest of the women.

Gasoline Crisis

You need to make a long circular automobile trip during a gasoline crisis. Inquiries have ascertained that the gas stations along the route contain just enough fuel to make it all the way around. If you have an empty tank but can start at a station of your choice, can you complete a clockwise round trip?

Solution: Yes. The trick is to imagine that you begin at station 1 (say) with *plenty* of fuel, then proceed around the route, emptying each station as you go. When you return to station 1, you will have the same amount of fuel in your tank as when you started.

As you do this, keep track of how much fuel you have left as you pull into each station; suppose that this quantity is minimized at station k. Then, if you start at station k with an empty tank, you will not run out of fuel between stations. ♡

Coins on the Table

One hundred quarters lie on a rectangular table, in such a way that no more can be added without overlapping. (We allow a quarter to extend over the edge, as long as its center is on the table.)

Prove that you can start all over again and cover the whole table with 400 quarters! (This time we allow overlap *and* overhang).

Solution: Let us observe first that if we double the radius (say, from $1''$ to $2''$) of each of the original coins, the result will be to cover the whole table. Why? Well, if a point P isn't covered, it must be $2''$ or more from any coin center, thus a (small) coin placed with its center at P would have fit into the original configuration. (See the first two figures below for an example of an original configuration, and what happens when the coins are expanded.)

Now, if we could replace each big coin by four small ones that cover the same area, we'd be done—but we can't.

But rectangles *do* have the property that they can be partitioned into four copies of themselves. So, let us shrink the whole picture (of big coins covering the table) by a factor of two in each dimension, and use four copies (as in the next figure) of the new picture to cover the original table!

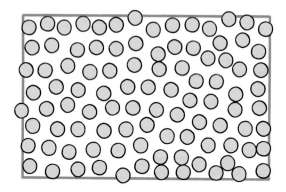

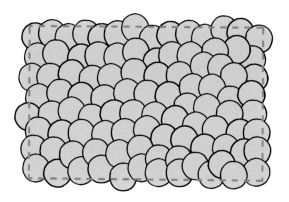

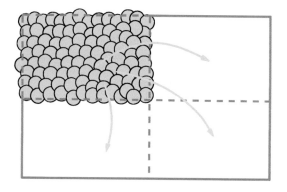

Surprisingly (perhaps), this lovely but seemingly rather crude argument gives the best possible factor: Replace the factor 4 by anything smaller, say 3.99, and the statement of the puzzle is no longer true.

To see this, we consider the limiting case where the table is very large and the coins numerous, so that boundary effects are negligible. Replace the table by a honeycomb-pattern bathroom floor with regular hexagonal tiles of diameter 2. Since each tile could then partitioned into six equilateral triangles of side 1 and thus area $\sqrt{3}/4$, the tile itself has area $6 \times \sqrt{3}/4 = 3\sqrt{3}/2$.

We can cover the floor now by covering each tile with a coin whose boundary is the tile's circumscribing circle (see next figure).

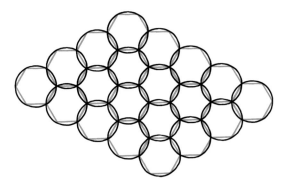

Each coin then has radius 1 and thus area π. If the floor has area A then, ignoring boundary effects, the total area of the coins will be $\pi A/(3\sqrt{3}/2) = 2\pi A/(3\sqrt{3})$.

Now, how thinly can we cover the floor without being able to add a non-overlapping coin? Let us use the same tiling, but this time we cover only every third tile (see next figure); and we cover it not with a circumscribed circle, but a coin placed over the center of the hexagon which is just a teensy bit bigger than the *inscribed* circle. This just barely prevents us from adding any more coins; how much is the coin area now?

Well, the coin radius is a tiny bit bigger than the altitude of one of the six equilateral triangles making up a hexagon—namely, $\sqrt{3}/2$. Hence the coin area just exceeds $\pi \times (\sqrt{3}/2)^2 = 3\pi/4$.

It follows that the total coin area on the floor is as close as we like to $(1/3) \times (3\pi/4) \times A/(3\sqrt{3}/2) = \pi A/(6\sqrt{3})$, one-fourth of what we had before!

The result of all this is that we have proved not only the statement of the puzzle, but two not-so-easy extremal properties of disks in the

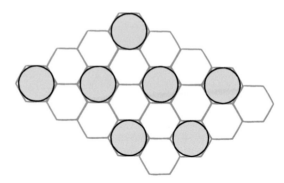

plane. The first says that there is no better way to cover the plane by unit disks than by circumscribing the tiles in a hexagonal tiling, as we did above; the second says that there is no more efficient way to *prevent* the addition of a non-overlapping unit disk than by centering on every third hexagonal tile a disk slightly bigger than the inscribed disk, again as we did above.

If you think these properties are obvious to begin with, consider the (seemingly) even more obvious fact that the densest way to *pack* unit disks in the plane is to use an inscribed disk in each hexagon. This was not rigorously proved until the great Hungarian geometer László Fejes Tóth (1915–2005) did it in 1972!

Coins in a Row

On a table is a row of 50 coins, of various denominations. Alix picks a coin from one of the ends and puts it in her pocket; then Bert chooses a coin from one of the (remaining) ends, and the alternation continues until Bert pockets the last coin.

Prove that Alix can play so as to guarantee at least as much money as Bert.

Solution: This puzzle resists the most obvious approaches. It's easy to check that Alix could do quite badly by always choosing the most valuable coin, or the coin that exposes the less valuable coin to Bert, or any combination of these. Basically, if she only looks a move or two ahead, she's in trouble.

In fact, for Alix to play *optimally*, she needs to analyze all the possible situations that may later arise. This can be done by a technique called "dynamic programming."

But we were not asked to provide an optimal strategy for Alix, just a strategy that guarantees her at least half the money. Experimenting with 4 or 6 coins instead of 50 might lead you to the following key observation.

Suppose the coins alternate quarter, penny, quarter, penny, and so forth, ending (since 50 is even) in a penny. Then Alix can get *all* the quarters! In fact, no matter what the coins are, if we number the coins from 1 to 50 left to right, Alix can take all the odd-numbered ones—or all the even-numbered ones.

But wait a minute—one of those two groups of coins must contain at least half the money! ♡

Powers of Roots

What is the first digit after the decimal point in the number $(\sqrt{2}+\sqrt{3})$ to the billionth power?

Solution: If you try entering $(\sqrt{2} + \sqrt{3})^{1,000,000,000}$ in your computer, you're likely to find that you get only the dozen or so most significant figures; that is, you don't get an accurate enough answer to see what happens after the decimal point.

But you can try smaller powers and see what happens. For example, the decimal expansion of $(\sqrt{2} + \sqrt{3})^{10}$ begins 95049.9999895. A bit of experimentation shows that each even power of $(\sqrt{2} + \sqrt{3})$ seems to be just a hair below some integer. Why? And by how much?

Let's try $(\sqrt{2} + \sqrt{3})^2$, which is about 9.9. If we play with $10 - (\sqrt{2} + \sqrt{3})^2$ we discover that it's equal to $(\sqrt{3} - \sqrt{2})^2$. Aha!

Yes, $(\sqrt{3}+\sqrt{2})^{2n}+(\sqrt{3}-\sqrt{2})^{2n}$ is always an integer, because when you expand it, the odd binomial coefficients cancel and the even ones are integers. But of course $(\sqrt{3} - \sqrt{2})^{2n}$ is very small, about 10^{-n}, so the first roughly n digits of $(\sqrt{3} + \sqrt{2})^{2n}$ after the decimal point are all 9's.

Coconut Classic

Five men and a monkey, marooned on an island, collect a pile of coconuts to be divided equally the next morning. During the night, however, one of the men decides he'd rather take his share now. He tosses one coconut to the monkey and removes exactly $\frac{1}{5}$ of the remaining coconuts for himself. A second man does the same thing, then a third, fourth, and fifth.

The following morning the men wake up together, toss one more coconut to the monkey, and divide the rest equally. What's the least original number of coconuts needed to make this whole scenario possible?

Solution: You can solve this by considering two men instead of five, then three, then guessing. But the following argument is irresistible, once found.

There's an elegant "solution" to the puzzle if you allow negative numbers of coconuts(!). The original pile has -4 coconuts; when the first man tosses the monkey a coconut, the pile is down to -5 but when he "takes" $\frac{1}{5}$ of this he is actually adding a coconut, restoring the pile to -4 coconuts. Continuing this way, come morning there are still -4 coconuts; the monkey takes one and the men split up the remaining -5.

It's not obvious that this observation does us any good, but let's consider what happens if there is no monkey; each man just takes $\frac{1}{5}$ of the pile he encounters, and in the morning there's a multiple of 5 coconuts left that the men can split. Since each man has reduced the pile by the fraction $\frac{4}{5}$, the original number of coconuts must have been a multiple of 5^6 (which shrinks to $4^5 \cdot 5$ by morning).

All we need to do now is add our two pseudo-solutions, by starting with $5^6 - 4 = 15{,}621$ coconuts. Then the pile reduces successively to $4 \cdot 5^5 - 4$ coconuts, $4^2 \cdot 5^4 - 4$, $4^3 \cdot 5^3 - 4$, $4^4 \cdot 5^2 - 4$, and $4^5 \cdot 5 - 4$. When the monkey gets his morning coconut, we have $4^5 \cdot 5 - 5$ coconuts, a multiple of 5, for the men to split. This is best possible because we needed $5^5 \cdot k - 4$ coconuts to start with, just to have an integer number come morning, and to get $4^5 \cdot k - 5$ to be a multiple of 5 we needed k to be a multiple of 5.

Doubtless, many theorems in mathematics were "discovered" when someone played around with small numbers and then saw a pattern that turned out to be a provable phenomenon.

Here's a theorem that could well have been found that way. Suppose you are running a dojo with an even number n of students. Each day you pair the students up for one-on-one sparring. Can you do this in such a way that over a period of days, each student spars with each other student exactly once?

Theorem. *For any even positive integer n there is a set of pairings ("perfect matchings") of the numbers $\{1, 2, \ldots, n\}$ such that every pair $\{i, j\}$ appears in exactly one pairing.*

A check of small numbers suggests that this seems to work: For $n = 2$, for instance, there is just the one pairing consisting of the pair $\{1, 2\}$, and for $n = 4$ we can (in fact, must) take the pairings $\{\{1, 2\}, \{3, 4\}\}$, $\{\{1, 3\}, \{2, 4\}\}$, and $\{\{1, 4\}, \{2, 3\}\}$.

For $n = 6$, though, we have choices to make. Is there a nice way to make them?

In fact, there are several; my favorite is the following. We know student n has to be paired with every other student; let's put her in the middle of a circle, with the rest of the students spaced equally around the circle. In the ith of our $n-1$ pairings, student n is paired with student i; draw a radius from n to i. The rest of the students are paired by line segments that run perpendicular to that radius (see the figure below).

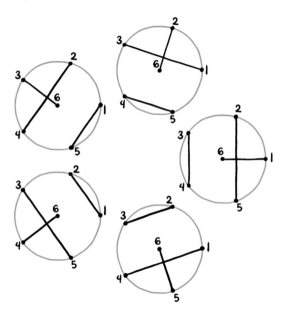

Since $n-1$ is odd, all the radii from n to other students are at different angles; it follows that no two students are paired twice, and since the eventual number of pairs is $(n-1) \times n/2 = \binom{n}{2}$, every pair is accounted for. ♡

8. Weighs and Means

You know how to compute the average of a set of numbers

$$\bar{x} := (x_1 + x_2 + \cdots + x_n)/n$$

or you wouldn't even have picked up this book. In solving puzzles, it's often useful to reverse this process, and compute the sum from the average instead.

For example, have you forgotten the formula for the sum of the integers from 1 to n? No problem. Their average is $(n+1)/2$ (you can pair them up to see that) and there are n of them, so their sum must be $n(n+1)/2$. Similarly, the sum of the integers from a to b must be $(b-a+1)(a+b)/2$.

Here's a theorem about averages that's helpful in problem-solving and is almost too obvious to call a theorem:

Theorem. *If $\bar{x}$ is the average of $\{x_1, x_2, \ldots, x_n\}$ then $\bar{x}$ lies between the minimum of the x_i's and the maximum. Moreover, the minimum, maximum and average are either all different or all the same.*

Another way of saying the first part is that *some* x_i must be larger than or equal to $\bar{x}$, and some x_i must be smaller than or equal to $\bar{x}$. Obvious, but useful!

The above theorem applies also in the more general case of *weighted averages*, also known as *convex combinations*. There, each element of the set to be averaged has some weight (a positive real number) w_i, and

$$\bar{x} := (w_1 x_1 + w_2 x_2 + \cdots + w_n x_n)/(w_1 + \cdots + w_n).$$

Often the weights already sum to 1, so you can skip the dividing. That happens, for example, when the w_i's constitute a probability distribution, in which case $\bar{x}$ becomes the *expected value* of the x_i's. (Much more about expectation is coming in a later chapter.)

The following puzzle is fairly typical of those in which the above theorem can come in handy.

Tipping the Scales

A balance scale sits on the desk of the science teacher, Ms. McGregor. There are weights on the scale, which is currently tipped to the right. On each weight is inscribed the name of at least one pupil.

On entering the classroom, each pupil moves every weight carrying his or her name to the opposite side of the scale. Must there be a set of pupils that Ms. McGregor can let in that will tip the scales to the left?

Solution: To set ourselves up for averaging, consider all subsets of students, including the empty set and the full set. Each weight will be on the left exactly half the time (pick any name on that weight, say Jay, and assume the decision of whether to include Jay in the set is the last decision to be made.) Thus the total weight on the left for all these subsets is the same as the total weight on the right. Another way to say it is that on the average, the scale balances. Since the empty set results in a tip to the right, some other set must tip to the left.

The "averaging" technique used in Tipping the Scales comes up often: Watch for it!

Here's a quickie.

Men with Sisters

On average, who has more sisters, men or women?

Solution: This puzzle can seem confusing. If, like me, you were raised in a family with both male and female children, you might have observed that in such a family the boys have more sisters (by one) than the girls have. So maybe men have more sisters than women on average?

Wait, in a family with several children, all girls, there are several girls with sisters; in a family of all boys, none of those boys has a sister. So maybe that cancels out the effect of the families with children of both sexes.

The real issue here is independence of gender. If you make the reasonable (and pretty accurate) assumption that the genders of siblings are not influenced by your own gender, then knowing the numbers of brothers and sisters a person has tells you nothing about whether that person is male or female. It follows that the reverse is true too, thus men and women have the same number of sisters on average.

If you want to get down to detail, there's actually a slight positive correlation of gender between siblings; that is, the sibling of a girl is slightly more likely to be a girl than the sibling of a boy is. This is caused mostly by the fact that identical twins are the same sex. As a result, women have *very slightly* more sisters on average than men do.

The next puzzle uses the continuum form of our theorem.

Watches on the Table

Fifty accurate watches and a tiny diamond lie on a table. Prove that there exists a moment in time when the sum of the distances from the diamond to the ends of the minute hands exceeds the sum of the distances from the diamond to the centers of the watches.

Solution: Considering just one watch, we claim that during the passing of an hour the *average* distance from the diamond D to the tip M of the watch's minute hand exceeds the distance from D to the center W of the watch. This is so because if we draw a line $\mathcal{L}$ through D perpendicular to the line from D to W, then the average distance from $\mathcal{L}$ to M is clearly equal to $\mathcal{L}W$ which is in turn equal to DW. But DM is at least equal to $\mathcal{L}M$ and usually bigger.

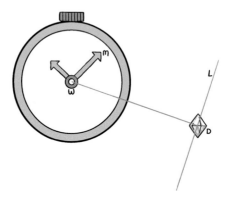

Of course, if we average over all the watches we reach the same conclusion, and it follows that sometime during the hour the desired inequality is achieved.

The requirement that the watches be accurate is to ensure that each minute hand moves at constant speed. It doesn't matter if some watches are slow and some fast (assuming each minute hand rotates at constant speed), unless our patience is limited to one hour.

One additional note: If the watches are accurate and you place and rotate them carefully, you can ensure that the sum of the distances from the center of the table to the ends of the minute hands is *always* strictly greater than the sum of the distances from the center of the table to the center of the watches.

Another nice property of averaging is that adding or subtracting one datum from a list has a predictable effect on the average: If the datum exceeds the average, then adding it will raise the average and deleting it will lower the average. If the datum is below average, the effect will be the opposite; and of course if the datum is exactly average, there is no effect either way. The next puzzle uses this idea to examine a frequently-heard type of insult.

Raising Art Value

Fans of Gallery A like to say that last year, when Gallery A sold to Gallery B "Still Life with Kumquats," the average value of the pictures in each gallery went up. Assuming that statement is correct, and that the value of each of the 400 pictures in the galleries' combined holdings is an integral number of dollars, what is the least possible difference between the average picture values of Gallery A and Gallery B before the sale?

Solution: It's a bit curious that we are asked for a dollar amount when no quantity of money is specified in the puzzle, but let's plunge in anyway. Say there were p pictures in Gallery A before the sale, thus $400 - p$ in Gallery B. Suppose the average value of A's pictures was a, and B's b; then the behavior of the averages says exactly that the value s of "Still Life with Kumquats" satisfies $a > s > b$.

We know that a is a fraction that can be written as m/p where m is a whole number (of dollars); similarly b can be written as $n/(400 - p)$ where n is a whole number. By how little can a exceed an integer? Easy: By $1/p$, which happens when m is one more than a multiple of p. Similarly, b's shortfall from an integer value can be as little as $1/(400 - p)$. Thus the least possible difference between a and b is $1/p + 1/(400 - p)$ which is minimized when the two denominators are equal, giving $a - b = 1/200 + 1/200 = 1/100$. The upshot is that the insult is not very damaging; we can conclude from it only that the average value of Gallery A's holdings exceeds the average value of Gallery B's holdings by at least one cent.

Averaging comes up a lot in probability theory—whole chapters on probability and on expectation lie ahead. But we can do the following two averaging puzzles without needing any special machinery.

Waiting for Heads

If you flip a fair coin repeatedly, how long do you have to wait, on average, for a run of five heads?

Solution: Since the probability of seeing HHHHH in a particular series of five coinflips is $1/32$, you might think it would take 32 flips on average to get HHHHH. Indeed, 32 flips is the average wait *between* occurrences of HHHHH, but this includes, for example, a wait of length 1 between the first five heads in HHHHHH and the last five.

Waits of length 1 can't help us, because we have no "head start" (OK, pun intended) when we begin flipping.

The real answer is much greater. Between runs, half the time you get the wait of 1 and the rest of the time $1+x$, where x is the desired quantity. Hence it is not x but the average of 1 and $1+x$ that is equal to 32, which gives us $x = 62$.

What are best patterns (of fixed length, say 5) to choose if you want to hit your pattern ASAP? HHHHT is one of them, because seeing an HHHHT gives you no head start toward seeing another. Therefore, starting fresh with your coinflips costs you nothing, and you expect to hit your target in just 32 steps on average.

Finding a Jack

In some poker games the right of first dealer is determined by dealing cards face-up (from a well-shuffled deck) until some player gets a jack. On average, how many cards are dealt during this procedure?

Solution: 10.6. If you tried to compute for each k the probability that the first jack would be the kth card, with the intent of computing the expected value of k directly, you might have found it tough going. But there's an easy way.

Each non-jack is equally likely to be in any of the five regions between successive jacks (to see this, insert a joker, permute circularly at random, then cut at the joker and remove it; or, get a shuffled deck by starting with the jacks, and inserting cards at random, so the next card is equally likely to be in any of the five slots.) Thus the expected number of cards before the first jack is $48/5 = 9.6$; add 1 for the jack itself.

Can we talk about the average of infinitely many things? Sometimes.

Sums of Two Squares

On average, how many ways are there to write a positive number n as the sum of two squares? In other words, suppose n is a random integer between 1 and a zillion. What is the expected number of ordered pairs (i, j) of integers such that $n = i^2 + j^2$?

Solution: Let $f(n)$ be the number of ways to write n as the sum of two squares. What we can't do is add up $f(n)$ for all integers n and then divide by (countable) infinity. But, as suggested in the puzzle, we can choose some huge number Z, and compute

$$\frac{1}{Z} \sum_{n=1}^{Z} f(n) \, .$$

If this value converges to some number r as $Z \to \infty$, it seems reasonable to call this r the average value of $f(n)$ over all n.

And in this case it does converge. Why? The sum $\sum_{n=1}^{Z} f(n)$ counts once each pair (i, j) of integers for which $i^2 + j^2 \le Z$; these are the integer points in the circle $x^2 + y^2 \le Z$. This circle has radius $\sqrt{Z}$ thus area πZ, and therefore the number q of integer points in it is about πZ (more precisely, $q/(\pi Z) \to 1$).

It follows that the number we want, q/Z, approaches π.

It is worth remarking that if we count just the *unordered* pairs of *positive* integers whose squares sum to n, the answer is only $\pi/8$. The reason is that most such pairs—in particular, pairs $\{i, j\}$ for which $i \ne j$—show up eight times in the circle, namely as (i, j), (j, i), $(i, -j)$, etc.

We might also ask for the probability that a random n can be written in at least one way as the sum of two squares. We can deduce from the above calculation that this probability cannot exceed $\pi/8$, which is what the answer would be if most numbers could be written as the sum of two squares in at most one way, up to signs and swapping. But this is not the case at all, and in fact the probability in question approaches zero! A famous theorem says that n is representable as the sum of two squares only if there is no prime which (a) appears an odd number of times in the prime factorization of n, and (b) gives a remainder of 3 when divided by 4. The probability that n passes this test is about 0.7642236535892206629906987312500092328116790541 divided by the square root of the natural log of n, and since $\log n$ grows without bound, we arrive at a curious fact: Although the average of our function $f(n)$ is π, its value is nearly always zero.

Increasing Routes

On the Isle of Sporgesi, each segment of road (between one intersection and the next) has its own name. Let d be the average number of road segments meeting at an intersection. Show that you can take a drive on the Isle of Sporgesi that covers at least d road segments, and hits those segments in strict alphabetical order!

Solution: This puzzle restates a very general, yet surprisingly little-known, theorem of graph theory. Recall that a graph is a (finite, for us) set of *vertices* together with some pairs of vertices, known as *edges*. The *degree* of a vertex is the number of edges it belongs to; if we sum up the degrees of all the vertices each edge is counted twice, thus the *average degree* of the graph is $2m/n$, where m is the number of edges and n is the number of vertices.

A *walk* of length k is a sequence of k edges, each one connecting the end-vertex of the previous edge (if there is one) to the start-vertex of the next one. In a walk, as opposed to a "path," a vertex may be used more than once. If the edges of the graph are ordered in some fashion, then a walk is *increasing* if the edges are encountered in strict increasing order—in particular, the edges of an increasing walk must be distinct. The figure below shows an increasing walk in a graph with edges ordered by letters.

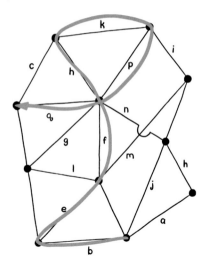

Here, then, is the claimed theorem.

Theorem. *Let G be an arbitrary graph with an arbitrary ordering of its edges. Then there is an increasing walk in G of length at least the average degree of G.*

Proof. We denote by d (as in the puzzle statement) the average degree of G. Since d is already an average, it's natural to look for some set of increasing walks whose average length is d; then we can apply our theorem to deduce that one of them has length at least d. But where do we get such a set of walks? The set of *all* walks doesn't work.

Here's an (admittedly brilliant) idea. Place a pedestrian at each vertex, and at time 1, let the pedestrians at each end of the first edge traverse that edge, thus changing places. At time 2, the pedestrians who now find themselves at the ends of the second edge change places, and so forth, until all m edges have been traversed.

Now, observe: Each pedestrian has taken an increasing walk! The total length of all these walks is $2m$ and there are n of them, so their average length is $d = 2m/n$ and we are done. $\heartsuit$

It's worth noting that this theorem is *tight*, that is, there are graphs with edges ordered so that no increasing walk exceeds d in length. For example, let K_{2t} denote the complete graph (all edges present) on $2t$ vertices. A *complete matching* in K_{2t} (or in any graph) is a set of edges whose endpoints hit every vertex exactly once. The theorem we proved in the last chapter shows that the edges of K_{2t} can be partitioned into disjoint complete matchings (necessarily, $2t-1$ of them, since every vertex is incident to $2t-1$ edges.)

Now, suppose we let the matchings be $M_1, \ldots, M_{2t-1}$ and we use them to order the edges as follows. First, the edges of M_1 are labeled 1 through t arbitrarily. Then the edges of M_2 are labeled with $t+1$ through $2t$, those of M_3 from $2t+1$ through $3t$, etc.

A walk in K_{2t} cannot step directly from one edge to another of the same matching, because no two matching edges meet at a vertex. But an *increasing* walk can't return later to the same matching, on account of the way we have numbered the edges. So an increasing walk hits each of the matchings $M_1, \ldots, M_{2t-1}$ at most once, thus can have length at most $2t-1 = d$.

9. The Power of Negative Thinking

Contradiction is a ridiculously simple, yet somehow incredibly powerful, tool. You want to prove statement A is true? Assume it isn't, and employ *reductio ad absurdum*: Derive an impossible conclusion. Voila!

The great mathematician Godfrey H. Hardy put it nicely: "*Reductio ad absurdum*, which Euclid loved so much, is one of a mathematician's finest weapons. It is a far finer gambit than any chess play: A chess player may offer the sacrifice of a pawn or even a piece, but a mathematician offers the game."

In fact, our first application of contradiction is to figure out who wins a certain game even though we didn't know how to play it well— and still don't.

Chomp

Alice and Bob take turns biting off pieces of an $m \times n$ rectangular chocolate bar marked into unit squares. Each bite consists of selecting a square and biting off that square plus every remaining square above and/or to its right. Each player wishes to avoid getting stuck with the poisonous lower-left square.

Show that, assuming the bar contains more than one square, Alice (the first to play) has a winning strategy.

Solution: It's easy to see that Alice can win if the bar is a square, as she can bite off all but the left-hand and bottom edges, then when Bob bites off a piece of one leg she bites of the same amount of the other. But how does she win in general? The surprising answer is: No one knows!

But we do know that either the first player (Alice) or the second (Bob) must have a winning strategy; suppose it is Bob. Then, in particular, Bob must have a winning answer to Alice's opening move when she merely nibbles off the upper right-hand square.

However, whatever Bob's reply is could have been made by Alice as her own opening move, after which she could follow Bob's allegedly winning strategy and chalk up the game. This of course contradicts the assumption that Bob can always win. Hence, it must be Alice that has a winning strategy.

This kind of proof is known as a *strategy-stealing argument* and does not, unfortunately, tell you how Alice actually wins the game. A similar argument shows that the first player must have a winning strategy in the game of Hex (look it up if you haven't heard of it), yet we don't even know a winning first move.

If you know how to play dots and boxes, you can try this one with the same idea in mind.

Dots and Boxes Variation

Suppose you are playing Dots and Boxes, but with each player having an *option* of whether to go again after making a box. Suppose the board has an odd number of cells (thus, someone must win). Show that the first player to make a box has a winning strategy.

Solution: This just amounts to noticing that by choosing to play again, you are putting yourself in the position your opponent would be in if you *didn't* play again. One of these choices must be winning!

Big Pairs in a Matrix

Suppose that the sum of the greatest two numbers in each row of a certain square matrix is r, and the sum of the greatest two numbers in each column is c. Show that $r = c$.

Solution: Suppose otherwise; by symmetry we may assume $r > c$. Circle the greatest number in each row, and draw a square around the next greatest. The circled numbers are all at least $r/2$, so they must all be in different columns. Suppose the highest number with a square around it, say x, is in column j along with column j's circled number y. Let z be the square-encased number in y's row; then $y + z = r$ but $y + x = c$, impossible because $c < r$ but $x \geq z$.

Bugging a Disk

Elspeth, a new FBI recruit, has been asked to bug a circular room using seven ceiling microphones. If the room is 40 feet in diameter and the bugs are placed so as to minimize the maximum distance from any point in the room to the nearest bug, where should the bugs be located?

Solution: Assuming the ceiling is flat, the problem becomes two-dimensional: Locate seven bugs in a disk of diameter 40 feet so that every point in the disk is within distance d of some bug, where d is as small as possible.

This means that the seven disks of radius d centered at the bugs cover the room. Imagine a regular hexagon inscribed in the big circle; each side will be 20 feet in length, and the centers of the sides, together with the center of the room, will be within 10 feet of any point in the room. That gives $d = 10$; can we do any better?

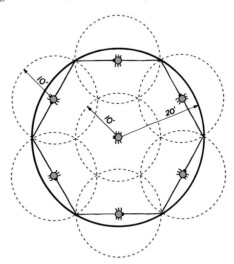

Suppose we could. Since the circles of radius 10 ft about the previous outer six points just barely cover the room's circumference, each using its full diameter, covering the room's circumference with circles of radius *less* than 10 would require at least seven circles. But that would leave the center of the room uncovered, and that is our contradiction.

Often using contradiction is tantamount to using mathematical induction: You assume the statement you wish is false, and consider a *minimal* counterexample.

Pairs at Maximum Distance

Let X be a finite set of points on the plane. Suppose X contains n points, and the maximum distance between any two of them is d. Prove that at most n pairs of points of X are at distance d.

Solution: To solve this puzzle, it's useful to observe that if A, B and C, D are two "max pairs" (pairs of points from X at distance d), then the line segments AB and CD must cross (else one of the diagonals of the quadrilateral $ABDC$ would exceed d in length).

Now assume the statement of the puzzle is false, and let n be the size of a smallest counterexample. Since there are more than n max pairs and each has two points, there must be a point P which participates in three max pairs (say, with points A, B, and C). Every two of the segments PA, PB, and PC must make at most a $60°$ angle at P, and one of them, say B, must lie between the others.

But this makes it pretty tough for B to be in any other max pair, since if BQ were a max pair, it would have to intersect both PA and PC—an impossibility. Thus we can drop B out of X altogether, losing only one max pair and obtaining a smaller counterexample. This contradiction completes the proof.

Lemming on a Chessboard

On each square in an $n \times n$ chessboard is an arrow pointing to one of its eight neighbors (or off the board, if it's an edge square). However, arrows in neighboring squares (diagonal neighbors included) may not differ in direction by more than 45 degrees.

A lemming begins in a center square, following the arrows from square to square. Is he doomed to fall off the board?

Solution: The lemming is indeed doomed. One way to see this is to imagine that the lemming can move to *any* neighboring square, but must turn to face in the direction of the arrow found there. Then the lemming cannot turn 360 degrees around, because if he could you could shrink the cycle on which he does that until it collapses into a contradiction. But the real lemming, if he is to stay on the board, must eventually cycle and when he does so, he will have to make that $360°$ turn.

The conclusion can also be reached using induction (thus also contradiction). If the lemming stays on the board, he must, as we have already noted, eventually settle into a cycle. Let C be the smallest-area cycle (on any board) on which this can happen, and suppose it's a clockwise cycle. Cut the whole board down to C and its interior, then rotate all the arrows $45°$ clockwise to force a smaller cycle!

Curve on a Sphere

Prove that if a closed curve on the unit sphere has length less than 2π, then it is contained in some hemisphere.

Solution: Pick any point P on the curve, travel half way around the curve to the point Q, and let N (standing for North Pole) be the point half-way between P and Q. (Since the distance $d(P,Q)$ from P to Q is less than π, N is uniquely defined). Thinking of N as the North Pole provides us with an equator, and if the curve lies entirely on N's side, that is, in the northern hemisphere, we are done. Otherwise, the curve crosses the equator, and let E be one of the points at which it does so. Then, we observe that $d(E,P) + d(E,Q) = \pi$, since if you poke P through the equatorial plane to P' on the other side, P' is antipodal to Q; hence, $d(E,P') + d(E,Q) = \pi$.

However, for any point X on the curve, $d(P,X) + d(X,Q)$ must be less than π, and this provides the desired contradiction. ♡

Soldiers in a Field

An odd number of soldiers are stationed in a large field. No two soldiers are exactly the same distance apart as any other two soldiers. Their commanding officer radios instructions to each soldier to keep an eye on his nearest neighbor.

Is it possible that every soldier is being watched?

Solution: The problem is most easily solved by considering the two soldiers at shortest distance from each other. Each of these watches the other; if anyone else is watching one of them, then we have a soldier being watched twice and therefore another not being watched at all. Otherwise, these two can be removed without affecting the others. Since the number of soldiers is odd, this procedure would eventually reduce to one soldier not watching anyone, a contradiction.

Our next puzzle is a more serious, and indeed classical, mathematical conundrum.

Alternating Powers

Since the series $1 - 1 + 1 - 1 + 1 - \cdots$ does not converge, the function $f(x) = x - x^2 + x^4 - x^8 + x^{16} - x^{32} + \cdots$ makes no sense when $x = 1$. However, $f(x)$ does converge for all positive real numbers $x < 1$. If we wanted to give $f(1)$ a value, it might make sense to let it be the limit of $f(x)$ as x approaches 1 from below. Does that limit exist? If so, what is it?

Solution: This question began life in a paper by G. H. Hardy, the same guy whose appreciation for the power of contradiction was noted at the beginning of this chapter. Ironically, although Hardy answers his question correctly, he comments that no completely elementary proof seems to be known; a century or so later, we have one, and it involves contradiction.

If you try to determine the answer to this puzzle by computing $f(x)$ for various x near 1, you might reach the wrong conclusion: It *appears* to converge to $\frac{1}{2}$. But appearances can be deceptive, and in fact the limit does not exist!

Suppose $f(x)$ does have a limit, say c, as x approaches 1 from below. Since $f(x) = x - f(x^2)$, that limit can only be $\frac{1}{2}$ (because, taking x as close to 1 as needed, we get $c = 1 - c$). But notice that $f(x^4) - f(x) = x - x^2 > 0$, for all x strictly between 0 and 1. From this we conclude that the sequence $f(x), f(x^4), f(x^{16}), \ldots$ is strictly increasing.

It follows that if there is *any* $x < 1$ for which $f(x) > \frac{1}{2}$, there can be no limit. In fact, $f(0.995)$, for instance, exceeds 0.50088.

What actually happens is that $f(x)$ oscillates more and more rapidly inside an interval of length about 0.0055 centered at $\frac{1}{2}$. Seems a bit capricious, no? The function $g(x) = 1 - x + x^2 - x^3 + x^4 - \cdots$ is also defined for each positive $x < 1$ and has the same problem at

$x = 1$ that f did. But this one is equal to $1/(x+1)$, as you can check by adding $xg(x)$ to $g(x)$, thus it docilely approaches $\frac{1}{2}$ as $x \to 1$.

Halfway Points

Let S be a finite set of points in the unit interval $[0,1]$, and suppose that every point $x \in S$ lies either halfway between two other points in S (not necessarily the nearest ones) or halfway between another point in S and an endpoint.

Show that all the points in S are rational.

Solution: Suppose S has n points, $x_1 < x_2 < \cdots < x_n$. By assumption, they satisfy n linear equations with rational coefficients; for example, if point x_j lies halfway between x_i and x_k, it satisfies $2x_j = x_i + x_k$. If instead x_j lies between x_i and 1, it satisfies $2x_j = x_i + 1$.

If this set of equations has just the one solution, they must be rational. Why? You may already know from linear algebra that when a set of linear equations over a field that has a unique solution, the values in the solution lie in the field. We can prove that easily by induction: Take any equation and solve it for one variable, say x_n. Then substitute in the remaining equations to get a uniquely solvable system with one fewer variable.

So we need only prove that *our* system has only one solution; suppose it has a second solution, $y_1, \ldots, y_n$. Letting $z_i = y_i - x_i$ and subtracting corresponding equations, we find that the z_i's satisfy similar conditions except that (1) they may not all be distinct (but they're not all 0); (2) they lie between -1 and 1; and (3) the endpoint 1 is no longer in play, that is, each z_j either lies halfway between two other z_i's or halfway between another z_i and zero.

We may assume one of the z_i's is positive; let z_j have the largest value and among those with that maximum value, the largest index. Obviously z_j can't be halfway between 0 and some other z_i, so it must be between z_i and z_k where $z_i = z_j = z_k$. By choice of j, i, and k must both be smaller indices than j, but that can't be because in the original solution x_j must have lain halfway between x_i and x_k, contradicting the fact that we indexed the original solution in size order.

Our theorem is quite a famous one, and the proof given below was often cited by Paul Erdős as an example of a "book proof."

(Although Erdős claimed not to believe in God, he spoke often of a book maintained by God that contained the best proof of every theorem.)

Theorem. *Let X be a finite set of points on the plane, not all on one line. Then there is a line passing through exactly two points of X.*

Proof. Assume that every line through two or more points in X in fact contains at least three points of X. The idea is to find such a line L, and a point P not on L, such that the distance from P to L is minimized.

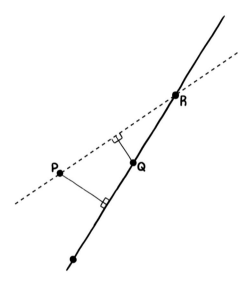

Since L contains at least three points of X, two of them, say Q and R, lie on the same side of the perpendicular to L from P. But then, if R is the farther one, the point Q is closer to the line through P and R than P is to L—contradiction! ♡

10. In All Probability

Dealing with probabilities is intuitive for some folks, less so for others. But there are just a few basic ideas, which, if mastered, convey a lot of power.

A thing that may or may not happen is called an *event*, and its probability measures the likelihood that it will happen; if the "trial" that could make the event happen is repeatable (like rolling a die), then the event's probability is the fraction of time that it occurs in many trials.

If the trial has n possible outcomes that are equally likely (perhaps because of symmetry) then the probability of the event A is the number of outcomes that result in A occurring, divided by n. For example, if we roll a die and A is the event that an even number appears on top, then the probability $\mathbb{P}(A)$ that A occurs is $3/6 = 1/2$, since three of the outcomes are even numbers. The fact that the outcomes are equally likely relies on the assumption that the die behaves like a true, homogeneous, *symmetrical*, cube.

If we roll a pair of dice, there are 36 equally likely outcomes; note that (2,3) is different from (3,2) because even if *you* can't distinguish between the two dice, Tyche (the goddess of chance) can. Thus, for example, there are five ways to roll a sum of 6—(1,5), (2,4), (3,3), (4,2), and (5,1)—so $\mathbb{P}$(two dice roll a sum of 6) = 5/36.

If two events A and B are *independent*, that is, the occurrence of one does not affect the probability of the other, then the probability $\mathbb{P}(A \wedge B)$ that they both occur is equal to the product $\mathbb{P}(A)\mathbb{P}(B)$ of the probabilities of the two events. Without the assumption of independence, we can only say $\mathbb{P}(A \wedge B) = \mathbb{P}(A)\mathbb{P}(B|A)$ where $\mathbb{P}(B|A)$, read "the probability of B given A," has intuitive meaning but is often defined simply as the quantity that makes the equation true.

The probability $\mathbb{P}(A \vee B)$ that at least one of A and B occurs is always equal to $\mathbb{P}(A) + \mathbb{P}(B) - \mathbb{P}(A \wedge B)$, as you can see from a Venn diagram. This reduces to $\mathbb{P}(A \vee B) = \mathbb{P}(A) + \mathbb{P}(B)$ only in the case where A and B are mutually exclusive—that is, they can't both occur.

Let's try a few amusing applications of these ideas.

Winning at Wimbledon

As a result of temporary magical powers (which might need to be able to change your gender), you have made it to the women's singles tennis finals at Wimbledon and are playing Serena Williams for all the marbles. However, your powers cannot last the whole match. What score do you want it to be when they disappear, to maximize your chances of notching an upset win?

Solution: Naturally, you want to have won the first set of the best-of-three-sets match, and be far ahead in the second. Your first thought might be that you'd like to be up five games to love and serving at 40-love (or, if your serve resembles mine, up 5-0 in games and *receiving* at love-40). Either gives you three bites at the apple, that is, if your probability of winning a point is some small quantity ε then your probability of winning is about 3ε.

(Technically, if the results of the next three points are independent, and you discount the possibility of winning the match despite losing them all, your success probability is $1 - (1 - \varepsilon)^3 = 3\varepsilon - 3\varepsilon^2 + \varepsilon^3$.)

But you can do better. Let the game score be 6-6, and suppose you are ahead 6-0 in the seven-point tiebreaker. Then you have six chances to win, almost doubling your probability when ε is small.

As I write this, you can do even better in the Australian Open, where the final-set tiebreaker is played to 10 points. The lesson, either way, is this: If you need a miracle, try to arrange for as many shots at that miracle as possible.

Birthday Match

You are on a cruise where you don't know anyone else. The ship announces a contest, the upshot of which is that if you can find someone who has the same birthday as yours, you (both) win a beef Wellington dinner.

How many people do you have to compare birthdays with in order to have a better than 50% chance of success?

Solution: Let's assume that (*a*) neither you (important!) nor anyone else on board (less important) was born on February 29, (*b*) other dates are equally likely to be a given shipmate's birthday, and (*c*) there are no sets of twins (or triplets etc.) on board that will cause you to waste queries. Then the probability that a given shipmate *fails* to match your birthday is 364/365, thus if you ask n people, the probability that you have no luck is $(364/365)^n$. You want this number to dip *below* 50%, which it does when n reaches 253; in fact $(364/365)^{253}$ = 0.49952284596... .

Is 253 higher than you guessed? You'd only need 183 queries to get your success probability over $\frac{1}{2}$, if the people you asked were guaranteed to have *different* birthdays. The trouble is, you will probably hear a lot of duplicated answers that don't help you. (Well, it could help you a bit if you demand a bite of their beef Wellington in return for telling each about the other).

If you thought the answer to the puzzle was 23, you confused the question with the more famous problem of how many people you need in a room to have a better than 50% probability that *some* pair of people share a birthday. In fact, you can get the answer "23" from our answer; the number of pairs of people in that room is $\binom{23}{2} = 253$. (The pairs are not precisely independent, but they're close enough to make this work.)

By the way, how do you determine that 253 is the least n for which $(364/365)^n < \frac{1}{2}$ without a lot of trial and error? My favorite way is to use the approximation $(1 - 1/m)^m \sim e^{-m}$ for large m. If you take this as an equality and solve $(1 - 1/365)^n = 1/2$ you get $n = \log_e(2) \cdot 365 \sim 252.9987$.

When the event A is decided before the event B, the conditional probability of B given A is often easy to deduce but $\mathbb{P}(A|B)$ is less intuitive. The laws of probability don't care about the order of events, though.

Other Side of the Coin

A two-headed coin, a two-tailed coin and an ordinary coin are placed in a bag. One of the coins is drawn at random and flipped; it comes up "heads." What is the probability that there is a head on the other side of this coin?

Solution: The temptation is to say that the drawn coin is either the two-headed coin or the fair coin, so the probability that its other side is heads is $\frac{1}{2}$. But is it? Imagine that the coin is flipped many times, and comes up heads every time. It could still be either coin, but you can't help thinking it's more likely to be the two-headed coin. That kind of reasoning is basically how we learn from statistical observations.

In fact, the inference that the drawn coin is probably two-headed exists already after one flip. We want to compute $\mathbb{P}(A|B)$ where A is the event that the two-headed coin was drawn, and B the event that heads was flipped. For any two events A and B, we calculate

$$\mathbb{P}(A|B) = \frac{\mathbb{P}(A \wedge B)}{\mathbb{P}(B)} = \frac{\mathbb{P}(A)\mathbb{P}(B|A)}{\mathbb{P}((A \wedge B) \vee (\overline{A} \wedge B))}$$

$$= \frac{\mathbb{P}(A)\mathbb{P}(B|A)}{\mathbb{P}(A \wedge B) + \mathbb{P}(\overline{A} \wedge B)}$$

$$= \frac{\mathbb{P}(A)\mathbb{P}(B|A)}{\mathbb{P}(A)\mathbb{P}(B|A) + \mathbb{P}(\overline{A})\mathbb{P}(B|\overline{A})}$$

where $\overline{A}$ is the event that A does *not* occur. This equation is known as Bayes' Rule, and if we plug in the values from our experiment, we get

$$\mathbb{P}(A|B) = \frac{\frac{1}{3} \cdot 1}{\frac{1}{3} \cdot 1 + \frac{2}{3} \cdot \frac{1}{4}} = \frac{2}{3}.$$

If all the above fractions leave you unconvinced, here's another way to get the answer. Steal a crayon or marker from your kid and label each of the six coin-faces as follows: "1" and "2" on the two sides of the two-headed coin, "3" on the heads side of the fair coin, "4" on its tails side, and finally "5" and "6" on the two sides of the two-tailed coin.

You will agree, I hope, that drawing a coin and flipping it has the same effect as rolling a fair die: You are equally likely to come up with any number from 1 to 6. Given that you flipped a head, then, you are equally likely to have seen 1, 2, or 3. Two of those three faces have a head on the other side.

"Reverse" conditional probabilities are crucial in statistics, where you need to compute $\Pr(A|B)$ where A is the hypothesis you are testing, and B the result of your experiment. Here are some more puzzles involving that kind of calculation.

Boy Born on Tuesday

Mrs. Chance has two children of different ages. At least one of them is a boy born on a Tuesday. What is the probability that both of them are boys?

Solution: We consider first the classic version, where the information is that at least one of Mrs. Chance's children is a boy. Here, the solution seems simple: *a priori*, the children (in age order) are equally likely to be boy–boy, boy–girl, girl–boy, or girl–girl. The last of these is ruled out, so the probability both children are boys is $\frac{1}{3}$.

The difficulty is that in *almost* every way that you are likely to find out that at least one of Mrs. Chance's kids is a boy, there is an additional inference which raises the probability of two boys from $\frac{1}{3}$ to $\frac{1}{2}$.

For example, if you find that the older child is a boy, then the girl–boy combination above is ruled out and it's clear that the probability that the younger child is a boy is $\frac{1}{2}$. But that reasoning also applies if your information is that the taller child is a boy, or the one with darker hair is a boy, or the child you saw Mrs. Chance with yesterday is a boy, or even if you happen to catch Mrs. Chance in the store buying boys' clothes.

Why the latter? Because if Mrs. Chance's other child were a girl, you'd be less likely to find her buying boys' clothes. It's quite similar to the previous puzzle; the fact that if the coin were fair it might not have come up "heads" affords an inference that the coin is the two-headed coin, raising the probability of two heads from $\frac{1}{2}$ to $\frac{2}{3}$. Here the effect is to raise the probability that the children are both boys from $\frac{1}{3}$ to $\frac{1}{2}$.

Why "almost"? Is there *any* way you could find out that at least one of Mrs. Chance's kids is a boy, that would lead to the conclusion that the probability they are both boys is $\frac{1}{3}$? As it happens, I have a friend whose wife became pregnant with fraternal twins, and an amniotic test revealed the presence of Y chromosomes in the placenta. The conclusion was that at least one of the twins was a boy, and naturally my friend and his wife hoped that the other was a girl—but assumed that the probability of this event was $\frac{1}{2}$. In fact, it really was $\frac{2}{3}$; I inquired and was told that the test always finds Y chromosomes as long as one or more fetuses is male. (In fact, the outcome was mixed twins.)

Now back to the boy born on a Tuesday. This is a very delicate question and you need to assume that the information has come to you in with no additional inferences—for example, you actually chose

Mrs. Chance randomly among many mothers of two (non-twin) children and asked her if at least one was a boy born on a Tuesday, and she said yes.

Compared to the classic version, here there is a substantial built-in inference that both kids are boys because that *nearly* doubles the probability that some boy was born on a Tuesday. But not quite, because if they were both born on Tuesday they count only once. Thus you'd expect the answer to be nearly $\frac{1}{2}$, and that is indeed the case. The calculation is straightforward. If we classify all two-kid families according to gender and birth-day-of-the-week we get $2 \times 7 \times 2 \times 7 = 196$ possibilities, of which $1 \times 2 \times 7 + 2 \times 7 \times 1 - 1 \times 1 = 27$ contain a boy born on a Tuesday (the -1 comes from over-counting the case of two boys born on Tuesdays). Of these, $1 \times 1 \times 7 + 1 \times 7 \times 1 - 1 \times 1 = 13$ involve two boys, so the probability we seek is $13/27$.

Whose Bullet?

Two marksmen, one of whom ("A") hits a certain small target 75% of the time and the other ("B") only 25%, aim simultaneously at that target. One bullet hits. What's the probability that it came from A?

Solution: You might think that since A is three times as good a shot as B, he is three times as likely to have been the one whose bullet hit; in other words, the probability that he deserves the credit is 75%.

But there are two things happening here: hitting and missing. Since the probability that A hits and B misses is $3/4 \times 3/4 = 9/16$ while the probability that B hits but A misses is only $1/4 \times 1/4 = 1/16$, the probability that the successful bullet was A's is actually a full 90%.

Conditional probabilities can sometimes defy your intuition. Another example:

Second Ace

What is the probability that a poker hand (five cards dealt at random from a 52-card deck) contains at least two aces, given that it has at least one? What is the probability that it contains two aces, given that it has the Ace of Spades? If your answers are different, then: What's so special about the Ace of Spades?

Solution: There are $\binom{52}{5}$ poker hands of which $\binom{52}{5} - \binom{48}{5}$ have at least one ace, and $\binom{52}{5} - \binom{48}{5} - \binom{4}{1} \cdot \binom{48}{4}$ have at least two aces;

dividing the last of these expressions by the second gives our first answer, 0.12218492854.

There are $\binom{51}{4}$ hands containing the Ace of Spades and $\binom{51}{4}$ − $\binom{48}{4}$ containing another ace as well, giving our second answer, 0.22136854741, quite a lot larger than the first answer.

Of course, there's nothing special about the Ace of Spades; specifying the ace held changes the conditional probabilities.

Here's a simpler example that you don't need a calculator for. Suppose a third of the families in town have two children, a third have one, and the remaining third none. Then the probability that a family has a second child, given that it has at least one, is $\frac{1}{2}$. But the probability that a family has a second child given that it has a *girl* is $\frac{3}{5}$, because only half the families with one child have a girl, but $\frac{3}{4}$ of those with two children have a girl.

Stopping After the Boy

In a certain country, a law is passed forbidding families to have another child after having a boy. Thus, families might consist of one boy, one girl and one boy, five girls and one boy... How will this law affect the ratio of boys to girls?

Solution: It might seem like limiting families to one boy should cut down the boy–girl ratio, but assuming that gender is independent among siblings, the child whose birth is prevented by having an older brother is as likely to be a girl as a boy. Thus the sex ratio should not be affected by this law.

One could make the case that, on account of the possibility of identical twins, gender among siblings is not precisely independent. However, in the statement of the puzzle, the law's effect when twin boys are born has not been made clear (and is perhaps a bit horrifying to think about). Thus we solvers are pretty much forced to discount that possibility.

Up until now we've been talking about *discrete* probability. Things get a bit more subtle when a random value is chosen from an interval of real numbers. Awkwardly, we have to assign probability 0 to the event that any particular number is chosen, even though some event of that kind is guaranteed to happen. Typically, we deal instead with the probability that the chosen point lies in a particular subinterval. For example, if we choose a point uniformly at random from the unit interval, the probability that it falls between 0.3 and 0.4 is 1/10 (regardless of whether we count 0.3 and 0.4 as part of the interval).

Points on a Circle

What is the probability that three uniformly random points on a circle will be contained in some semicircle?

Solution: It's not hard to see that no matter where we put the first two points, assuming they don't coincide, it is more likely that the third point will be in some semicircle with the first two, than not. Thus, in fact, the answer to the puzzle should be more than $\frac{1}{2}$. But how do we compute it, without the bother of trying to condition on the distance between the first two points?

The answer is (as is often the case in probability problems) to choose the points in a different way—but, of course, one that still results in uniformly random points. Here, let us choose three random *diameters* of our circle. Each hits the circle at two (antipodal) points, and we use three coinflips do decide which three points to use.

This may seem like an unnecessarily complicated way to choose our points—why in six steps, when we can do it in three steps? To see the answer, draw *any* three diameters and the six points where they hit the circle. Observe that of those six, if three *consecutive* points are chosen, they will certainly lie in a semicircle; otherwise, they won't!

Well, there are $2^3 = 8$ ways to pick the points, using our coinflips, and 6 of those result in consecutive points being chosen. So the probability of their being contained in some semicircle is 6/8 = 3/4. ♡

(If you found that puzzle too easy, try to determine the probability that four uniformly random points on a sphere are all contained in some hemisphere!)

Comparing Numbers, Version I

Paula (the perpetrator) takes two slips of paper and writes an integer on each. There are no restrictions on the two numbers except that they must be different. She then conceals one slip in each hand.

Victor (the victim) chooses one of Paula's hands, which Paula then opens, allowing Victor to see the number on that slip. Victor must now guess whether that number is the larger or the smaller of Paula's two numbers; if he guesses right, he wins $1, otherwise he loses $1.

Clearly, Victor can at least break even in this game, for example, by flipping a coin to decide whether to guess "larger" or "smaller"— or by always guessing "larger," but choosing the hand at random. The question is: Not knowing anything about Paula's psychology, is there any way he can do better than break even?

Solution: Amazingly, there is a strategy which guarantees Victor a better than 50% chance to win.

Before playing, Victor selects a probability distribution on the integers that assigns positive probability to each integer. (For example, he plans to flip a coin until a "head" appears. If he sees an even number $2k$ of tails, he will select the integer k; if he sees $2k-1$ tails, he will select the integer $-k$.)

If Victor is smart, he will conceal this distribution from Paula, but as you will see, Victor gets his guarantee even if Paula finds out. It is perhaps worth noting that although Victor might like the idea of assigning the *same* probability to every integer, this is not possible— there is no such probability distribution on the integers!

After Paula picks her numbers, Victor selects an integer from his probability distribution and adds $\frac{1}{2}$ to it; that becomes his "threshold" t. For example, using the distribution above, if he flips five tails before his first head, his random integer will be -3 and his threshold t will be $-2\frac{1}{2}$.

When Paula offers her two hands, Victor flips a fair coin to decide which hand to choose, then looks at the number in that hand. If it exceeds t, he guesses that it is the larger of Paula's numbers; if it is smaller than t, he guesses that it is the smaller of Paula's numbers.

So, why does this work? Well, suppose that t turns out to be larger than either of Paula's numbers; then Victor will guess "smaller" regardless of which number he gets, and thus will be right with probability exactly $\frac{1}{2}$. If t undercuts both of Paula's numbers, Victor will inevitably guess "larger" and will again be right with probability $\frac{1}{2}$.

But, *with positive probability*, Victor's threshold t will fall *between* Paula's two numbers; and then Victor wins regardless of which hand he picks. This possibility, then, gives Victor the edge which enables him to beat 50%. ♡

Neither this nor any other strategy enables Victor to guarantee, for some fixed $\varepsilon > 0$, a probability of winning greater than $50\% + \varepsilon$. A smart Paula can choose randomly two consecutive multidigit integers, and thereby reduce Victor's edge to an insignificant smidgen.

Comparing Numbers, Version II

Now let's make things more favorable to Victor: Instead of being chosen by Paula, the numbers are chosen independently at random from the uniform distribution on [0,1] (two outputs from a standard random number generator will do fine).

To compensate Paula, we allow her to examine the two random numbers and *to decide which one Victor will see*. Again, Victor must guess whether the number he sees is the larger or smaller of the two, with $1 at stake. Can he do better than break even? What are his and Paula's best (i.e., "equilibrium") strategies?

Solution: It looks like the ability to choose which number Victor sees is paltry compensation to Paula for not getting to pick the numbers, but in fact *this* version of the game is strictly fair: Paula can prevent Victor from getting any advantage at all.

Her strategy is simple: Look at the two random real numbers, then feed Victor the one which is closer to $\frac{1}{2}$.

To see that this reduces Victor to a pure guess, suppose that the number x revealed to him is between 0 and $\frac{1}{2}$. Then the unseen number is uniformly distributed in the set $[0, x] \cup [1-x, 1]$ and is, therefore, equally likely to be smaller or greater than x. If $x > \frac{1}{2}$, then the set is $[0, 1-x] \cup [x, 1]$ and the argument is the same.

Of course, Victor can guarantee probability $\frac{1}{2}$ against any strategy by ignoring his number and flipping a coin, so the game is completely fair. ♡

This amusing game was brought to my attention at a restaurant in Atlanta. Lots of smart people were present and were stymied, so if you failed to spot this nice strategy of Paula's, you're in good company.

Biased Betting

Alice and Bob each have $100 and a biased coin that comes up heads with probability 51%. At a signal, each begins flipping his or her coin once a minute and bets $1 (at even odds) on each outcome, against a bank with unlimited funds. Alice bets on heads, Bob on tails. Suppose both eventually go broke. Who is more likely to have gone broke first?

Suppose now that Alice and Bob are flipping the *same* coin, so that when the first one goes broke the second one's stack will be at $200. Same question: Given that they both go broke, who is more likely to have gone broke first?

Solution: In the first problem, Alice and Bob are equally likely to have gone broke first. In fact, conditioned on Alice's having gone broke, they have exactly the same probability of having gone broke at any particular time t.

To see this, pick any win–loss sequence s that ends by going broke; suppose s has n wins, thus $n + 100$ losses. It will have probability $p^n q^{n+100}$ for Alice, but $p^{n+100} q^n$ for Bob, where here $p = 51\%$ and $q = 49\%$. The ratio of these probabilities is the constant $(q/p)^{100}$, thus $\mathbb{P}(\text{Alice goes broke}) = (q/p)^{100}$ (which happens to be about 0.0183058708; call it 2%). Once we divide Alice's probability of encountering s by this number, her probability is the same as Bob's.

For the second problem, where Alice and Bob were flipping the same coin, your intuition might tell you that Alice probably went broke first. The reasoning is that if Bob went broke first, Alice would have to have gone broke after building up to $200, an unlikely occurrence. But is this true? Can't we argue as above, comparing win–lose sequences that result in both going broke?

No. The proof above doesn't work because here, Alice's going broke also affects Bob's longevity. In fact, the same ratio $(q/p)^{100}$ applies when comparing a both-broke win–loss sequence and the conclusion is that the probability that Bob goes broke first is $(q/p)^{100}$ times the probability that Alice goes broke first. Intuition was correct; Alice is more than 50 times more likely to have gone broke first!

It's worth noting that, putting Bob aside, we can deduce from the first calculation how long it takes, on average, for Alice to go broke *given that she goes broke*. The reason is that under this condition, we have shown Alice's expected time to ruin is the same as Bob's. But Bob *always* goes broke and since he loses 2 cents per toss on average, his expected time to ruin is $100/0.02 = 5000$ tosses. At 60 tosses per hour that comes to $83\frac{1}{3}$ hours, or about $3\frac{1}{2}$ days.

Home-Field Advantage

Every year, the Elkton Earlies and the Linthicum Lates face off in a series of baseball games, the winner being the first to win four games. The teams are evenly matched but each has a small edge (say, a 51% chance of winning) when playing at home.

Every year, the first three games are played in Elkton, the rest in Linthicum.

Which team has the advantage?

Solution: The series will last 4, 5, 6, or 7 games; it seems that the average should be somewhere between 5 and 6 games, in which case more games are played on average in Elkton than in Linthicum; therefore, Elkton has the advantage.

First issue: Are more games really played in Elkton, on average? To verify that we'll need to compute the probabilities that the series will last 4, 5, 6, or 7 games. If all the games were 50-50, the probabilities that a series will last 4, 5, 6, or 7 games would be, respectively, 1/8, 1/4, 5/16, and 5/16. Thus the average number of games played would be

$$4 \cdot \frac{1}{8} + 5 \cdot \frac{1}{4} + 6 \cdot \frac{5}{16} + 7 \cdot \frac{5}{16} = \frac{93}{16} = 5\frac{13}{16} = 5.8125$$

and since three games are always played in Elkton, on average only 2 13/16 would be played in Linthicum.

As one might expect, re-calculating the above odds taking into account the 1% home-field advantage changes the numbers very little. The average number of games played shifts upward a teensy bit to 5.81267507002.

So, indeed, more games are played in Elkton, on average. Over a very long period of time, this will be reflected in the game statistics; you can expect Elkton to have won about 50.03222697428% of the games over a million-year period.

However, shockingly, you can expect Linthicum to have won most of the series! Think of it this way: Imagine that all seven games are played (with equal ferocity) regardless of the outcomes. This doesn't change the probability of winning the series (if it did, we would never be able to quit after fewer than seven games). Those seven consist of four games in Linthicum and only three in Elkton, giving the advantage to Linthicum: Their probability of winning the series works out to 50.31257503002%.

If it seems odd to you that the home-field advantage in games 5, 6, and 7, which might not get played, counts as much as home-field advantage in earlier games, that's understandable. But the point is that the late home-field advantage *is there when Linthicum needs it*, and something that's there when you need it is just as good as its being there all the time.

Have you ever wondered about the effect of different rules, from sport to sport, concerning which player or team serves next? Maybe the next puzzle will answer your questions.

Service Options

You are challenged to a short tennis match, with the winner to be the first player to win four games. You get to serve first. But there are options for determining the sequence in which the two of you serve:

1. Standard: Serve alternates (you, her, you, her, you, her, you).

2. Volleyball Style: The winner of the previous game serves the next one.

3. Reverse Volleyball Style: The winner of the previous game receives in the next one.

Which option should you choose? You may assume it is to your advantage to serve. You may also assume that the outcome of any game is independent of when the game is played and of the outcome of any previous game.

Solution: Again using the idea (from the previous puzzle's solution) that it doesn't hurt to assume extra games are played, assume that you play lots of games—maybe more than are needed to determine the match winner. Let A be the event that of the first four served by you and the first three served by your opponent, at least four are won by you. Then, it is easily checked that no matter which service option you choose, you will win if A occurs and lose otherwise. Thus, your choice makes no difference. Notice that the independence assumptions mean that the probability of the event A, since it always involves four particular service games and three particular returning games, does not depend on when the games are played, or in what order.

If the game outcomes were not independent, the service option could make a difference. For example, if your opponent is easily discouraged when losing, you might benefit by using the volleyball scheme, in which you keep serving if you win.

The lesson learned above from *Home-Field Advantage* will again come in handy in the next puzzle.

Who Won the Series?

Two evenly-matched teams meet to play a best-of-seven World Series of baseball games. Each team has the same small advantage when playing at home. As is customary for this event, one team (say,

Team A) plays games 1 and 2 at home, and, if necessary, plays games 6 and 7 at home. Team B plays games 3, 4 and, if needed, 5 at home.

You go to a conference in Europe and return to find that the series is over, and six games were played. Which team is more likely to have won the series?

Solution: You could work this about by assigning a variable $p > 1/2$ to the probability of winning a game at home, then doing a bunch of calculations. But no calculations are actually necessary, if you remember that home-field advantage in *potential* games is as good as it is in games that are always played.

It follows that if you know that *at most* six games were played, you can conclude that teams A and B are equally likely to have won. On the other hand, if you know that at most 5 had been played, then (as host of 3 games) Team B is more likely to have won.

From these two facts we deduce that if *exactly* six games were played, Team A is the more likely winner.

It sometimes happens that there's alternative way to solve a puzzle if you trust the poser, and this is the case here. Since the degree of advantage experienced by the home team is not specified, you might choose to believe that the answer should be the same even if that degree is huge—in other words, if it requires a virtual miracle to win an away game.

In that case, with high probability there was just one upset (game won by the visitor) in the six games that were played, but that upset cannot have occurred in Game 1 or Game 2 else Team B would have won in only five games. Thus it happened in Game 3, 4, 5, or 6 and in the first three of those cases it is Team A who benefitted.

Next is the first of several probability-comparison puzzles.

Dishwashing Game

You and your spouse flip a coin to see who washes the dishes each evening. "Heads" she washes, "tails" you wash.

Tonight she tells you she is imposing a different scheme. You flip the coin 13 times, then she flips it 12 times. If you get more heads than she does, she washes; if you get the same number of heads or fewer, you wash.

Should you be happy?

Solution: If "should you be happy?" means anything at all, it should mean "are you better off than you were before?"

Here, the easy route is to imagine that first you and your spouse flip just 12 times each. If you get different numbers of heads, the one with fewer will be washing dishes regardless of the outcome of the next flip; so those scenarios cancel. The rest of the time, when you tie, the final flip will determine the washer. So it's still a 50-50 proposition, and you should be indifferent to the change in procedure (unless you dislike flipping coins).

Now suppose that some serious experimentation with the coin in question shows that it is actually slightly biased, and comes up heads 51% of the time. Should you still be indifferent to the new decision procedure? It might seem that you would welcome it now: Heads are good for you, so, the more flipping, the better.

But re-examination of the above argument shows the opposite. When you and your spouse flip 12 times each, it's still equally likely that you flip more heads than she does, or the reverse. Only if the two of you flip the same number of heads do you get your 51% advantage from the final flip. Thus, overall, your probability of ducking the dishwashing falls to somewhere between 50% and 51%.

The analysis suggests a third procedure: You and your spouse alternate flipping the coin until you reach a point where you have both flipped the same number of times but one of you (the winner) has flipped more heads. The advantage of this scheme? It's fair even if the coin is biased!

Many puzzles that involve comparing two probabilities can be approached by a technique called *coupling*, in which two events associated with different conditions are somehow built into a single experiment for which both events make sense.

Then, if you want to determine which of the two events A and B is more probable, you can ignore outcomes where they both occur or neither occurs, and just compare the probability of A happening without B with the probability of B happening without A. This amounts to comparing the areas of the two crescents in the Venn diagram below.

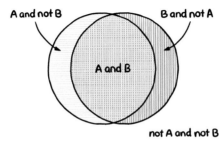

OK, that's a lot of abstract theory. Let's see some examples.

Random Judge

After a wild night of shore leave, you are about to be tried by your U.S. Navy superiors for unseemly behavior. You have a choice between accepting a "summary" court-martial with just one judge, or a "special" court-martial with three judges who decide by majority vote.

Each possible judge has the same (independent) probability—65.43%—of deciding in your favor, except that one officer who would be judging your special court-martial (but not the summary) is notorious for flipping a coin to make his decisions.

Which type of court-martial is more likely to keep you out of the brig?

Solution: It's not that hard to "do the math" on this one and simply compute your chance of getting off in the special court-martial, then compare it to your 65.43% chance of surviving the summary court-martial. (You might save yourself some arithmetic if you replace 65.43% by a variable, say p, and put the number back in later if necessary.)

But let's try coupling. You may as well suppose officer A would be on both panels and would vote the same way on each; officer C is the random one. If you go with the special court-martial you will regret it precisely when A votes innocent and B and C vote guilty, but thank your stars if A votes guilty while B and C vote innocent.

Without any calculations you know that these events have the same probability, just by exchanging the roles of A and B and reversing C's coinflip. So your two choices are equally good.

Suppose there's also an option of a "general" court-martial with five judges, two of whom are coin-flippers. You can apply the coupling method again, but this time it tells you to choose the five judges if p (as here) is greater than $\frac{1}{2}$.

In fact, the Law of Large Numbers tells you that even if 90% of the judges were coin-flippers, the remaining 10% voting in your favor with some fixed probability $p > 1/2$, you could boost your chances to 99% with enough judges.

Wins in a Row

You want to join a certain chess club, but admission requires that you play three games against Ioana, the current club champion, and win two in a row.

Since it is an advantage to play the white pieces (which move first), you alternate playing white and black.

A coin is flipped, and the result is that you will be white in the first and third games, black in the second.

Should you be happy?

Solution: You can answer this question by assigning a variable w to the probability of winning a game as White, and another variable $b < w$ to the probability of winning as Black; then doing some algebra.

Using coupling, you can get the answer without algebra—even if the problem is modified so that you have to win two in a row out of *seventeen* games, or m in a row out of n. (If n is even, it doesn't matter who plays white first; if n is odd, you want to be black first when m is odd, white first when m is even.)

The coupling argument in the original 2-out-of-3 puzzle goes like this. Imagine that you are going to play *four* games against Ioana, playing white, then black, then white, then black. You still need to win two in a row, but you must decide *in advance* whether to discount the first game, or the last.

Obviously turning the first game into a "practice game" is equivalent to playing BWB in the original problem, and failing to count the last game is equivalent to playing WBW, so the new problem is equivalent to the old one.

But now the events are in the same experiment. For it to make a difference which game you discounted, the results must be either WWLX or XLWW. In words: If you win the first two games, and lose (or draw) the third, you will wish that you had discounted the last game; if you lose the second but win the last two, you will wish that you had discounted the first game.

But it is easy to see that XLWW is more likely than WWLX. The two wins in each case are one with white and one with black, so those cases balance; but the loss in XLWW is with black, more likely than the loss in WWLX with white. So you want to discount the first game, that is, start as black in the original problem.

A slightly more challenging version of this argument is needed if you change the number of games played, and/or the number of wins needed in a row.

Split Games

You are a rabid baseball fan and, miraculously, your team has won the pennant—thus, it gets to play in the World Series. Unfortunately, the opposition is a superior team whose probability of winning any given game against your team is 60%.

Sure enough, your team loses the first game in the best-of-seven series and you are so unhappy that you drink yourself into a stupor. When you regain consciousness, you discover that two more games have been played.

You run out into the street and grab the first passer-by. "What happened in games 2 and 3 of the World Series?"

"They were split," he says. "One game each."

Should you be happy?

Solution: In my experience, presented with this puzzle, about half of respondents answer "Yes—if those games *hadn't* been split, your team would probably have lost them both."

The other half argue: "No—if your team keeps splitting games, they will lose the series. They have to do better."

Which argument is correct—and how do you verify the answer without a messy computation?

The task at hand is to determine, after hearing that games 2 and 3 were split, whether you are better or worse off than before. In other words, is your team's probability of winning the series better now, when it needs three of the next four games, than it was before, when it needed four out of six?

Computing and comparing tails of binomial distributions is messy but not difficult. Before the news, your team needed to win 4, 5, or 6 of the next six games. (Wait, what if fewer than seven games are played? As in both *Home-Field Advantage* and *Who Won the Series?*, it costs nothing to imagine that all seven games are played no matter what.)

The probability that your team wins exactly four of six games is "6 choose 4" (the number of ways that can happen) times 0.4^4 (the probability that your team wins a particular four games) times 0.6^2 (the probability that the other guys win the other two). Altogether, the probability that your team wins at least four of six is

$$\binom{6}{4} \cdot 0.4^4 \cdot 0.6^2 + \binom{6}{5} \cdot 0.4^5 \cdot 0.6 + \binom{6}{6} \cdot 0.4^6$$

$$= 15 \cdot 2^4 \cdot 3^2/5^6 + 6 \cdot 2^5 \cdot 3/5^6 + 2^6/5^6 = 112/625.$$

After the second and third games are split, your team needs at least three of the remaining four. The probability of winning is now:

$$\binom{4}{3} \cdot 0.4^3 \cdot 0.6 + \binom{4}{4} \cdot 0.4^4$$

$$= 4 \cdot 2^3 \cdot 3/5^4 + 2^4/5^4 = 112/625(!)$$

So you should be exactly indifferent to the news! The two arguments (one backward-looking, suggesting that you should be happy, the other forward-looking, suggesting that you should be unhappy) seem to have canceled each other precisely. Can this be a coincidence? Is there any way to get this result "in your head"?

Of course there is—by coupling. To set this up a little imagination helps. Suppose that, after game 3, it is discovered that an umpire who participated in games 2 and 3 had lied on his application. There is a movement to void those two games, and a counter-movement to keep them. The commissioner of baseball, wise man that he is, appoints a committee to decide what to do with the results of games 2 and 3; and in the meantime, he tells the teams to keep playing.

Of course, the commissioner hopes—and so do we puzzle-solvers—that *by the time the committee makes its decision, the question will be moot.*

Suppose five(!) more games are played before the committee is ready to report. If your team wins four or five of them, it has won the series regardless of the disposition of the second and third game results. On the other hand, if the opposition has won three or more, they have won the series regardless. The only case where the committee can make a difference is where your team won exactly three of those five new games.

In that case, if the results of games 2 and 3 are voided, one more game needs to be played; your team wins the series if they win that game, which happens with probability 2/5.

If, on the other hand, the committee decides to count games 2 and 3, the series ended before the last game, and whoever *lost* that game is the World Series winner. Sounds good for your team, no? Oops, remember that your team won three of those five new games, so the probability that the last game is among the two they lost is again 2/5. ♡

Angry Baseball

As in *Split Games*, your team is the underdog and wins any given game in the best-of-seven World Series with probability 40%. But, hold on: This time, whenever your team is behind in the series, the players get angry and play better, raising your team's probability of winning that game to 60%.

Before it all begins, what is your team's probability of winning the World Series?

Solution: It would be a pain in the neck to work out all the possible outcomes and their probabilities, but coupling again comes to our rescue. Let k be the number of the first game played *after the last time the teams were tied.* Since the teams were tied 0-0 at the start, k could be 1, but it could also be 3, 5, or 7. (It turns out, you don't care!)

Suppose team X wins game k. Let us represent the winners of the rest of the games by a sequence of X's and Y's. Since there are no more ties, X must have won the series, each win by X having had probability 40% and each win by Y, 60%.

If it was Team Y that won game k, the same reasoning applies and the same probability attaches to the subsequent win-sequence except with X's and Y's exchanged. In effect, we couple win-sequences after X wins game k with the complementary win-sequences after Y wins game k.

We conclude that the probability of winning the series for Team X is exactly the probability that Team X wins game k. That probability is 40% when Team X is your team. Since that does not depend on the value of k, your team's probability of winning the series is 40% from the start.

Note that unlike *Split Games*, this puzzle works with any game-winning probability p (changing to $1-p$ when the team is behind). The answer is then p.

Let's move on to (American) football. Are you ready to use probability theory to be an effective coach?

Two-point Conversion

You, coach of the Hoboken Hominids football team, were 14 points behind your rivals (the Gloucester Great Apes) until, with just a minute to go in the game, you scored a touchdown. You have a choice of kicking an extra point (95% success rate) or going for two (45% success rate). Which should you do?

Solution: You must assume the Hominids will be able to score a second touchdown; and it's reasonable to postulate that if the game goes to overtime, either team is equally likely to win it.

Whatever you do, if it doesn't work you'll need to go for two points after the second TD, winning with probability $0.45 \cdot \frac{1}{2} = 0.225$.

If you do succeed with the first conversion, you'll want to go for just one the next time, winning with probability $\frac{1}{2} \cdot 0.95 = 0.475$ if you had kicked last time but a full 0.95 if you had gotten two extra points last time.

So, altogether (assuming that you did get that second touchdown) your probability of winning with the "kick after first TD" option is

$$0.95 \cdot 0.475 + 0.05 \cdot 0.225 = 0.4625,$$

while your probability of winning if you went for two after the first TD is

$$0.45 \cdot 0.95 + 0.55 \cdot 0.225 = 0.55125,$$

so going for two is much better.

It might be easier to see why this is so if you instead imagine that the kick is a sure thing while the two-point conversion has probability of success $\frac{1}{2}$. Then if you kick, you are headed for overtime for sure (given, of course, the second touchdown) where you win with probability $\frac{1}{2}$. If you go for two and succeed you win for sure, so already you get probability $\frac{1}{2}$ of winning the game; but if you miss it ain't over. Altogether the two-point conversion strategy wins 5/8 of the time.

So why do coaches frequently try to kick the extra point in these situations? Do they think missing the conversion will demoralize their team and make the second touchdown unlikely?

My own theory is that coaches tend to make conservative decisions, making it harder for their managers to point at a coach's decision that cost the game. In this situation, for example, if the team misses both two-point conversions, blame is likely to fall on the coach; but if they kick both extra points and lose in overtime, most of the blame will fall on whichever player screwed up in overtime.

Random Chord

What is the probability that a random chord of a circle is longer than a side of an equilateral triangle inscribed in the circle?

Solution: Call a chord "long" if it is longer than the side of our inscribed triangle.

Suppose we choose our random chord by fixing some direction—horizontal is as good as any. Then we can draw a vertical diameter in our circle, pick a point on it uniformly at random, and draw the horizontal chord through that point.

In the left-hand third of the figure below, the inscribed triangle has been drawn with its base down, so that the diameter we just drew is perpendicular to the base. We claim the triangle's base is exactly halfway between the circle's center and the bottom point of the diameter; one way to see this is to draw another inscribed triangle, this one upside-down, so that the two triangles make a "star of David." The hexagon in the middle of the star can be broken into six equilateral triangles and once we do that, it is evident that the part of our diameter that yields a long chord is the same length as the rest. Thus the probability that the random chord is long is $\frac{1}{2}$.

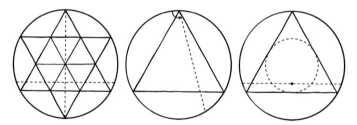

But ... we could instead fix one end of the chord, say at the triangle's apex, and choose its angle uniformly at random, as in the middle third of the figure. Since the angle made by the triangle is $60°$ and our random angle is between $0°$ and $180°$, we deduce that the probability the chord is long is $\frac{1}{3}$.

Wait, here's yet another way to solve the problem. Except for diameters (which have probability zero), every chord has a different midpoint and every point inside the circle, except for the center, is the midpoint of a unique chord. So let's pick a point inside the circle uniformly at random, and use the chord with that midpoint. It's easy to see that the chord will be long exactly if our point is closer to the center of the circle than the midpoint of a side of the triangle (see right-hand third of figure). Thus, we get a long chord if and only if the chosen point lies inside the circle that's inscribed *in* our triangle. But this circle has half the radius, thus one-quarter the area, of the original circle; therefore the probability that our chord is long is $\frac{1}{4}$.

What gives? What't the right answer, $\frac{1}{2}$, $\frac{1}{3}$, $\frac{1}{4}$ or something else?

The answer is that the concept of "uniform randomness" for an object chosen from an infinite set is not automatically defined. True, for a point chosen from a geometric shape (e.g., an interval or a polygon) we generally take uniformity to mean that the probability that the point lands in a particular subset is proportional to the measure (e.g., length or area) of that subset.

For chords, we don't have a standard notion of uniform randomness and as we've just seen, different attempts at defining uniformity can lead to different answers.

Random Bias

Suppose you choose a real number p between 0 and 1 uniformly at random, then bend a coin so that its probability of coming up heads when you flip it is precisely p. Finally you flip your bent coin 100 times. What is the probability that after all this, you end up flipping exactly 50 heads?

Solution: This puzzle again calls for a form of coupling; the idea in this case is to couple the coinflips even though they involve different coins, depending on the choice of p. To do this we choose, in advance, a uniformly random threshold $p_i \in [0, 1]$ for the ith coinflip, independently for each i. Then, after p is chosen, the ith coinflip is deemed to be heads just when $p_i < p$.

Now we need only observe that getting 50 heads is tantamount to p being the middle number, that is, the 51st largest, of the 101 values $p_1, p_2, \ldots, p_{100}, p$. Since these numbers are all uniform and independent, their order is uniformly random and the probability that p ends up in the middle position is exactly $1/101$.

Coin Testing

The Unfair Advantage Magic Company has supplied you, a magician, with a special penny and a special nickel. One of these is supposed to flip "Heads" with probability $\frac{1}{3}$, the other $\frac{1}{4}$, but UAMCO has not bothered to tell you which is which.

Having limited patience, you plan to try to identify the biases by flipping the nickel and penny one by one until one of them comes up Heads, at which point that one will be declared to be the 1/3-Heads coin.

In what order should you flip the coins to maximize the probability that you get the correct answer, and at the same time to be fair, that is, to give the penny and the nickel the same chance to be designated the 1/3-Heads coin?

Solution: From the point of view of accuracy, at any point when you've flipped each coin the same number of times, it doesn't matter which one you flip next. At any point when you've flipped the penny more times than the nickel, you need to flip the nickel next, and vice-versa.

Thus, for the purpose of accuracy, the order doesn't matter as long as for every i, flips $2i-1$ and $2i$ are of different coins.

However, if you want a *fair* test—one equally likely to damn either coin—you need to be more careful. Imagine that you make an extra flip when the first Head turns up at an odd flip, and declare the result to be inconclusive if the succeeding flip (necessarily, of the other coin) is again a Head. This would then certainly be fair to the two coins, but it may mean more flipping. And if you're reading this book, you're likely to be the kind of person that would rather spend an hour figuring something out then an extra minute actually doing it.

It follows from considering this tie scenario that the bias toward the penny (say) is the probability that the "tie" happens with the penny having flipped Heads first, and the nickel immediately after; and vice-versa.

To arrange these biases so that they cancel, we need to order the flips so that if a tie "would have" occurred, it is equally like to have been with the penny or the nickel coming up Heads first.

Suppose we flip the penny first, that is, we begin "PN". The probability of getting Heads already on the first two flips is $\frac{1}{3} \cdot \frac{1}{4} = \frac{1}{12}$.

What's the probability that the modified game ends in a tie? You will be flipping in pairs—odd flip, then even flip—until you either get HT, TH, or HH, and the tie occurs in the last of these cases, which has probability $(\frac{1}{3} \cdot \frac{1}{4})/(\frac{2}{3} \cdot \frac{1}{4} + \frac{3}{4} \cdot \frac{1}{3} + \frac{1}{3} \cdot \frac{1}{4}) = (\frac{1}{12})/(\frac{6}{12}) = \frac{1}{6}$.

Therefore, in order to cancel the penny's first-flip advantage, *all* the rest of the odd flips must be performed by the nickel!

In conclusion, the flip order should either be PNNPNPNPNPNP... or NPPNPNPNPNPN... .

Coin Game

You and a friend each pick a different heads–tails sequence of length 4 and a fair coin is flipped until one sequence or the other appears; the owner of that sequence wins the game.

For example, if you pick HHHH and she picks TTTT, you win if a run of four heads occurs before a run of four tails.

Do you want to pick first or second? If you pick first, what should you pick? If your friend picks first, how should you respond?

Solution: At first it seems as if this must be a fair game; after all, any particular sequence of four flips has the same probability of appearing (namely, $\frac{1}{16}$) as any other. Indeed, if the coin is flipped four times and then the game begins anew, nothing of interest arises. But because the winning sequence could start with any flip, including the second, third or fourth, significant imbalances appear.

For example, if you pick HHHH (a miserable choice) as your sequence and your friend counters with THHH, you will lose unless the game begins with four heads in a row (only a $\frac{1}{16}$ probability). Why? Because otherwise the first occurrence of HHHH will be preceded by a tail, thus your friend will get there first with her THHH.

How can you determine how to play optimally? For that you need a formula for the probability that string B beats string A. There is, in fact, a very nice one, discovered by the late, indomitable John Horton Conway. Here's how it works.

We use the expression "A·B" to denote the degree to which A and B can overlap when A begins first (or they begin simultaneously). We measure this with a four-digit binary number $x_4x_3x_2x_1$ where x_i is 1 if the first i letters of B match the last i of A, otherwise 0. For example, HHTH·HTHT = 0101(binary) = 5 since H[HTH]T gives an overlap of 3 while HHT[H]THT gives an overlap of 1. The left-most digit, x_4, can only be 1 if $A = B$; thus A·B is equal to or greater than 8 if and only if $A = B$.

Now if your friend has chosen A as her sequence, how should you choose B? You want B to have low waiting time, thus B·B should be as low as possible. You want B·A to be high, so that you maximize your probability of sneaking in ahead of your friend; and you want A·B to be low to minimize the probability that she sneaks in ahead of you. The formula says that the odds that B beats A are (A·A − A·B):(B·B − B·A).

For example, suppose your friend picks A = THHT as her sequence (then A·A = 9). Your best choice is (always) to tack an H or T to the front of her choice and drop the last letter; this ensures that B·A is at least 4. If you pick B = HTHH you get B·B = 9, B·A = 4, and A·B = 2, so Conway's formula says the odds in your favor are $(9 − 2):(9−4) = 7:5$. If instead you pick B = TTHH you get B·B = 8, B·A = 4, and A·B = 1, and the formula says the odds in your favor are $(9 − 1):(8 − 4) = 2:1$. So in this case B = TTHH is the better choice.

Was A = THHT a poor choice by your friend? Actually, it was among the best. Anything your friend chose would leave her at a 2:1 disadvantage or worse, with best play by you.

Conway's formula continues to work beautifully for strings of length k, for any integer $k \geq 2$, and also to non-binary randomizers such as dice. In the case of dice, the overlap vectors still consist of zeroes and ones, but are now interpreted base 6.

Sleeping Beauty

Sleeping Beauty agrees to the following experiment. On Sunday she is put to sleep and a fair coin is flipped. If it comes up Heads, she is awakened on Monday morning; if Tails, she is awakened on Monday morning and again on Tuesday morning. In all cases, she is not told the day of the week, is put back to sleep shortly after, and will have no memory of any Monday or Tuesday awakenings.

When Sleeping Beauty is awakened on Monday or Tuesday, what—to her—is the probability that the coin came up Heads?

Solution: This puzzle, introduced by philosopher Adam Elga in 2001, became the subject of major controversy in the philosophy community over the next 20 years. Many were persuaded by the following argument: On Sunday SB (Sleeping Beauty) knows that the probability of Heads is $\frac{1}{2}$. She knows she will be awakened, so when she *is* awakened (on Monday or Tuesday), she has no new information and thus no reason to change her mind. So the answer is $\frac{1}{2}$.

The flaw in this reasoning is that there is indeed new information: that she is awake *now*. Many arguments show that the correct answer is $\frac{1}{3}$; one very nice example, offered by philosophers Cian Dorr and Frank Arntzenius, runs as follows.

Suppose that SB is awakened both days, regardless of the coin, but if the coin comes up Heads, then on Tuesday, 15 minutes after she is awakened, she will be told "It's Tuesday and the coin came up Heads." She knows all this in advance.

When she is awakened, it's equally likely to be Monday or Tuesday (by symmetry) and equally likely to be Heads or Tails (again by symmetry). Thus, the four events "Monday, Tails," "Monday, Heads," "Tuesday, Tails," and "Tuesday, Heads" are all equally likely.

When, 15 minutes later, SB is *not* told that it's "Tuesday, Heads" she rules out that last option and the rest remain equally likely. ♡

You might, on the other hand, be more persuaded by a frequency argument. If the experiment is run 100 times, you'd expect about 150 Monday/Tuesday awakenings, about $\frac{1}{3}$ of which would be after Heads was flipped.

The last puzzle will lead us to our theorem for this chapter.

Leading All the Way

Running for local office against Bob, Alice wins with 105 votes to Bob's 95. What is the probability that as the votes were counted (in random order), Alice led the whole way?

Solution: Of course, what we'd really like to do is solve this problem in general (when Alice has won with a votes to Bob's b, where a and b are arbitrary positive integers with $a > b$).

The answer is given by what is often known as Bertrand's Ballot Theorem (although it was apparently first proved by W. A. Whitworth in 1878, not J. L. F. Bertrand in 1887).

Theorem. *Let $a > b > 0$ and let $x = (x_1, x_2, \ldots, x_{a+b})$ be a uniformly random string of 1's and -1's, containing a 1's and b -1's. For each k between 1 and $a+b$, let s_k be the sum of the first k elements of x, that is,*

$$s_k = \sum_{i=1}^{k} x_k.$$

Then the probability that $s_1, s_2, \ldots, s_{a+b}$ are all positive is $(a-b)/(a+b)$.

Let's first check that this statement tells us what we want to know. The string x codes the ballot-counting process; each 1 represents a vote for Alice, each -1 a vote for Bob. The sum s_k tells us by how much Alice leads after the kth vote has been counted; for example, if $s_9 = -3$ we conclude that Alice was actually *behind* by three votes after the ninth vote was counted.

If we apply the theorem to the puzzle above, we get the answer $(105 - 95)/(105 + 95) = 1/20 = 5\%$. So how can we prove the theorem?

There's a famous and clever proof that uses reflection of random walk, but doesn't really explain why the probability turns out to be the vote difference divided by the total vote. Instead, consider the following argument.

One way to choose the random order in which the ballots are counted is first to place the ballots randomly in a circle, then to choose a random ballot in the circle and start counting (say, clockwise) from there. Since there are $a+b$ ballots, there are $a+b$ places to start, and the claim is that no matter what the circular order is, exactly $a-b$ of those starting points are "good" in that they result in Alice leading all the way.

To see this we observe that any occurrence of $1, -1$ (i.e., a vote for Alice followed immediately by a vote for Bob in clockwise order) can be deleted without changing the number of good starting positions. Why? First, you can't start with *that* 1 or -1, since if you start with the 1 the candidates are tied after the second vote, and if you start with the -1, Alice actually falls behind. Secondly, the $1, -1$ pair has no influence upon the the eligibility of any other starting point, since all it does is raise the sum (Alice's lead) by one and then lower it again.

So we can keep deleting $1, -1$'s until all that's left is $a-b$ 1's, each of which is obviously a good starting point. We conclude that all along there were $a-b$ good starting points, and we are done! ♡

11. Working for the System

There's no getting around it—many puzzles respond only to trying out ideas until you find one that works. Even so, there are good ways and bad ways to go about this process. Some people have a tendency to circle back and try the same answer many times.

How can you avoid doing that? By classifying your attempts. Make some choice and stick with it, until it either yields a solution or runs out of steam; if the latter, you can then rule out one possibility, and progress has been made.

There are other benefits to this besides avoiding redundancy: For example, you might actually see what's going on and jump to the solution!

No Twins Today

It was the first day of class and Mrs. O'Connor had two identical-looking pupils, Donald and Ronald Featheringstonehaugh (pronounced "Fanshaw"), sitting together in the first row.

"You two are twins, I take it?" she asked.

"No," they replied in unison.

But a check of their records showed that they had the same parents and were born on the same day. How could this be?

Solution: We have begun with a puzzle that might not seem to respond to systematic search; you either get it or you don't. But these days many people train themselves to "think out of the box" and forget to first try thinking *in* the box. Accept that the boys are genetically identical and born to the same parents at the same time, and ask how this can happen if they are not twins. Then, at least, it *might* occur to you that there was an additional simultaneous sibling (or more!).

The most likely possibility is that Donald and Ronald are triplets; the third (Arnold, perhaps?) is in another class.

Natives in a Circle

An anthropologist is surrounded by a circle of natives, each of whom either always tells the truth or always lies. She asks each native whether the native to his right is a truth-teller or a liar, and from their answers, she is able to deduce the fraction of liars in the circle.

What fraction is it?

Solution: Observe that if all truth-tellers are changed to liars and vice-versa, none of answers would change. Therefore, if the fraction x can be determined, it must have the property that $x = 1-x$, thus $x = 1/2$.

It remains to check that there is a set of answers from which the anthropologist can deduce that half the natives are truth-tellers. That requires, of course, that there be an even number of natives in the circle. One possibility is that every native says his right-hand neighbor is a liar, in which case natives must alternate between liar and truth-teller.

Poker Quickie

What is the best full house?

(You may assume that you have five cards, and you have just one opponent who has five random other cards. There are no wild cards. As a result of the Goddess of Chance owing you a favor, you are entitled to a full house, and you get to choose whatever full house you want.)

Solution: All full houses with three aces are equally good, because there can only be one such hand in a deal from a single deck. But there are *other* hands that can beat them: any four-of-a-kind, of which there are always 11 varieties possible, and more relevantly any straight flush. Since AAA99, AAA88, AAA77, and AAA66 kill the most straight flushes (16—each ace kills only two, but each spot card kills five) they are the best full houses. Note that AAA55 doesn't quite make the list, because one of the 5's must be in the same suit as an ace and the A2345 straight flush in that suit is thus doubly covered.

If you greedily insist on AAAKK, there are $40 - 9 = 31$ possible straight flushes that can beat you—even more if you haven't got the four suits covered—instead of just $40 - 16 = 24$.

Although the argument above only works for head-to-head poker (two players), the result has in fact been verified by computer for any number of players.

Fewest Slopes

If you pick n random points in, say, a disk, the pairs of points among them will with probability 1 determine $n(n-1)/2$ distinct slopes. Suppose you get to pick the n points deliberately, subject to no three being collinear. What's the smallest number of distinct slopes they can determine?

Solution: Three points, since they are not permitted to lie on one line, determine a triangle with three slopes. A little experimentation will convince you that with four, you can't have any fewer than the four slopes you get from a square.

In fact the vertices of a regular n-sided polygon determine just n different slopes. If n is odd, the polygon's sides already have n different slopes, but every diagonal is parallel to some side. If n is even, the sides exhibit only $n/2$ different slopes but now the diagonals that connect two vertices an even number of edges apart are not parallel to any side, but they form $n/2$ parallel classes; so again we get n different diagonals.

Can you do any better than the vertices of a regular n-gon? No. Let X be any configuration of n points in the plane, no three on a line, and let P be the southernmost point of X. Then you already have $n-1$ lines through P, each determining a different angle between $-90°$ and $+90°$ (relative to north $= 0°$). The line through the two points that give the smallest and largest such angles gives you your nth slope.

Two Different Distances

Find all configurations of four points in the plane that determine only two different distances. (Note: There are more of these than you probably think!)

Solution: It may pay to think of the points as vertices of a graph, all of whose edges are the same length, and all of whose *non*-edges are the same length. If there is a "3-clique" in the graph, that is, three vertices of which any two are adjacent, then three of our points—call them, say, A, B, and C—form an equilateral triangle.

The fourth point, D, can't be connected to all the others, since in the plane we can't have four points determining only *one* distance. If it is adjacent to two of the others, it forms another equilateral triangle and we get the upper-left configuration in the figure.

If D is connected to one of the others, say A, it must be equidistant from B and C; thus it must either stick out from A directly away from the triangle, or directly toward it, giving the other two configurations on the top row. Finally, if D is adjacent to none of the three it must be equidistant from all three, thus in the middle of the triangle; this gives the lower-left configuration.

What if there's no triangle of edges (or non-edges) in the graph? If there's a 4-cycle (but no triangle), its diagonals must be of equal length, thus the points are the corners of a square.

If there's no triangle or 4-cycle in either the graph or its complement, the graph must be precisely a path three edges in length (hence its complement is, too). This is the graph you get by deleting one edge of a 5-cycle, and corresponds to the configuration you get by deleting one vertex of a regular pentagon.

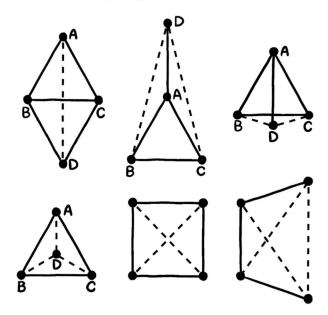

First Odd Number

What's the first odd number in the dictionary? More specifically, suppose that every whole number from 1 to, say, 10^{10} is written out in formal English (e.g., "two hundred eleven," "one thousand forty-two") and then listed in dictionary order, that is, alphabetical order with spaces and hyphens ignored. What's the first odd number in the list?

Solution: In dictionary order, one word precedes another precisely when the first *letter* in which the two words differ is, in the first word, earlier in the alphabet. Spaces and punctuation are ignored, and if there is no differing letter, the shorter word appears first.

So finding the dictionary-first word in a set of words is just a matter of carefully and systematically considering the successive words involved in the description of a number.

The earliest actual digit is "eight," and the earliest odd digit "five." We don't need to consider any other digits, but other possibly useful words that appear in numbers include "billion," "eighteen," "eighty," "hundred," "million," and "thousand." Our earliest odd number must begin with a digit so should start with "eight billion." After that, "eighteen" is the best we can do, and proceeding along these lines, we eventually get the answer 8,018,018,885: "eight billion, eighteen million, eighteen thousand, eight hundred eighty-five."

A little more work will get you to the first prime number in the dictionary, 8,018,018,851.

Annoyingly, "dictionary order" is (according to Wikipedia) subtly different from "alphabetical order," and the difference makes a difference. The issue is that in alphabetical order, which is used with people's names, spaces typically are *not* ignored, but instead precede all other characters. Thus, alphabetically, "eight hundred" precedes "eighteen," and the first odd number in *alphabetical* order becomes 8,808,808,885. The first prime would in that case be the very next odd number in alphabetical order, namely, 8,808,808,889.

Measuring with Fuses

You have on hand two fuses (lengths of string), each of which will burn for exactly 1 minute. They do not, however, burn uniformly along their lengths. Can you nevertheless use them to measure 45 seconds?

Solution: Simultaneously light both ends of one fuse and one end of the other; when the first fuse burns out (after half a minute), light the other end of the second. When it finishes, 45 seconds have passed. ♡

If you allow midfuse ignition and arbitrary dexterity, you can do quite a lot with fuses. For example, you can get 10 seconds from a single 60-second fuse by lighting at both ends and at two internal points, then lighting a new internal point every time a segment finishes; thus, at all times, three segments are burning at both ends and the fuse material is being consumed at six times the intended rate.

Bit of a mad scramble at the end, though. You'll need infinitely many matches to get perfect precision.

King's Salary

Democracy has come to the little kingdom of Zirconia, in which the king and each of the other 65 citizens has a salary of one zircon. The king can not vote, but he has power to suggest changes—in particular, redistribution of salaries. Each person's salary must be a whole number of zircons, and the salaries must sum to 66. Each suggestion is voted on, and carried if there are more votes for than against. Each voter can be counted on to vote "yes" if his or her salary is to be increased, "no" if decreased, and otherwise not to bother voting.

The king is both selfish and clever. What is the maximum salary he can obtain for himself, and how many referenda does he need to get it?

Solution: There are two key observations: (1) that the king must temporarily give up his own salary to get things started, and (2) that the game is to reduce the number of salaried citizens at each stage.

The king begins by proposing that 33 citizens have their salaries doubled to 2 zircons, at the expense of the remaining 33 citizens (himself included). Next, he increases the salaries of 17 of the 33 salaried voters (to 3 or 4 zircons) while reducing the remaining 16 to no salary at all. In successive turns, the number of salaried voters falls to 9, 5, 3, and 2. Finally, the king bribes three paupers with one zircon each to help him turn over the two big salaries to himself, thus finishing with a royal salary of 63 zircons.

It is not difficult to see that the king can do no better at any stage than to reduce the number of salaried voters to just over half the previous number; in particular, he can never achieve a unique salaried voter. Thus, he can do no better than 63 zircons for himself, and the seven rounds above are optimal. ♡

More generally, if the original number of citizens is n, the king can achieve a salary of $n-3$ zircons in k rounds, where k is the least integer greater than or equal to $\log_2(2n-4)$.

Packing Slashes

Given a 5×5 square grid, on how many of the squares can you draw diagonals (slashes or backslashes) in such a way that no two of the diagonals meet?

Solution: You can get 15 easily (see figure below) using rows (or columns) 1, 3, and 5, with all slashes or all backslashes on each line, or by nested Ls. Is it possible to do any better than that?

There are $6 \times 6 = 36$ vertices (cell corners) in the grid and each diagonal accounts for two, so you certainly will not be able to fit in more than 18 diagonals. But you can see that, for example, it won't be possible to use all 12 corners of the top row of squares.

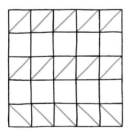 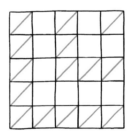

In fact, come to think of it, every diagonal touching one of the 20 outer vertices must also touch one of the 12 outer vertices of the inside 3×3 grid—except for those in a corner cell that miss the corner vertex of the big square, for example, a backslash in the upper right cell. But there can't be backslashes in both cells neighboring the upper right corner cell, so putting a backslash in the upper right corner will cause at least one other outside vertex to be missed. In conclusion, no matter what we do, at least four outside vertices will be unmatched, and we're down to an upper bound of $(36-4)/2 = 16$ diagonals.

When we try to achieve 16, we notice that using up all the outer vertices of the inside 3×3 grid to take care of outer vertices won't work—that leaves the four corners of the center cell unmatched, and a diagonal in that cell takes care of only two, taking us back to 15 diagonals. We'll need to leave at least two of the four corners of the big square unmatched, and use two or four unmatched outer vertices of the inside 3×3 grid to cover the corners of the central cell. A little experimentation shows that this works only if you miss *all* the corners of the big square, producing the picture below—the unique solution to the puzzle, up to reflection.

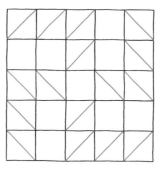

Unbroken Lines

Can you re-arrange the 16 square tiles below into a 4×4 square—no rotation allowed!—in such a way that no line ends short of the boundary of the big square?

Solution: Notice that there are $2^4 = 16$ different tiles that can be made by choosing, for each line-type, whether to include that type or not. Each tile is represented once in the puzzle.

This is, again, a matter of intelligent search. The horizontal lines must all be in two rows, the vertical lines in two columns; it costs nothing to assume those are the top two rows and the leftmost two columns, and already you have severe constraints.

The lengths of the SW–NE diagonals must sum to 8 squares worth, and thus there are only a few possibilities: a 4 and two 2s, a 4, a 3 and a 1, etc.

There are in fact several ways to do it of which one is illustrated below. Curiously, all ways of doing it work on the torus as well as on the square—in other words, if you imagine that the square wraps around top-to-bottom and left-to-right, no line ends *at all*; all are, ultimately, loops.

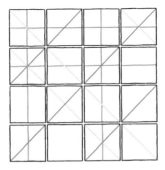

[See *Notes & Sources* for the composer and inspiration behind this and the next puzzle.]

Similar, but harder:

Unbroken Curves

Can you re-arrange the 16 square tiles below into a 4 × 4 square— no rotation allowed!—in such a way that no curve ends short of the boundary of the big square?

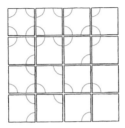

Solution: Again, there are $2^4 = 16$ different tiles that can be made by choosing, for each corner of the tile, whether to include a quarter-circle at that corner or not. Each tile is represented once in the puzzle.

For this puzzle, the best route to a solution seems to be to guess how the squares with the most quarter-circles fit together and (systematically!) chase down the possibilities.

Not all solutions to this one work on the torus as well as the square, but the one illustrated below does work on both.

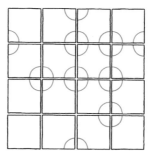

Conway's Immobilizer

Three cards, an ace, a deuce, and a trey, lie face-up on a desk in some or all of three marked positions ("left," "middle," and "right"). If they are all in the same position, you see only the top card of the stack; if they are in two positions, you see only two cards and do not know which of them is concealing the third card.

Your objective is to get the cards stacked on the left with ace on top, then deuce, then trey on bottom. You do this by moving one card at a time, always from the top of one stack to the top of another (possibly empty) stack.

The problem is, you have no short-term memory and must, therefore, devise an algorithm in which each move is based entirely on what you see, and not on what you last saw or did, or on how many moves have transpired. An observer will tell you when you've won. Can you devise an algorithm that will succeed in a bounded number of steps, regardless of the initial configuration?

Solution: Since there aren't that many things you can do, it's tempting to try some arbitrary set of rules and see if they work. Actually, there's a surprising number of different algorithms possible with the above constraints—more than a quintillion. So some reasoning will be needed. But it's tricky to design an algorithm that makes progress, avoids cycling, and doesn't do something stupid when it's about to win.

For example, it looks obvious to try to win by putting the A on the 2 if you see 2,A,– or 2,–,A. Then if you see three cards with the 3 on the left, you can win in two moves by putting the 2 on the 3. So far, so good.

Suppose the 2 is exposed on the left but no other card is seen; then we may as well move the 2 to the center. Now, if you see 3,2,–, it's tempting to put the 2 on the 3. But if the A is hiding under the 3, you will get caught in a loop, stuck forever going back and forth between 2,–,– and 3,2,–. So you'll need to move the 2 to the right from 3,2,–. But then you can't put 2 on 3 from 3,–,2 without getting caught in a loop of order 3. So you'll have to move the 3 from that position. Darn!

It's beginning to look like we should be filling holes by moving cards in one direction, say to the right, to avoid loops. So let's agree that when we see only 2 cards, we move a card to the right (around the corner from R to L, if necessary) to an open slot; the only exceptions will be 2,A,– and 2,–,A where we place A on 2 to try to win.

Moving cards to the right will spread out the cards in two moves at most, so we can see them. Seeing only one stack, sending the top card to the *left* will then also spread the cards in two moves. It remains only to decide what to do when we see all three cards, but the 3 is not on the left. Clearly we can't move a card left; it would come back immediately. So we move a card to the right; which one? After two moves whichever card we covered will be re-exposed, with the other two switched.

Having got this far, several successful algorithms can now be described. Here are the rules for one of them, in priority order:

1. Seeing 2,1,–, place 1 on 2;

2. Otherwise, seeing two cards, move a card right (around the corner if necessary) to the open place;

3. Seeing just one card, move that card to the left;

4. Seeing three cards, move the middle-rank card to the right.

This same algorithm works for any number of cards, numbered 1 through n. It takes quadratic time in the worst or random case, that is, approximately a constant times n^2 steps. A computer scientist would conclude that a list of items can be sorted with just three LIFO (last-in, first-out) stacks and no memory.

Here's a second solution that's slower but has a compensating feature:

1. Seeing just one card, move a card to the right.

2. Seeing two cards:

   (a) If the left is empty, move the center card to the right stack;
   (b) If the center is empty, move the right card to the left stack;
   (c) If the right is empty, move the left card to the right stack.

3. Seeing three cards, let the numbers be x, y, and z from left to right.

   (a) If z is the largest or smallest, move the smaller of x and y onto z;
   (b) If z is the middle, move the larger of x and y onto z.

This algorithm is cubic, requiring $(2/3)n^3 + (1/2)n^2 - (7/6)n$ steps if it is started with the cards stacked in reverse order on the left. Once it gets started, it never has more than one card in the center. Thus, we see that sorting can be done with only two LIFO stacks, a single register, and no memory! If you can beat that, let me know.

Seven Cities of Gold

In 1539 Friar Marcos de Niza returned to Mexico from his travels in what is now Arizona, famously reporting his finding of "Seven Cities of Gold." Coronado did not believe the "liar friar" and when subsequent expeditions came up empty, Coronado gave up the quest.

My (unreliable) sources claim that Coronado's reason for disbelieving de Niza is the latter's claim that the cities were laid out in the desert in a manner such that of any three of the cities, at least one pair were exactly 100 furlongs apart.

Coronado's advisors told him no such pattern of points existed. Were they right?

Solution: Putting ourselves in the position of Coronado's advisors, we need to decide whether there's a way to position seven points in the plane in such a way that among any three, there are at least two that are a specified distance apart.

What could we try? What we want is a seven-vertex graph drawn on the plane, with the property that every edge is a line segment of unit length (the unit being 100 furlongs = $12\frac{1}{2}$ miles), having the following property: Every three vertices contain an edge.

If there were only three vertices, we could of course arrange them as an equilateral triangle of side 1, and get more: an edge between every pair of vertices. In fact, two of these equilateral triangles (say, ABC and DEF), no matter their relative position, would get us a six-vertex solution. Why? Because any three vertices would include at least two in one of the two triangles, and there's our edge.

So let's see if we can add a seventh vertex, call it G, to our two triangles. Then we only have to worry about sets of three vertices that contain vertex G.

We can't connect G to all three vertices of one of our triangles, but we can connect it to two (say, A and B from the first triangle, and D and E from the second) by making diamonds $AGBC$ and $DGEF$ that share vertex G. That leaves just one set of three vertices without and edge: the set $\{G, C, F\}$. No problem: We just angle the two diamonds, hinged at G, so that their far ends are a unit distance apart. Done! Here's what it looks like:

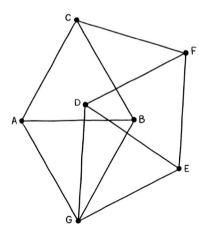

This graph is known as the "Moser spindle," and it's the only graph with the desired property.

Theorem. *Suppose seven distinct points in the plane have the property that among any three of them, there are two that are exactly a unit distance apart. Then the graph on those seven points, defined by inserting an edge between each pair that are a unit distance apart, is the Moser spindle.*

Proof. To prove this we again need to be systematic. We assume the graph H satisfies the requisite properties, and gradually learn more and more about it until we can show it actually is the Moser spindle.

First step: We observe that for any vertex v, the vertices not equal to or adjacent to v must form a "clique," that is, they must all be connected to one another. Why? Because if two (say, x and y) are not, then the set $\{v, x, y\}$ would violate the condition. But four vertices cannot all be connected to each other by unit-distance edges, so the number of vertices not equal or adjacent to v is at most three; it follows that v, and thus every vertex of H, has at least three neighbors.

Second step: No vertex can have as many as five neighbors. Reason: If vertex u, say, has five neighbors, they all lie in a circle of unit radius with center at u and cannot be connected among themselves any better than by a path of four edges. Such a path has three vertices no two of which are mutually adjacent, and that's not allowed.

Third step: We now know all vertices have either three or four neighbors, but they can't all have just three neighbors—for then, the sum of the degrees of the vertices of H would be $3 \times 7 = 21$ which is an odd number, and each edge contributes 2 to that sum. Hence, there is at least one vertex, say w, that has four neighbors.

Fourth step: How are w's neighbors connected among themselves? At least in two pairs, say p with q and r with s, because otherwise there'd be three with no edge. The two remaining vertices (call them x and y) must be adjacent to each other, else $\{w, x, y\}$ would be a bad set. Neither x nor y can be adjacent to three of p, q, r, and s because there's already a point (w) with that property and there can't be two.

Wrap-up: Of p, q, r, and s, the ones x is *not* adjacent to must form a clique, otherwise we'd have another bad set; and the same for y. That is possible only if one of x and y is adjacent to precisely p and q, and the other to r and s.

And that's the Moser graph! ♡

12. The Pigeonhole Principle

The *pigeonhole principle* is nothing more than the evident fact that if n pigeons are to be housed in m holes, and $n > m$, then some hole must contain more than one pigeon. Yet this principle has wide-ranging application, and is the key to proving some surprising statements and some important mathematical truths.

First, a warm-up.

Shoes, Socks, and Gloves

You need to pack for a midnight flight to Iceland but the power is out. In your closet are six pairs of shoes, six black socks, six gray socks, six pairs of brown gloves, and six pairs of tan gloves. Unfortunately, it's too dark to match shoes or to see any colors.

How many of each of these items do you need to take to be sure of getting a matched pair of shoes, two socks of the same color, and matching gloves?

Solution: The easiest way to solve such puzzles is to figure out the maximum number of items you could have *without* getting what you want, then adding one. In this way you are implicitly invoking the pigeonhole principle.

For the shoes, you could have one shoe from each pair, six in all; so you need seven to guarantee a pair.

For the socks, one black and one gray will not work but the third sock gets you a pair.

The gloves are yet another issue: You need a right- and a left-hander, so the biggest disaster would be six brown all left or all right, and six tan, also all left or all right. So 13 gloves are required, although you would have to count yourself very unlucky indeed to take only 12 and not get a match!

The same idea is easily employed in certain poker calculations. A standard deck of 52 playing cards contains 13 cards of each suit, and within each suit, an ace, deuce, 3, 4, etc. up to 9, 10, jack, queen, king. Thus there are four cards of each rank.

A poker hand is five cards. Two cards of the same rank constitute a *pair*, three, *trips* (or "three of a kind"). Five cards of the same suit make a *flush*; five cards of consecutive ranks (including 5432A as well as AKQJ10) make a *straight*.

How many cards do you need to hold to guarantee a pair—that is, to be sure that among your cards is a 5-card poker hand containing two cards of the same rank? Easy: The worst holding (for this problem!) is 13 cards consisting of one card of each rank, therefore the answer is 14 (but note that 12 cards are enough to guarantee a pair *or better*).

If it's trips you seek, the worst holding is two cards of each rank, so you need $2 \times 13 + 1 = 27$ for your guarantee. For a flush, four of each suit will stump you, so you need just 17 cards. The straight is the trickiest: You must have a 5 or a 10 to make a straight, so you can hold $52 - 2 \times 4 = 44$ cards without making a straight, hence it takes a stunning 45 cards to guarantee one. Note the contrast with relative probabilities of getting trips, flush or straight when dealt five cards: The flush is the hardest, and indeed outscores the straight and trips in poker, even though it takes the fewest cards to guarantee.

For the next puzzle, one preliminary bit of cleverness is needed.

Polyhedron Faces

Prove that any convex polyhedron has two faces with the same number of edges.

Solution: The key is to pick a face with the maximum number of edges, say E. Since it has at least E neighboring faces, the number of faces of the polyhedron is at least $E+1$. But their numbers of edges range only from 3 to E, comprising only $E-2$ different numbers, so two must have the same number of edges. (In fact, since we have some overkill here, you can deduce that there are either *three* pairs of faces with the same number of edges, or a triple and a pair, or four faces with the same number of edges.)

Sometimes you have to look around a bit to find the right items to use for your pigeonholes, as in this geometric example.

Lines Through a Grid

If you want to cover all the vertices in a 10×10 square grid by lines none of which is parallel to a side of the square, how many lines are needed?

Solution: Your first idea here might be to use 19 diagonals running $45°$ running (say) north–west to south–east, but you immediately see that the two extreme ones can be replaced by a single SW–NE diagonal covering the SW and NE corners, as in the figure below. This reduces the count to 18; can you do any better?

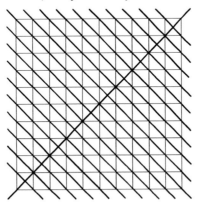

No, because there are 36 vertices on the boundary of the grid, and no non-vertical, non-horizontal line can cover more than two of them.

Now let us move on to an oddly counterintuitive fact about numbers.

Same Sum Subsets

Amy asks Brad to pick ten different numbers between 1 and 100, and to write them down secretly on a piece of paper. She now tells him she's willing to bet $100 to $1 that his numbers contain two nonempty disjoint subsets with the same sum! Is she nuts?

Solution: Of course, the answer is that Amy is perfectly sane and in fact the bet is not merely favorable to her; it's a sure thing. How can you prove it?

There's no unique maximum set-size Brad can pick that *doesn't* have the desired property, so the shoe-sock-glove technique won't work. But whenever you are asked to show that there are two objects that share a property, you can try to show that there are more objects than properties; then, by pigeonhole, you're done.

Here, the number of nonempty subsets of Brad's set of 10 numbers is $2^{10} - 1 = 1023$. The possible sums range only from 1 to $100 + 99 + \cdots + 91 < 1000$, so there are indeed fewer sums than subsets; by the pigeonhole principle, there must be two different subsets with the same sum.

But wait—the problem asked for *disjoint* subsets. That was a red herring: Throw away the common elements of two sets with equal sums to get disjoint sets with equal sums. $\heartsuit$

The next problem looks geometric, but is really number-theoretic; nor is it obvious at first glance that the pigeonhole principle has a role to play here.

Lattice Points and Line Segments

How many lattice points (that is, points with integer coordinates) can you have in 3-space with the property that no two of them have a third lattice point on the line segment between them?

Solution: The first issue to be addressed here is: When do two lattice points, say (a, b, c) and (d, e, f), have another lattice point on the line segment between them? A moment's thought will convince you that this happens when the numbers $a-d$, $b-e$, and $c-f$ have a common divisor. The "easiest" common divisor for them to have is 2, and this will happen if $a-d \equiv b-e \equiv c-f \equiv 0 \mod 2$, that is, if the two vectors have the same parity coordinatewise. When that happens, the midpoint of the two vectors is a lattice point.

But there are only $2^3 = 8$ parities available, so by pigeonhole, we can't have nine points in our set. Can we have eight? Yes, just take the corners of a unit cube.

Here's another number problem.

Adding, Multiplying, and Grouping

Forty-two positive integers (not necessarily distinct) are written in a row. Show that you can put plus signs, times signs, and parentheses between the integers in such a way that the value of the resulting expression is evenly divisible by one million.

Solution: This puzzle has a pigeonhole flavor to it and indeed the pigeonhole principle is useful in it, but not perhaps where you might expect. Since a million is $2^6 \cdot 5^6$, and $42 = 6 \cdot 2 + 6 \cdot 5$, a logical idea is to break up the numbers into six contiguous groups of size 2 and six more of size 5, with the multiplication symbols between the groups; then all that is needed is to "persuade" each group of size 2 to represent an even number, and each group of size 5 to represent a multiple of 5.

The groups of size 2 are easily handled: If the two numbers inside a group are both even or both odd, add them; otherwise, multiply them.

The groups of size 5 are only slightly trickier. If we can get a contiguous substring of the five numbers that sums to a multiple of 5, we're done; we can add those and multiply the sum by the product of the remaining numbers.

Here's where pigeonhole comes in. Let the numbers be $n_1, \ldots, n_5$ and examine the partial sums $s_1, \ldots, s_5$ where $s_i = n_1 + \cdots + n_i$. Let r_i be $s_i \mod 5$, that is, r_i is the remainder when s_i is divided by 5. If any r_i is 0, we use the substring $n_1, \ldots, n_i$; otherwise, the five s_i's all lie in the set $\{1, 2, 3, 4\}$. Thus two of them, say r_i and r_j with $i < j$, are equal. Then $s_j - s_i = n_{i+1} + n_{i+2} + \cdots + n_j$ is a multiple of 5 and we are done. ♡

Notice that this argument works for 5 replaced by any n. We have in effect proven a nice lemma which says that every string of n numbers has a nonempty substring that sums to a multiple of n.

Line Up by Height

Yankees manager Casey Stengel famously once told his players to "line up alphabetically by height." Suppose 26 players, no two exactly the same height, are lined up alphabetically. Prove that there are at least six who are also in height order—either tallest to shortest, or shortest to tallest.

Solution: Let's try small numbers to see what's going on. Two players are, obviously, enough to get two that are also in height order one way or the other. Three or four players won't get you any more: For example, their heights could be $73''$, $75''$, $70''$, and $72''$ respectively. But five players will get you three in height order; by the time you reach that fifth player (say, Rizzuto) alphabetically, there's a previous player (say, Berra) who's got both a taller and shorter player before him. Then, whether Rizzuto is taller or shorter than Berra, he ends a height-order subsequence of length 3.

This reasoning suggests a sort of "dynamic programming" approach to the 26-player problem. Give each player two numbers, an "up" number and a "down" number. The up number records the length of the longest height-increasing subsequence that ends with that player, and the down number is defined analagously. For example, in the height sequence $73''$, $75''$, $70''$, $72''$ from above, the up-down number pairs would be $(1,1)$, $(2,1)$, $(1,2)$, and $(2,2)$.

Notice these are all different—they have to be! Rizzuto and Berra, for instance, can't have the same pair because if the alphabetically later player (Rizzuto) is taller his up number would be higher, and if shorter, his down number would be higher.

Now we see why five players guarantees a subsequence of three in height order: Only four different up-down pairs can be made of numbers less than 3. Similarly, if there were no height-order subsequence of 26 players of length 6, all the up-down pairs would have to be made up of numbers from 1 to 5 and there are only 25 of those. By the pigeonhole principle, there must be a player with an up-number or down-number of 6.

The key relationship is that if a subsequence of length $k+1$ is sought, you need the number of players to be greater than k^2. Still more generally, if the number of players is at least $jk+1$, you are guaranteed *either* a height-increasing subsequence of length $j+1$ or a height-decreasing subsequence of length $k+1$. The argument is the same: If you didn't have either such subsequence, the number of up-down pairs would be limited to jk, and since the pairs are all different, you get a pigeonhole contradiction.

In fact, this last statement is known as the Erdős–Szekeres Theorem, and can be even further generalized (with almost the same pigeonhole proof) to a statement about height and width of partially ordered sets—for those readers who know what they are. The Erdős–Szekeres Theorem is "tight" in the sense that if there are only jk players, you might indeed be limited to length j for height-increasing subsequences and k for height-decreasing. An example: If, say, $j = 4$ and $k = 3$, the heights (in inches) could be $76, 77, 78, 79, 72, 73, 74, 75, 68, 69, 70, 71$.

Here's a related problem where we can use a similar idea.

Ascending and Descending

Fix a number n and call a list of the numbers from 1 to n "good" if there is no descending subsequence of length 10. Show that there are at most 81^n good lists.

Solution: Let $x_1, \ldots, x_n$ be a good list and let d_i be the length of the longest descending subsequence ending at x_i. Then d_i is a number between 1 and 9, which we call the "color" of the position i, and also of the number x_i. Essentially, d_i is the "down-number" from the solution to the previous puzzle.

The numbers x_i for which d_i is some particular color, say color 7, form an increasing subsequence. Suppose we know the positions with color 7 and also the numbers of color 7. Then we can fill in all the numbers that go in those positions, because we know they are in increasing order.

For example, if the list begins $42, 68, 35, 50$, then positions 1 and 2 get color 1, as do the numbers 42 and 68. Knowing that $d_1 = d_2 = 1$ and the set of numbers that got color 1, we can conclude that $x_1 = 42$ and $x_2 = 68$ because the first two positions are colored 1 and the two lowest numbers that got the first color are 42 and 68.

Now, since there are at most 9^n ways to assign colors to positions, and 9^n ways to assign colors to numbers, we get (using, again, the Pigeonhole Principle) a maximum of $(9^n)^2 = 81^n$ different lists.

Note that this bound of 81^n, though exponentially large, is a vanishingly small fraction of all the $n!$ ways to list the numbers from 1 to n.

Sometimes the pigeonhole principle can be applied even though a puzzle asks only for one item, not two similar ones. The trick is to use subtraction to turn two similar items into one useful one.

Zeroes and Ones

Show that every natural number has a non-zero multiple which, when written out (base 10), contains only zeroes and ones. Even better, prove that your telephone number, if it ends with a 1, 3, 7, or 9, has a multiple whose decimal representation is all ones!

Solution: There are infinitely many numbers whose decimal representations contain only ones and zeroes—or, for that matter, only ones. So some of them ought to be multiples of n unless there's some reason why they can't be. But how can we prove it?

Perhaps we can make use of the fact that if we subtract two numbers each of which has the same remainder when divided by n, the result will be a multiple of n. There are only n different remainders possible when you divide by n, so pigeonhole tells us that $n+1$ numbers are enough to guarantee two with the same remainder.

Now we only have to find $n+1$ numbers with the property that the difference between any two has only zeroes and ones as its digits. That's not hard: Take numbers of the form $111\ldots111$, where the number of digits runs from 1 to $n+1$. Formally speaking, these numbers are $(10^k - 1)/9$ for $1 \leq k \leq n+1$. (Actually, we only need n of them because if all their remainders modulo n differ, one of them already has remainder 0, and we can use it directly.)

To wrap up, we find two distinct numbers all of whose digits are ones, that have the same remainder when divided by n; subtracting the smaller from the larger gives a number of the form $111\ldots111000\ldots000$.

This means that we have actually proved a stronger statement than required, namely that any n has a multiple of the form $111\ldots111000\ldots000$. But, as often happens, asking for more would have made the puzzle easier. We'll see a more extreme example of that phenomenon in the next puzzle.

Wait, what about your telephone number? We know now that it has a multiple of the form $N = 111\ldots111000\ldots000$, ending in, say, k zeroes. But if your number ends in 1, 3, 7, or 9, it is not a multiple of 5 or 2. It follows that $N/(5 \cdot 2)^k$, which is all ones, is still a multiple of your phone number.

The next puzzle looks like one in which pigeonhole refuses to work. It takes some serious insight to see how it ultimately does.

Same Sum Dice

You roll a set of n red n-sided dice, and a set of n black n-sided dice. Each die is labelled with the numbers from 1 to n. Show that there must always be a nonempty subset of the red dice, and a nonempty subset of the black dice, showing the same sum.

Solution: This seems like an ideal pigeonhole application: There are lots of nonempty subsets ($2^n - 1$ of each color) and sums can only go from 1 to n^2. But it doesn't do any good to have, say, two red subsets with the same sum, and two black subsets with some other sum.

Come to think of it, there may be very few *distinct* sums among, say, the red dice; for example, they could all be showing the same number of pips, say j. In the meantime the black dice could all be agreeing on some other number, k. That does not give us a counterexample, because then k of the reds and j of the blacks both sum to jk, but that coincidence seems like something the pigeonhole principle would be hard pressed to lay claim to.

Nonetheless, with help from a subtraction trick similar to the one we used in *Zeroes and Ones*, pigeonhole saves us. The idea is the following: Suppose the subsets R (of the red dice) and B (of the black) don't work, because the sum of the dice in R exceeds that of B by some non-zero number d. Let's add some more red dice to R to get a bigger set R^+, and similarly add some black dice to B to get B^+. Now we check the sums for R^+ and B^+ and they still don't agree, but suppose that the discrepancy is the same as it was before: The dice in R^+ sum to exactly d more than the dice in B^+. Then we're done, because the dice we *added* to R must have had the same sum as those we added to B.

To get this idea to work we put each set of dice in some fixed left-to-right order. Amazingly, it doesn't matter a bit what orders we choose. Let r_i be the sum of the first i red dice, reading the dice from left to right, and similarly let b_j be the sum of the leftmost j black dice. We claim that there are *contiguous* subsets of the red and of the black dice with the same sum.

A subset is "contiguous" if it is a substring of the dice as laid out, for example, all the red dice from the third to the seventh. There are $\binom{n+1}{2} = n(n+1)/2$ such subsets of each color, because we can get one by choosing two of the $n+1$ spaces between or at the ends of the dice in which to put down supermarket dividers. It's hard to believe that we can get away with limiting ourselves, in this seemingly arbitrary manner, to just these few of the exponentially many subsets of each color. Can this really work no matter how the dice are rolled, *and* no matter how we order them?

We can assume $r_n < b_n$; if they are equal we're done, and if the inequality goes the other way we can just switch the roles of the colors. Let us find, for each i, the j for which b_j comes closest to r_i without exceeding it. More formally, let

$$j(i) = \max\{j : b_j \le r_i\}$$

with the understanding that $b_0 = 0$, and thus $j(i)$ may be zero for some small i's if the red sequence starts with smaller numbers than the black sequence.

Now let's look at the discrepancies $r_i - b_{j(i)}$. All are non-negative by definition of $j(i)$; moreover, all are less than n, because if $r_i - b_{j(i)}$ were equal to n or more, we could have thrown in the next black die without exceeding r_i. (Since $r_n < b_n$, there always *is* a next black die.)

If any of the discrepancies $r_i - b_{j(i)}$ is zero, we're done, so we can assume all the discrepancies are among the numbers $1, 2, \ldots, n-1$. There are n such discrepancies, one for each i, so there most be two that are equal; in other words, there are two distinct indices $i < i'$ with

$$r_{i'} - b_{j(i')} = r_i - b_{j(i)}.$$

Now we play the subtraction game, concluding that the red dice from number $i+1$ up to number i' must add up to the same value as the black dice from number $j(i)+1$ up to number $j(i')$. Done!

Our last puzzle looks perhaps less like a pigeonhole problem than its predecessors, but now that you've seen the subtraction trick, you won't be fooled. Nonetheless, it's not easy.

Zero-sum Vectors

On a piece of paper, you have (for some reason) made an array whose rows consist of all 2^n of the n-dimensional vectors with coordinates in $\{+1, -1\}^n$—that is, all possible strings of $+1$'s and -1's of length n.

Notice that there are lots of nonempty subsets of your rows which sum to the zero vector, for example any vector and its complement; or the whole array, for that matter.

However, your 2-year-old nephew has got hold of the paper and has changed some of the entries in the array to zeroes.

Prove that no matter what your nephew did, you can find a nonempty subset of the rows in the new array that sums to zero.

Solution: The technique used in *Same Sum Dice* suggests that to get a set of new rows that sum to zero, we might try to build a sequence of new rows and use pigeonhole to argue that two of its partial sums coincide; then the portion of the sequence between those two partial sums will add to the zero vector.

To make that work, the sequence will have to be designed so that the partial sums belong to some small set. How small can we make it? Would you believe that we can force all the partial sums to be $\{0, 1\}$-vectors?

We can certainly start the list with a $\{0, 1\}$-vector: We just take the row u' that *used to be* $u = \langle 1, 1, 1, \dots, 1 \rangle$. If our nephew happened to have changed *all* of u's coordinates to zeroes, we're done. Otherwise, there will be some 1's in u', and we need to choose the next row v' in our sequence in such a way that $u' + v'$ has all of its coordinates in $\{0, 1\}$. But we can do that by choosing v to be the vector that has -1 in the places where u' is 1, and $+1$ in the places where u' is zero. Then no matter how v was altered by your nephew, $u' + v'$ will have the desired property.

We continue this procedure as follows: At time t, t altered rows have been put on our sequence and all partial sums, including the sum x of all t of them, are $\{0, 1\}$-vectors. Let z be the original row with -1 wherever t has a 1, and $+1$ wherever t has a 0; add z's altered form z' to the sequence.

Since there are only finitely many $\{0, 1\}$-vectors of length n, we will eventually hit the same sum twice; let us stop when this happens for the first time. (We consider the 0th partial sum to be the zero vector, so one way this can happen is for the zero vector to arise again as a partial sum.) Call the vectors $w_1, w_2, \dots$, so that our critical time is the least k such that $w_1 + w_2 + \cdots + w_j = x = w_1 + w_2 + \cdots + w_k$, with $0 \le j < k$.

Then $w_{j+1} + w_{j+2} + \cdots + w_k$ is the zero vector, and we are done. Note, very importantly, that we have not used any altered vector too many times; the reason is that since up to time k all the partial sums have been different, all the *original* vectors z whose altered forms we appended to our sequence were different.

Here's a more serious use of the pigeonhole principle in mathematical analysis. Suppose we fix some real number r and let $\{nr\}$ be the fractional part of nr, that is, the unique number in the half-open interval $[0, 1)$ that can be expressed as nr minus an integer.

If r is a fraction (say, a/b) written in lowest terms, then the values $\{r\}, \{2r\}, \{3r\}, \ldots$ will cycle among the multiples of $1/b$ in $[0, 1)$. What if r is irrational? Then the $\{nr\}$'s will not repeat and it's natural to conjecture that they are in fact dense in the unit interval; in other words, for any $\varepsilon > 0$ and any real $x \in [0, 1)$, there is some positive integer n for which $|\{nr\} - x| < \varepsilon$.

Theorem. *For any irrational number r, if $\{nr\}$ is the fractional part of nr, then the numbers $\{r\}, \{2r\}, \{3r\}, \ldots$ are dense in the unit interval.*

Proof. Fix some irrational r. Of course $\{nr\}$ is never zero for any positive integer n, but we can borrow an idea from the *Zeroes and Ones* puzzle above to show that we can choose n to get nr is as close as we want to an integer.

Let $1 \le i < j$. If $\{ir\} < \{jr\}$, , then $\{(j-i)r\} = \{jr\} - \{ir\}$; if $\{ir\} > \{jr\}$, then $\{(j-i)r\} = 1 - (\{ir\}-\{jr\})$. Pick any $\varepsilon > 0$, and let m be an integer bigger than $1/\varepsilon$. Divide the unit interval into m equal intervals; by pigeonhole, one of those little intervals (each of length $< \varepsilon$) contains two of the values $\{r\}, \{2r\}, \ldots, \{(m+1)r\}$. Let those be $\{ir\}$ and $\{jr\}$ with $i < j$; then if $k = j-i$, we conclude that either $\{kr\} < \varepsilon$ or $1-\{kr\} < \varepsilon$.

Suppose that the former inequality holds, that is, $\{kr\} < \varepsilon$. Given any real value $x \in [0, 1]$, let p be the greatest integer such that $p\{nr\} < x$. Then $\{pkn\} = p\{kn\} < x \le (p+1)\{kr\}$ but the outside quatities differ by $\{kr\}$ which is less than ε, so in particular $|x-\{pkr\}| < \varepsilon$ and we have our approximation.

If instead $1-\{kr\} < \varepsilon$, we take p to be the greatest integer such that $p(1-\{kr\}) > 1-x$, and a similar argument again shows $|x-\{pkr\}| < \varepsilon$. $\heartsuit$

13. Information, Please

The basis of this chapter is simple: If you want to distinguish among n possibilities, you have to do something that results in at least n different possible outcomes. A common puzzle application is to balance-scale problems.

Each weighing by a balance scale can produce at most three outcomes: tilt left (L), tilt right (R), or balance (B). If you are permitted two weighings, you get $3 \times 3 = 9$ possible outcomes: LL, LR, LB, RL, RR, RB, BL, BR, and BB. Similarly, with w weighings you could have as many as 3^w outcomes, but no more.

So, if a puzzle gives you a bunch of objects to test, k possible answers, and allows w weighings on a balance scale, the first thing to ask yourself is whether $k \le 3^w$. If not, the task is impossible (or you have missed some trick in the wording). If you do have $k \le 3^w$, the task may or may not be possible, and you can use the idea above to help you construct a solution or to prove impossibility.

Here's a classic example.

Finding the Counterfeit

You have a balance scale and 12 coins, 11 of which are genuine and identical in weight; but one is counterfeit, and is either lighter or heavier than the others. Can you determine, in three weighings with a balance scale, which coin is counterfeit and whether it is heavy or light?

Solution: There are 24 possible outcomes, less then $3^3 = 27$, so the task may be possible but needs to be pretty efficient—in other words, most weighings will need to have all three possible outcomes. We certainly want to start by weighing k coins against k others, where $k \le 6$. That would leave $12 - 2k$ suspect coins if the initial weighing balances, so it better be possible to test that many in two weighings. Two weighings have at most nine possible outcomes, so we can't afford to leave more than four coins out of the first weighing; thus k is at least 4.

In fact k can't be more than 4 because then "tilt left" would lead to $10 > 3^2$ possible outcomes. So we have reached a firm conclusion: If the task can be done, it must be right to put 4 coins against 4 (say, coins ABCD against EFGH) in the first weighing.

Suppose they balance. Then the temptation is to put 1 against 1 or 2 against 2 of the remaining coins, but either is information-theoretically deficient: If 1 balances against 1, four (so, more than three) possibilities remain; if 2 tilts against 2, again four possibilities remain. Either way one more weighing is insufficient. So we must use the known-genuine coins A through H in the second weighing, and the easiest thing to do is put three of them against three of the suspect coins (say, IJK). If they balance, coin L is counterfeit and we can weigh it against A to determine whether it is light or heavy. If IJK are light or heavy, test I against J.

If the initial 4-against-4 weighing tilts left, so that either one of ABCD is heavy or one of EFGH is light, put (say) ABE against CDF. If they balance, the culprit is a light G or H and you can test either against the genuine L. If they tilt left, you're left with A or B heavy or F light and you can determine which by putting A and F against two genuine coins, say K and L; the solution is similar if CDF outweigh ABE.

Now suppose you are given the same problem but with 13 coins. There are now 26 possible outcomes, which is still less than or equal to 27, but our previous reasoning already showed that nonetheless three weighings are not enough. Why? Because we saw that 5-against-5 in the first weighing is no good, while 4-against-4, when they balance, leaves five coins—too many—untested.

Three Natives at the Crossroads

A logician is visiting the South Seas and as is usual for logicians in puzzles, she is at a fork, wanting to know which of two roads leads to the village. Present this time are three willing natives, one each from a tribe of invariable truth-tellers, a tribe of invariable liars, and a tribe of random answerers. Of course, the logician doesn't know which native is from which tribe. Moreover, she is permitted to ask only two yes-or-no questions, each question being directed to just one native. Can she get the information she needs? How about if she can ask only *one* yes-or-no question?

Solution: We can dispose of the one-question case easily: If the question is directed to the random answerer, the logician gets no information, thus can never guarantee to identify the right road. (This argument does not apply if you assume the random answerer first flips a mental coin to determine whether to lie or tell the truth; you could then gain information with a well-chosen self-referential statement, for example, "Out of the other two guys, if I pick the one whose response's truthfulness will least likely match your response's truthfulness and ask him if Road 1 goes to the village, will he answer 'Yes'?" But we assume the random answerer just randomly answers "yes" or "no" regardless of the question, so no information is imparted.)

Similarly, if the logician doesn't know the right road after one question, *and her second question is directed to the random answerer*, she is in trouble. It follows that after the first answer, she must be able to identify a native who is not the random answerer.

If she can do that, she's in business, because she can then use a traditional one-native query as her second question: For example, something like "If I were to ask you whether Road 1 goes to the village, would you say yes?"

To attain the objective, she'll need to ask Native A something about Native B or Native C, then use the answer to choose between B and C. Here's one that works: "Is B more likely than C to tell the truth?"

Curiously, if A says "yes" she picks C, and if he says "no" she picks B! If A is the truth-teller she wants next to query the companion who is *less* likely to tell the truth, namely, the liar. If A is the liar she queries the *more* truthful of his companions, namely the truth-teller.

Of course, if A is the random answerer it doesn't matter which of B and C she turns to next. ♡

If in a puzzle the information seems insufficient, we may have to find extra sources.

Attic Lamp

An old-fashioned incandescent lamp in the attic is controlled by one of three on–off switches downstairs—but which one? Your mission is to do something with the switches, then determine after *one* trip to the attic which switch is connected to the attic lamp.

Solution: This seems not to be doable: There are three possible answers (as to which of three switches controls the bulb), but when you get to the attic either the bulb is on or off. So only one bit of information is available, and we need a "trit."

Is there anything that can be observed about the bulb other than whether it is on or off? There's a hint in the puzzle wording: It's an incandescent bulb, whose efficiency is limited by its habit of putting out quite a lot of heat as well as light. So you could feel the bulb and see if it's hot; can that help?

Yes. Flip switches A and B on, wait 10 minutes or so, then flip switch B off. Go quickly up to the attic: If the bulb is on, it's switch A; if off but warm, switch B; off but cold, switch C.

You could even do four switches: Flip A and B on, wait 10 minutes, then flip B off and C on and run up the stairs. The four possible states of the bulb (on/warm, off/warm, on/cool, or off/cool) will tell you what you need to know.

Information is customarily measured in bits; k bits gives you 2^k possibilities. The thriving field of "communication complexity" attempts to determine the minimum number of bits of communication needed between two or more parties, in order for them to accomplish some task. Here is a toy example.

Players and Winners

Tristan and Isolde expect to be in a situation of severely limited communication, at which time Tristan will know which two of 16 basketball teams played a game, and Isolde will know who won. How many bits must be communicated between Tristan and Isolde in order for the former to find out who won?

Solution: Tristan and Isolde can, in advance, label the 16 teams (in alphabetic order, perhaps) by the binary numbers 0000 through 1111. Then, when the time comes, Tristan can send Isolde eight bits: the four bits corresponding to one team, and the four bits corresponding to the other one. Isolde can then send back a 0 if the winning team was the one with the smaller number, and a 1 otherwise. That comes to nine bits, and we can trim that to 8 by noting that the number of unordered pairs of teams is only $16 \times 15/2 = 120 < 2^7$ and thus Tristan can code the matchup with only seven bits.

But an easier and faster solution is to just have Isolde send Tristan the four bits corresponding to the winning team. Surely, we can't do any better than four bits, right?

Amazingly, we can! We can code those four positions (leftmost, next-to-left, next-to-right, rightmost) by, say, 00, 01, 10, and 11 respectively. The codes of the two playing teams must differ in at least one of the four positions; Tristan picks such a position and sends its code (two bits) to Isolde. She now only needs to look at the value of the bit in that position of the winning team and sent it to Tristan. Three bits total!

For example, suppose the playing teams are 0010 and 0110, and 0110 won. The only position where the playing teams differ is the next-to-left, code 01, so Tristan sends "01" to Isolde. She checks the next-to-left position of the winning team—it is a "1"—and sends it back to Tristan. Done.

It is not hard to show three bits is unbeatable. Note that in the most efficient communication scheme, more information is passed from the student to the teacher than vice-versa. Is there a lesson to be learned from that?

Here's another example where we need to think about information, and also to be clever.

Missing Card

Yola and Zela have devised a clever card trick. While Yola is out of the room, audience members pull out five cards from a bridge deck and hand them to Zela. She looks them over, pulls one out, and calls Yola into the room. Yola is handed the four remaining cards and proceeds to guess correctly the identity of the pulled card.

How do they do it? And once you've figured that out, compute the size of the biggest deck of cards they could use and still perform the trick reliably.

Solution: Let's try to look at this trick information-theoretically. The obvious (and correct, for this trick) way for Zela to communicate information to Yola when handing her the four remaining cards is to *order* the cards in some way. There are $4! = 24$ ways to order four objects, but that doesn't seem like enough—there are 48 possible values of the missing card. We could, of course, try to sneak in another bit of information, for example, whether the cards are passed face up or face down.

We could do that pretty easily and we would in fact need to, if the card to be removed were chosen by an innocent bystander. But Zela gets to decide which card is removed, and thus can limit the possibilities to much less than 48.

Since Zela gets to pick the card, the right information-theoretic way to look at the problem is to ask whether seeing four cards and their order is enough to deduce the original *five* cards. The number of ways to pick five cards from a deck is

$$\binom{52}{5} = \frac{52 \cdot 51 \cdot 50 \cdot 49 \cdot 48}{5 \cdot 4 \cdot 3 \cdot 2 \cdot 1} = 2{,}598{,}960,$$

while the number of ways to pick four cards *where the order counts* is

$$52 \cdot 51 \cdot 50 \cdot 49 = 6{,}497{,}400,$$

so we are OK, at least from an information-theoretic standpoint. What is needed is a human-implementable one-to-one map from a subset of the ordered foursomes to all the unordered fivesomes, with the property that each foursome is contained in the fivesome it is mapped to.

That requires some cleverness; here's how magician William Fitch Cheney did it back in the 1930's. Making use of the pigeonhole principle, he noted that every five cards contain at least two of some suit. One of those will be removed and a second of that suit left on top of the pile of four, so already Yola will know the suit of the removed card. The remaining three of the four can be ordered six ways, and Yola will use that info to determine the rank of the missing card.

Here's how. Yola and Zela fix an ordering of all the cards, perhaps A,2,3,... up to 10,J,Q,K, with ties broken by suit (the traditional magician's ordering of suits is the "CHaSeD" order, namely Clubs, Hearts, Spades, Diamonds). Let the three cards in question be called A, B, and C, according to the ordering above. If they are handed to Yola in order ABC, that codes the number 1; if ACB, the number 2; BAC = 3, BCA = 4, CAB = 5, and CBA = 6.

Yola looks at the rank of the top card and adds her code number to it, going "around the corner" if necessary. For example, suppose the top card is the queen of spades, and the order of the other three cards is CAB. Since the code of CAB is 5 and the queen is rank 12 of 13, Yola adds 5 to 12: 13=K, 14=1=A, 2, 3, 4. She concludes that the missing card is the 4 of spades.

Wait, suppose the missing card had been the 6 of spades; then Yola would have had to add 8 to the Q, and there's no code 8. Not to worry: In that case, Zela would have removed not the 6 of spades but the Q of spades; and adding 6 gets you from 6 to queen. In other words, given two cards of the same suit, we can always pick one so that you can get to the other by adding a number from 1 to 6. ♡

Yes, this will take some practice on the part of Yola and Zela.

There's actually quite a bit of "room" in this trick. Information-theoretically, it can be done with a deck of any size n satisfying $\binom{n}{5} \leq 4! \times \binom{n}{4}$ which is true up to $n = 124$. In fact, you really can do it with a deck of size 124, but we will leave it to you, the reader, to figure out how.

What is the value of information? We can quantify it the following way: Subtract your expectation without the information from your expectation with. Since you don't *have to* act on the information, this quantity can never be negative. Conversely, if the information can't or shouldn't change your actions, it's worthless.

Peek Advantage

You are about to bet $100 on the color of the top card of a well-shuffled deck of cards. You get to pick the color; if you're right you win $100, otherwise you lose the same.

How much is it worth to you to get a peek at the *bottom* card of the deck? How much more for a peek at the bottom *two* cards?

Solution: Without the peek, the bet is fair, that is, your expected gain is $0. If your peek reveals a red card on the bottom, you will of course bet on black and be right with probability 26/51; then your expectation will be

$$\frac{26}{51} \cdot \$100 + \frac{25}{51} \cdot (-\$100) \sim \$1.96.$$

Since the calculation is the same if the card on the bottom is black, the value of the information is $1.96.

We can do a similar calculation for peeking at two cards, but why bother? The second peek is worthless! Suppose the bottom card is red. If the next-to-bottom card is red, then of course you still bet that the top card is black. If the next-to-bottom card is black, then you're back to a fair game, so you may as well keep your plan of betting on black. Since the second peek does not have the power to change your strategy, its additional value to you is exactly $0.

The same kind of reasoning is useful in the next puzzle.

Bias Test

Before you are two coins; one is a fair coin, and the other is biased toward heads. You'd like to try to figure out which is which, and to do so you are permitted two flips. Should you flip each coin once, or one coin twice?

Solution: Suppose you only get one flip. If you flip coin A and it comes up heads, you will of course guess that it is the biased coin; if it comes up tails, that it is the fair coin.

Now if you flip coin B and get the opposite face, you will be even happier with your previous decision. If you get the same face, you will be reduced to no information and may as well stick with the same choice. Thus, as with the card-peeking, flipping the second coin is worthless.

Could flipping the same coin twice change your mind? Yes. If you get heads the first time you are inclined to guess that coin is biased, but seeing tails the next time will change your best guess to "unbiased." We conclude that flipping one coin twice is strictly better than flipping each coin once.

As it turns out, it's a theorem that in trying to determine which is which of two known probability distributions, it's better to draw twice from one than once from each. However, unlike in the puzzle above, it may happen that *after* seeing the result of your first draw you prefer to draw from the other distribution. We'll leave it to the reader to concoct an example.

Information about information—typically involving a chain of deductions to the effect that seeing A, person X would have deduced B—is a common puzzle theme. The following example is perhaps the best variation of a classic.

Dot-Town Exodus

Each resident of Dot-town carries a red or blue dot on his (or her) forehead, but if he ever thinks he knows what color it is he leaves town immediately and permanently. Each day the residents gather; one day a stranger comes and tells them something—*anything*—nontrivial about the number of blue dots. Prove that eventually Dot-town becomes a ghost town, *even if everyone can see that the stranger's statement is patently false.*

Solution: This puzzle, often in non-PC versions involving unfaithful spouses, murder, or suicide, has been around for at least a century. Usually the stranger is assumed to have made a true statement, but as we shall see, that stipulation is not necessary. The condition of "nontriviality" means that, *a priori*, the statement could be true or false; in other words, there is some number of blue dots that would make it true, and some other that would falsify it.

Example: All the dots are blue but the stranger says they are all red. Then everyone knows that the stranger is lying and moreover, that everyone else knows that the stranger is lying. How can the stranger's statement have any effect?

It's easiest to see when the population of Dot-town is small, for example, 3. Then you, if you are a Dot-town resident, know that the stranger is lying. But if your own dot is red, then your friend Fred is looking at a red dot and a blue dot and wondering if the third townsperson, Emily, is seeing two red dots. If so, Emily will believe the stranger and leave town. When she doesn't, Fred will correctly conclude that his own dot is blue, and will leave town the second night. When this fails to happen, you—and Fred and Emily, as well— can all conclude that your hats are blue, and all leave town on the third night.

To do a proof of the general case, we need some notation. Let $S \subset \{0, 1, \ldots, n\}$ be the set of numbers x with the property that if there are x blue dots among the n residents of Dot-town, then the stranger's statement is true. In other words, the stranger's statement is equivalent to "The number of blue dots is in S." Our non-triviality assumption tells us that S is a proper, non-empty subset. Let b be the actual number of blue dots, which may or may not be in S.

For resident i, let B_i be the set consisting of the possible numbers of blue dots, from i's point of view. Prior to the stranger's visit, $B_i = \{b_i, b_i + 1\}$ where b_i is the number of blue dots resident i sees among his compatriots.

If at any point B_i drops to one value, resident i is cooked. This will happen immediately if $|B_i \cap S| = 1$, but it will also happen the night after any departures occur. To see this, we observe first that all residents with the same dot color will behave identically, since they all see the same number of dots. Thus if resident i sees that anyone has left town, he deduces (correctly) that that person's dot color is different from his own; he therefore knows his own color and is doomed to exile.

Given S and b, let d_b be the number of steps (increments or decrements by 1) needed to get from b across the border of S; in other words, d_b is the least k such that $b + k$ or $b - k$ is in S if b is not, and out of S (but still in $\{0, 1, \ldots, n\}$) if b is in S. You can think of d_b as the distance from the stranger's statement to the truth, if the statement is false; and the distance to falsehood, if it is true.

Example: If $n = 10$ and $S = \{0, 1, 2, 9, 10\}$ then $d_0 = 3$, $d_1 = 2$, $d_2 = d_3 = 1$, $d_4 = 2$, $d_5 = d_6 = 3$, $d_7 = 2$, $d_8 = d_9 = 1$, and $d_{10} = 2$.

We claim that the first departures will occur exactly on night d_b.

The proof is by induction on d_b. If $d_b = 1$, some residents will find the stranger immediately credible, and will leave the first night. Suppose $t > 1$ and assume that our claim is true whenever $d_b < t$. The day after night $t-1$, since no departures have yet occurred, everyone will deduce that $d_b \geq t$. However, if $d_b = t$ then either d_{b-1} or d_{b+1} must be equal to $t-1$. If the former, then those residents with blue dots, who see that the number of blue dots is either b (the actual number) or $b-1$, can rule out $b-1$ and are toast. If the former, it is the red-dotted folks who can rule out $b+1$ and must pack their things. Finally, if $d_{b-1} = d_{b+1} = t-1$ then nobody stays the night.

Since d_b is at most n, the proof tells us that everyone will have exited by the $n+1$st night. But in the $d_b = n$ case all residents have the same dot-color so they all leave together on the nth night. It follows that Dot-town is always deserted within n days.

It is perhaps worth noting also that the definition of d_b makes no distinction between S and its complement; from this it follows that it makes no difference whether the stranger says "X" or "not X," the residents of Dot-town will behave exactly the same way in either case.

You might reasonably wonder whether the Dot-town residents, knowing that a stranger is coming and liable to break the manifestly justifiable no-talk-about-dot-colors taboo, can organize some defense. For example, everyone who knows the stranger is lying jumps up and says so. Alas, a little thought will show you that neither this nor any similar strategy can save the town. A fragile lot, these Dot-towners.

A whole class of puzzles is built around conversations in which the speakers take several rounds to make a deduction. Your job is to deduce something of your own from the conversation, even though you have less information than the speakers!

My favorite of these is the following. Part of the beauty of this one is that no numbers are named in the puzzle statement.

Conversation on a Bus

Ephraim and Fatima, two colleagues in the Mathematics Department at Zorn University, wind up seated together on the bus to campus.

Ephraim begins a conversation with "So, Fatima, how are your kids doing? How old are they now, anyway?"

"It turns out," says Fatima as she is putting the $1 change from the bus driver into her wallet, "that the sum of their ages is the number of this bus, and the product is the number of dollars that happen to be in my purse at the moment."

"Aha!" replies Ephraim. "So, if I remembered how many kids you have, and you told me how much money you are carrying, I could deduce their ages?"

"Actually, no," says Fatima.

"In that case," says Ephraim, "I know how much money is in your purse."

What is the bus number?

Solution: We will assume all numbers in the puzzle are positive integers—and that Ephraim knows the number of the bus he's on. That number (call it n) must allow two different sets of kids' ages, with the same number of kids, the same sum n, and the same product;

otherwise Ephraim would have been able to deduce the kids' ages from sum, product and number of kids.

One would expect that large values of n would tend to have this property; what prevents any large number from being the answer? Ah, the last line of the conversation shows that there can't be *two* pairs of kids'-age-sets with sum n, one pair sharing one product, the other sharing a different product. If there were two such pairs, Ephraim would not have been able to deduce the amount of money in Fatima's purse at the conversation's end.

Bus number $n = 13$ is such an example. Fatima could have five kids of ages 6, 2, 2, 2, and 1, or of ages 4, 4, 3, 1, and 1; either would put $48 in her purse. But she could instead have three kids of ages 9, 2, and 2, or of ages 6, 6, and 1; those possibilities would both give her purse $36. So even with the information that knowing the number of kids and purse contents (in addition to the bus number) would not have been sufficient to deduce the kids' ages, Ephraim would still not be in a position to deduce the purse contents if the bus number had been 13.

What about higher numbers? Here's the crucial observation: Once you've hit a number with 13's "ambiguity" property, every higher number has it too! Why? Because you can take those four sets of kids' ages and add one kid of age 1 to each one of them. That changes no products but raises all the sums by 1, so if n is ambiguous, so is $n+1$.

Now it's time to do the math, checking values of n *below* 13 to try to find one that boasts exactly one pair of kids'-age-sets with the same number of kids, sum n, and the same product. That's not as time consuming as it sounds, because (a) you only need consider sets of size at least three (one or two numbers can always be deduced from their sum and product), and (b) many products yield the kids' ages readily on account of unique prime factorization.

Turns out the puzzle-composer has not misled you: There is just one bus number that qualifies, $n = 12$. The critical kids'-age-sets are $\{4, 4, 3, 1\}$ and $\{6, 2, 2, 2\}$.

The next example illustrates the complexities of trying to communicate and play a game at the same time.

Matching Coins

Sonny and Cher play the following game. In each round, a fair coin is tossed. Before the coin is tossed, Sonny and Cher *simultaneously* declare their guess for the result of the coin toss. They win the round if both guessed correctly. The goal is to maximize the fraction of rounds won, when the game is played for many rounds.

So far, the answer is obviously 50%: Sonny and Cher agree on a sequence of guesses (e.g., they decide to always declare "heads"), and they can't do any better than that. However, before the game begins, the players are informed that just prior to the first toss, Cher will be given the results of all the coin tosses in advance! She has a chance now to discuss strategy with Sonny, but once she gets the coinflip information, there will be no further opportunity to conspire. Show how Sonny and Cher can guarantee to get at least 6 wins in 10 flips.

(If that's too easy for you, show how they can guarantee at least six wins in only nine flips!)

Solution: Guaranteeing 5 wins in 10 flips is a cinch: Cher chooses her first five guesses to match the last five coinflips. Then Sonny and Cher both repeat that sequence and collect their five wins. With luck they might catch one or two additional wins among the first five flips, but that can't be guaranteed.

Is there a better way? Suppose Cher tries to communicate three flips at a time to Bob. The nice thing about three flips is that they are either mostly heads or mostly tails. If flips 2 through 4 are mostly heads, Cher calls "heads" at flip 1 and this tells Sonny to guess "heads" at flips 2, 3, and 4. Suppose flips 2 and 4 are heads but flip 3 is a tail. Then Cher will call "heads" at flips 2 and 4, garnering two wins, and will use flip 3 to tell Sonny what to do in flips 5 through 7.

What if flips 2, 3, and 4 are all tails? Then, by agreement, we just use Cher's last call (on flip 4) to tell Sonny what to guess at the next three flips.

Flips 5 through 7 are treated similarly, with the minority flip (or the last flip, if all flips are the same) used to tell Sonny which way to guess on flips 7 through 9.

In this way Sonny and Cher are guaranteed to get at least two wins out of three in each of the flip-groups 2-4, 5-7, and 8-10.

Readers who like to imagine life in asymptopia will want to know what fraction of correct flips can be guaranteed when the total number of flips is large. It turns out Sonny and Cher can do much better than 2/3, in fact they can beat 80%(!) if the number of flips is huge and they work hard to squeeze the last drop of blood out of their scheme.

As for 6 out of 9, that takes some rather careful work and I wouldn't want to deprive the ambitious reader of the opportunity to devise a successful scheme. (Advice: Do 3 out of 5 first.) Eventually, I'll post a solution on the web.

In the science of cryptography, communication must be done with an eye to keeping information from falling into the wrong hands. This is most easily done if the parties that wish to communicate have a common secret.

For example, suppose Alice is expecting to see Bob at a party and hoping to tell him "yes" or "no" to a certain business proposition, privately, even though others at the party may be nearby and interested. No problem; she and Bob arrange beforehand that "Hi" means yes and "Hello" means no. This *shared secret* enables them to communicate secretly.

The last puzzle in this chapter was designed to show that a shared secret can sometimes be created over an open channel, provided that the participants have some shared knowledge—that is, when put together they could deduce things that an eavesdropper couldn't know.

Two Sheriffs

Two sheriffs in neighboring towns are on the track of a killer, in a case involving eight suspects. By virtue of independent, reliable detective work, each has narrowed his list to only two. Now they are engaged in a telephone call; their object is to compare information, and if their pairs overlap in just one suspect, to identify the killer.

The difficulty is that their telephone line has been tapped by the local lynch mob, who know the original list of suspects but not which pairs the sheriffs have arrived at. If they are able to identify the killer with certainty as a result of the phone call, he will be lynched before he can be arrested.

Can the sheriffs, who have never met, conduct their conversation in such a way that they both end up knowing who the killer is (when possible), yet the lynch mob is still left in the dark?

Solution: Label the suspects $ABCDEFGH$ and organize them into lists so that each list consists of four disjoint (unordered) pairs, and each pair appears in exactly one list. For example:

$$
\begin{array}{cccc}
AB & CD & EF & GH \\
AC & BD & EG & FH \\
AD & BC & EH & FG \\
AE & BF & CG & DH \\
AF & BE & CH & DG \\
AG & BH & CE & DF \\
AH & BG & CF & DE
\end{array}
$$

Let's name the sheriffs Lew and Ralph. Lew can share the above array with Ralph over the phone and then tell Ralph which row his pair (i.e., the pair of suspects to which Lew has narrowed down his search) lies in. If Ralph's pair lies in the same row, he tells Lew; that means the two sheriffs have arrived at the same pair, so they may as well hang up.

Otherwise, Ralph announces a partition of the row into sets of two pairs each, with the property that Ralph's two final suspects are in the same part.

For example, suppose Lew's pair is EF. Then he announces that his pair is in the first row. If Ralph's pair is FG, he partitions that row into one part consisting of the pairs AB and CD, and another containing the pairs EF and GH. (At this point, Lew and Ralph already have established a shared secret: They both know which part the killer is in.)

Lew now reveals whether his pair is the leftmost or rightmost in its part of the row (in this case, it's the leftmost). That tells Ralph which pair is Lew's, and hence who the killer is (here, F). To communicate that safely to Lew, he just tells Lew whether the killer is the left or the right member of Lew's pair (here, the rightmost).

The conversation would be identical if the pairs had been AB and BC, making B the killer instead of F. We hope that the mob either refuses to chance lynching an innocent person, or doesn't have enough rope for two.

Cryptography is the study of means for sending private messages over an open channel. Classically, the sender (Alice, say) and receiver (Bob) are equipped in advance with a "key" that enables Alice to *encrypt* her message and Bob to *decrypt* it, while eavesdropper Eve, not possessing the key, can make no sense of the intercepted stream.

Suppose Eve has the time and computing power to test every possible key, looking for one which, when used to decrypt the message she intercepted, yields a credible possibility for the message Alice intended Bob to get. When might this work?

If the number of possible keys is small, Eve can expect that only one will yield an intelligible message; but if it's large, many might do so. Information theory can give us an idea of how many possible keys there should be, in order for Alice and Bob to be safe.

Suppose that for every choice of message Alice might have wanted to send Bob on this occasion, and every "cyphertext" $x_1, \ldots, x_n$ that Eve might intercept, there is a unique key that would encrypt that message as $x_1, \ldots, x_n$. Then, assuming the key was chosen uniformly at random and Eve doesn't know it, we have an easy conclusion.

Theorem. *In the above situation, Eve can gain no information whatever about Alice's intended message.*

Proof. Since any message could have been intended, each associated with one key, and all keys are equally likely, Eve learns nothing. ♡

There actually is a cryptographic scheme that achieves this goal—assuming accurate random generation and perfect secrecy in the key choice. It's called a "one-time pad," and here's an example of how it could work. Alice translates her intended message into a stream of n 0s and 1s (by some means that Bob needs to know and Eve may also know). Then she extracts the next n bits from the one-time pad that she and Bob share, and adds those bits one by one, modulo 2, to the message bits. For example, if the message is 00101110 and the bits from the one-time pad are 01001001, then the encrypted message is 01100111 and this is what Alice sends (and Eve intercepts).

Bob, who knows where he and Alice are in the one-time pad, takes his received message 01100111 and, one by one, adds the one-time pad bits 01001001 to it to get the original message, 00101110, intended by Alice. (This works nicely because modulo 2, addition and subtraction are equivalent. Much more about this useful kind of arithmetic is found in Chapter 17.)

Eve is kept completely in the dark, because looking only at 01100111, any eight-bit message could have been intended by Alice.

As its name suggests, the security of the one-time pad is heavily dependent on each bit of the pad being used only once, and in only one message. Otherwise a "depth" is created and Eve may be able to crack the code.

The difficulty with one-time pads is that if Alice and Bob are planning any long conversations, they need to share (and keep private) a great many bits. In most applications, keys are much shorter than messages, and the security of encryption relies on Eve not having the time and/or computing power to test enough keys.

14. Great Expectation

The *expectation* or *expected value* of a random number is the average value that is anticipated (via the Law of Large Numbers) when the random number is "redrawn" independently many times. If the random number can take only finitely or countably many values (e.g., if it takes only integer values), we can define its expectation as the average of the values, weighted by their probabilities.

Numbers whose values are determined by chance are called by probabilists *random variables*, and denoted typically by boldface capital letters from the end of the Latin alphabet, such as $\mathbf{X}$. If $\mathbf{X}$'s possible values are $x_1, x_2, \ldots$ with corresponding probabilities $p_1, p_2, \ldots$, then the expectation of $\mathbf{X}$ is

$$\mathbb{E}\mathbf{X} := \sum_i p_i x_i.$$

The terminology is a bit misleading; for example, the expectation when you roll a fair die is $3\frac{1}{2}$, but you don't "expect" the result $3\frac{1}{2}$. It would be more accurate to call $3\frac{1}{2}$ the "expected average" rather than the "expectation" or "expected value." But convenience triumphs.

Expectation is often easier to work with than probability. Suppose $\mathbf{X}$ and $\mathbf{Y}$ are two random variables. If we want to know the probability that $\mathbf{X} + \mathbf{Y} > 5$ (say) it is not enough to know everything there is to know about each of $\mathbf{X}$ and $\mathbf{Y}$ *individually*; you will also need to know how $\mathbf{X}$ and $\mathbf{Y}$ interact with each other.

But if you only want the *expectation* of $\mathbf{X} + \mathbf{Y}$, it is guaranteed to satisfy

$$\mathbb{E}(\mathbf{X} + \mathbf{Y}) = \mathbb{E}\mathbf{X} + \mathbb{E}\mathbf{Y}$$

so all you need to know are the expectation of $\mathbf{X}$ and the expectation of $\mathbf{Y}$. For example, suppose $\mathbf{X}$ and $\mathbf{Y}$ each take values 1, 2, 3, 4, 5, or 6 with equal probability, so that each has expectation $3\frac{1}{2}$. What does the random variable $\mathbf{X} + \mathbf{Y}$ look like? Well, if $\mathbf{X}$ and $\mathbf{Y}$ are independent—for example, if $\mathbf{X}$ is the first roll of a die and $\mathbf{Y}$ the second—then $\mathbf{X} + \mathbf{Y}$ takes on values 2, 3, $\ldots$, 12 with probabilities

$1/36,\ 2/36, \ldots, 5/36,\ 6/36,\ 5/36, \ldots, 1/36$. But if $\mathbf{X}$ and $\mathbf{Y}$ are both the outcome of the first roll, that is, if $\mathbf{X} = \mathbf{Y}$, then $\mathbf{X} + \mathbf{Y} = 2\mathbf{X}$ takes on only even values from 2 to 12, each with probability $\frac{1}{6}$. And if $\mathbf{Y}$ happens to be defined as $7 - \mathbf{X}$, we find that $\mathbf{X} + \mathbf{Y}$ is the constant value 7. All very different, but notice that in every case $\mathbb{E}(\mathbf{X} + \mathbf{Y}) = 3\frac{1}{2} + 3\frac{1}{2} = 7$.

This miracle is a special case of what mathematicians call "linearity of expectation," whose general formulation is

$$\mathbb{E} \sum_i c_i \mathbf{X}_i = \sum_i c_i \mathbb{E}\mathbf{X}_i,$$

that is, the expectation of a weighted sum of random variables is the weighted sum of their expectations.

Let's start with a calculation of expectation in the service of making a strategic decision.

Bidding in the Dark

You have the opportunity to make one bid on a widget whose value to its owner is, as far as you know, uniformly random between \$0 and \$100. What you do know is that you are so much better at operating the widget than he is, that its value to *you* is 80% greater than its value to him.

If you offer more than the widget is worth to the owner, he will sell it. But you only get one shot. How much should you bid?

Solution: You should not bid. If you do bid \$$x$, then the expected value to the widget's owner, *given that he sells*, is \$$x/2$; thus, its expected value to you, if you get it, is $1.8 \cdot \$x/2 = \$0.9x$. Thus you lose money on average if you win, and of course, you gain or lose nothing when you do not, so it is foolish to bet. ♡

One common setting for expectation is waiting time; in particular, the average number of times you expect to have to try something until you succeed. For example, how many times, on average, do you need to roll a die until you get a 6? We encountered this sort of problem in *Odd Run of Heads* in Chapter 5. Let x be the expected number of rolls. "If at first you don't succeed" (probability $\frac{5}{6}$) you still have an average of x rolls to "try, try again," so $x = 1 + \frac{5}{6} \cdot x$ giving $x = 6$. The same argument shows that in general, if you conduct a series of "trials" each of which has probability of success p, then the average number of trials you need to get your first success is $1/p$.

Let's apply this observation to a slightly trickier dice puzzle.

Rolling All the Numbers

On average, how many times do you need to roll a die before all six different numbers have turned up?

Solution: Often the most important step in applying linearity of expectation is simply recognizing that what you are being asked for is the expectation of a sum of random variables. The key for this puzzle is to think of rolling all the numbers as a 6-stage process, where at Stage i you are going for your ith new number. In other words, let $\mathbf{X}_i$ be the number of rolls needed to get your ith new number, given that you have already seen $i-1$ different numbers.

Then the total number of rolls needed is $\mathbf{X} := \mathbf{X}_1 + \mathbf{X}_2 + \cdots + \mathbf{X}_6$ and what we are asked for is $\mathbb{E}\mathbf{X} = \mathbb{E}\mathbf{X}_1 + \mathbb{E}\mathbf{X}_2 + \cdots + \mathbb{E}\mathbf{X}_6$. We already know that $\mathbb{E}\mathbf{X}_1 = 1$ since it always takes exactly one roll to get the first new number. The second will come with probability $5/6$ from any given roll, so takes $6/5$ rolls on average. Continuing in this manner, we have

$$\mathbb{E}\mathbf{X} = 1 + \frac{6}{5} + \frac{6}{4} + \frac{6}{3} + \frac{6}{2} + \frac{6}{1} = 14.7$$

and we are done.

Warning: Some experimentation will remind you that $\mathbf{X}$ is by no means guaranteed to be near 14.7; the last new number, in particular, could take a frustratingly long time to roll. Later we'll be reminded that the Law of Large Numbers sometimes takes a while to kick in.

Next, see if you can set up a sum of random variables for this puzzle.

Spaghetti Loops

The 100 ends of 50 strands of cooked spaghetti are paired at random and tied together. How many pasta loops should you expect to result from this process, on average?

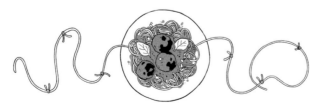

Solution: The key here is to recognize that you start with 100 spaghetti-ends and each tying operation reduces the number of spaghetti-ends by two. The operation results in a new loop only if the first end you pick up is matched to the other end of the same strand. At step i there are $100-2(i-1)$ ends, thus $100-2(i-1)-1 = 101-2i$ *other* ends you could tie any given end to. In other words, there are $101 - 2i$ choices of which only one makes a loop; the others just make two strands into one longer one.

It follows that the probability of forming a loop at step i is $1/(101-2i)$. If $\mathbf{X}_i$ is the number of loops (1 or 0) formed at step i, then the expected value of $\mathbf{X}_i$ is $1/(101 - 2i)$. Since $\mathbf{X} := \mathbf{X}_1 + \cdots + \mathbf{X}_{50}$ is the final number of loops, we apply linearity of expectation to get

$$\mathbb{E}\mathbf{X} = \mathbb{E}\mathbf{X}_1 + \cdots + \mathbb{E}\mathbf{X}_{50} = \frac{1}{99} + \frac{1}{97} + \frac{1}{95} + \cdots + \frac{1}{3} + \frac{1}{1} = 2.93777485\ldots,$$

less than three loops! ♡

For the last two puzzles, the summands were independent random variables—we didn't actually need to know that independence is not required for linearity of expectation. Independence of two events A and B means that the probability they both occur is the product of their individual probabilities, so, for example, if we wanted to know the probability of getting all six numbers in only six rolls of a die, we could have obtained it as

$$1 \cdot \frac{5}{6} \cdot \frac{4}{6} \cdot \frac{3}{6} \cdot \frac{2}{6} \cdot \frac{1}{6} = \frac{6!}{6^6} \sim 1.5\%.$$

For all the spaghetti to end up in one big loop, you must never tie a strand to its other end except at the last step; in other words, $\mathbf{X}_1, \ldots, \mathbf{X}_{49}$ must all be zero. This happens with probability

$$\frac{98}{99} \cdot \frac{96}{97} \cdot \frac{94}{95} \cdot \ldots \cdot \frac{2}{3}$$

$$= \frac{49! \cdot 2^{49}}{99!/(49! \cdot 2^{49})} = \frac{49!^2 \cdot 2^{98}}{99!} = 0.12564512901\ldots,$$

so about one time in eight.

Ping-Pong Progression

Alice and Bob play table tennis, with Bob's probability of winning any given point being 30%. They play until someone reaches a score of 21. What, approximately, is the expected number of points played?

Solution: With high probability, Alice will win, and thus the expected number of points won by Bob is $(3/7)21 = 9$ and therefore the expected total number of points is close to 30. Another way to make that calculation: Since Alice wins any given point with probability 70%, it takes on average $1/0.7 = 10/7$ rallies to earn Alice a point, thus $21 \times 10/7 = 30$ rallies on average for her to get to 21.

The actual answer is a bit less than 30, because playing until Alice scores 21 points is sufficient, but not always necessary, to finish the game. How much less?

To find out, we employ a bit of statistics (readers not familiar with variance and normal approximations are invited to skip to the next puzzle).

The probability of Bob winning is equal to the probability that he wins the majority of the first 41 points. Let $X_i = 1$ if Bob wins the ith point, 0 otherwise. The variance of X_i is $0.3 \times 0.7 = 0.21$, thus the standard deviation of the number of points Bob wins out of 41 is $\sqrt{41 \times 0.21} = 2.9343$. His expected score being $0.3 \times 41 = 12.3$, we apply the normal approximation with the intent of computing the probability that Bob's score exceeds 20.5; that amounts to $(20.5 - 12.3)/2.9343 \sim 2.795$ sigmas, which works out to a probability of about 0.0000387. If Bob does win, he'll win by about 2 points on average, leading to 40 points played instead of the approximately 43 points needed if we wait for Alice to hit 21. Thus, the cost to our 30-point estimate is around $3 \times 0.0000387 \sim 0.0001$.

So the result of all that work is revising our estimate to 29.9999.

Do you recall a puzzle called "Finding a Jack" from a Chapter 8? You can use its solution technique—twice!—for the next puzzle.

Drawing Socks

You have 60 red and 40 blue socks in a drawer, and you keep drawing a sock uniformly at random until you have drawn all the socks of one color. What is the expected number of socks left in the drawer?

Solution: It's useful to imagine that the sock-drawing is done by arranging all the socks in a uniformly random order, then removing them, starting from the beginning of the order, until all of one color have been removed. If the color of the last sock is blue (probability 0.4), then it must have been the red socks that were all removed, and the number of socks "left in the drawer" is the number of blue socks encountered starting from the *end* of the arrangement before any red sock is found.

If indeed the last sock is blue, there are 39 other blue socks distributed in 61 *slots* (before the first red sock, between the first and second red socks, and so forth, finally after the 60th red sock). Thus, we expect 39/61 blue socks in that last slot; adding the final blue sock gives on average of 100/61 blue socks left in the drawer. Similarly, if the last sock is red (probability 0.6), we have 59 other red socks in 41 slots so on average 59/41 in the last slot; that leaves $1 + 59/41 = 100/41$ red socks behind. Putting these together, we have that the expected number of socks left is $0.4 \times 100/61 + 0.6 \times 100/41 \sim 2.12$.

If there are m socks of one color and n of the other, the above argument gives you an average of $m/(n+1) + n/(m+1)$ socks left in the drawer.

Linearity of expectation applies even when there are continuum-many random variables. Modern mathematicians would replace the sum by an integral, but sometimes an Archimedean approach will do the trick.

Random Intersection

Two unit-radius balls are randomly positioned subject to intersecting. What is the expected volume of their intersection? For that matter, what is the expected *surface area* of their intersection?

Solution: We can assume that the center of the first ball is at the origin of a three-dimensional coordinate system; the first ball thus consists of all points in space within distance 1 of the origin. The center C of the second ball must be somewhere in the ball of radius *two* about the origin, in order for it to intersect the first ball. The conditions of the problem imply that C is uniformly random subject to this constraint.

(Note that this is not at all the same as taking the center of the second ball to be a random point between -2 and 2 on the x-axis. The latter assumption would give a much-too-big expected intersection size of $\pi/2$.)

Now consider a point P inside the first ball: What's the probability that P will be in the intersection of the two balls? That will happen just when C falls in the unit ball whose center is P. This ball is inside the ball of radius 2 about the origin that C is chosen from, with $1/2^3 = 1/8$ of the volume. Thus the probability that P lies in our intersection is $1/8$, and it follows from linearity of expectation that the expected volume of the intersection is $1/8$ times the volume of the first ball, that is, $1/8 \times 4\pi/3 = \pi/6$.

If it seems fishy to you to be adding up points P like this, good for you! Think of P not as a point but a small cell, or "voxel," inside the ball. If the voxel's volume is $(4\pi/3)/n$ with the ball divided into n voxels, n being some large integer, then an ordinary finite application of linearity of expectation gives approximately the same answer, $\pi/6$. It's not exact owing to the fact that a few voxels will lie partly inside and partly outside the intersection. As the voxels get smaller the approximation gets better.

For the surface area of the intersection, the point P—now representing a two-dimensional *pixel*—is taken on the surface of the unit ball centered about the origin. This will lie on the surface of the intersection if it's inside the second ball, that is, again, if C lies in the unit ball centered at P. We've already seen that this happens with probability $1/8$, but note that the intersection of the two spheres has *two* surfaces of the same area, one that was part of the surface of the first ball—that's the part that could contain P—and one that was part of the surface of the second ball. Thus the expected surface area of the intersection is $2 \times \frac{1}{8} \times 4\pi = \pi$.

You may have noticed, by the way, that all this can be done in any dimension; with disks on the plane, for example, you get expected area $\pi/4$ and perimeter π. You can even do it with (some) other shapes.

One place where linearity of expectation arises frequently is in gambling. A game is called "fair" if the expected return for each player is 0, and we can deduce that any combination of fair games is fair. This applies even if your choice of game, or strategy within a game, is made "as you play" (as long as you can't foresee the future).

Of course, in a gambling casino, the games are not fair; your expectation is negative, the casino's positive. For example, in (American) roulette there are 38 numbers (0, 00, and 1 through 36) on the wheel, but if you bet $1 on a single number and win, you earn only $36, not $38, in return for your dollar. The result is that your expectation is

$$\frac{1}{38} \cdot \$35 + \frac{37}{38} \cdot (-\$1) = -\$\frac{1}{19},$$

so that you lose, on average, about a nickel per dollar bet. This same negative expectation applies also to bets made on groups of numbers or colors.

However:

Roulette for the Unwary

Elwyn is in Las Vegas celebrating his 21st birthday, and his girlfriend has gifted him with twenty-one $5 bills to gamble with. He saunters over to the roulette table, noting that there are 38 numbers (0, 00, and 1 through 36) on the wheel. If he bets $1 on a single number, he will win with probability 1/38 and collect $36 (in return for his $1 stake, which still goes to the bank). Otherwise, of course, he just loses the dollar.

Elwyn decides to use his $105 to make 105 one-dollar bets on the number 21. What, approximately, is the probability that Elwyn will come out ahead? Is it better than, say, 10%?

Solution: We know from linearity of expectation that on average, Elwyn should expect to lose $\$\frac{105}{19} \sim \$5.53$. But this says very little about the probability that Elwyn comes out ahead. A random variable with expectation -5.53 could be anywhere from always negative, to almost never negative!

In this case, we need to work out the probability that Elwyn wins three or more of his 105 bets, because if he wins three times he pulls in $3 \cdot \$36 = \$108$ and comes out ahead.

That takes a little work. The probability that Elwyn loses every time is $(37/38)^{105} \sim 0.06079997242$. To compute the probability that he wins exactly once, we multiply the number of bets by the probability of winning a given bet and losing the rest; this gives us

$$105 \cdot \frac{1}{38} \cdot \left(\frac{37}{38}\right)^{104} \sim 0.17254046227.$$

The probability that Elwyn wins exactly twice is the number of *pairs* of bets times the probability of winning two specified bets and losing the rest, which works out to

$$\frac{105 \cdot 104}{2} \cdot \left(\frac{1}{38}\right)^2 \cdot \left(\frac{37}{38}\right)^{103} \sim 0.24248929833.$$

Adding these three numbers and subtracting from 1 gives the probability that Elwyn comes out ahead: 0.52417026698, better than 50%!

No, this does not mean that Elwyn has figured out how to beat 'Vegas. He comes out *slightly* ahead rather often but *far* behind quite often as well, since winning only twice leaves him with just $72 to show for his $105 stake. Elwyn's negative expectation is the more relevant statistic here.

If you want to go to Vegas and be able to tell your friends that you made a profit at roulette, here's a suggestion: Pick a number of dollars that's one less than a power of 2, and that you can afford to lose. Suppose it's $63. Then bet $1 on "red." (Half the positive numbers are red, so you get your dollar back and win an additional $1 with probability 18/38, but lose your dollar otherwise.) If you do win, go home. If you lose, bet $2 on red; if you win, go home having made a $1 profit. If you lose, bet $4 next time, etc., until you win or have lost your sixth straight bet ($32). At this point, you must give up or risk ruin; fortunately the probability that you will be this unlucky is only $(20/38)^6$, about 2%. Your *expectation* for the whole experiment is of course still negative.

An Attractive Game

You have an opportunity to bet $1 on a number between 1 and 6. Three dice are then rolled. If your number fails to appear, you lose your $1. If it appears once, you win $1; if twice, $2; if three times, $3.

Is this bet in your favor, fair, or against the odds? Is there a way to determine this without pencil and paper (or a computer)?

Solution: This game seems pretty decent, maybe even favorable. You have three chances to get your number; if you were guaranteed to roll three different numbers, the probability that one would be yours would be 50%, making the game perfectly fair. If you *don't* roll three different numbers, you could end up with a bonus. So what's not to like?

In fact this is a well known game, often called "Chuck-a-Luck" or "Birdcage" (the latter because the dice are typically kept inside a cage and rolled by shaking the cage). Like any gambling game it is designed to look attractive but to make money for the house, and indeed that is the case.

The easiest way to see this is to imagine that there are six players each of whom bets on a different number. The expectation for each player is of course the same, since the rules don't discriminate. If that expectation is x then the house's expectation must be $-6x$ by linearity of expectation—this is a "zero-sum game," since no money leaves or enters from outside.

But now if the three rolls are different, three players win and three lose so the house breaks even. If two numbers are the same, the house pays $2 to one player and $1 to another while picking up $1 from each of the remaining four players, so makes a $1 profit. If all three dice come up the same, the house pays out $3 and collects $5, making a $2 profit. So the house's expectation is positive and therefore the players' is negative.

(It's not hard to work out that the probability the three rolls are different is $(6 \cdot 5 \cdot 4)/6^3 = 5/9$, the probability that they are all the same is $1/6^2 = 1/36$, and thus the probability that just two of them match is the remaining $15/36 = 5/12$. Thus the house's expectation is $(5/12) \cdot \$1 + (1/36) \cdot \$2 = \$17/36$ and each player's expectation is thus $(-1/6) \cdot \$17/36 \sim -\$0.0787037037$, so a player loses on the average about 8 cents per dollar bet—worse even than American roulette, about which more later.)

Here's a game which is definitely favorable; the question is, to what extent can you turn positive expectation into a sure thing?

Next Card Bet

Cards are turned face up one at a time from the top of a well-shuffled deck. You begin with a bankroll of $1, and can bet any fraction of your current worth, prior to each revelation, on the color of the next card. You get even odds regardless of the current composition of the deck. Thus, for example, you can decline to bet until the last card, whose color you will of course know, then bet everything and be assured of going home with $2.

Is there any way you can *guarantee* to finish with more than $2? If so, what's the maximum amount you can assure yourself of winning?

Solution: With a little thought you can improve on $2 by waiting to bet until three cards remain. If they're all one color, great—you can bet everything for the remaining three rounds and go home with $8. If one is red and the others black, bet 1/3 of your money on black; if you're right you now have $1\frac{1}{3}$ which you can double at the last card for a final tally of $2\frac{2}{3}$. If you're wrong you're down to $\frac{2}{3}$ but the remaining cards are both black, so you can double this twice to again end with $2\frac{2}{3}$. Since a similar strategy works when the remaining cards are two red and one black, you have found a way to guarantee $2\frac{2}{3}$.

To extend that backward and guarantee even more starts to look messy, so let's look at the problem another way. Whatever betting strategy you use, you'll certainly want to bet everything at every round as soon as the remaining deck becomes monochromatic; let's call strategies with this property "sensible."

It is useful first to consider which of your strategies are optimal in the sense of expectation, that is, which maximize your expected return. It is easy to see that all such strategies are sensible.

Surprisingly, the converse is also true: No matter how crazy your betting is, as long as you come to your senses when the deck becomes monochromatic, your expectation is the same! To see this, consider first the following *pure* strategy: Imagine some fixed specific distribution of red and black in the deck, and bet everything you have on that distribution at every turn.

Of course, you will nearly always go broke with this strategy, but if you win you can buy the earth—your take-home is then $2^{52} \times \$1$, around 4.5 quadrillion dollars. Since there are $\binom{52}{26}$ ways the colors can be distributed in the deck, your *expected* return is $\$2^{52}/\binom{52}{26} = \$9.0813$.

Of course, this strategy is not realistic, but it is "sensible" by our definition, and, most importantly, *every sensible strategy is a combination of pure strategies of this type*. To see this, imagine that you have $\binom{52}{26}$ assistants working for you, each playing a different one of the pure strategies.

We claim that every sensible strategy amounts to distributing your original $1 stake among these assistants, in some way. If at some point your collective assistants bet x on red and y on black, that amounts to you yourself betting $x - y$ on red (when $x > y$) and $y - x$ on black (when $y > x$).

Each sensible strategy can be implemented by a distribution of money to the assistants, as follows. Say you want to bet $0.08 that the first card is red; this means that the assistants who are guessing "red" first get a total of $0.54 while the others get only $0.46. If, on winning, you plan next to bet $0.04 on black, you allot $0.04 more of the $0.54 total to the "red–black" assistants than to the "red–red" assistants. Proceeding in this manner, eventually each individual assistant has his assigned stake.

Now, any mix of strategies with the same expectation shares that expectation, hence every sensible strategy has the same expected return of $9.08 (yielding an expected profit of $8.08). In particular, all sensible strategies are optimal.

But one of these strategies *guarantees* $9.08; namely, the one in which the $1 stake is divided equally among the assistants. Since we can never guarantee more than the expected value, this is the best possible guarantee.

This strategy is actually quite easy to implement (assuming as we do that US currency is infinitely divisible). If there are b black cards and r red cards remaining in the deck, where $b \geq r$, you bet a fraction $(b - r)/(b + r)$ of your current worth on black; if $r > b$, you bet $(r - b)/(b + r)$ of your worth on red. You will be entirely indifferent to the outcome of each bet, and can relax and collect your $8.08 profit at the end.

Sometimes the notion of fair game comes in handy even when there is no mention of a game.

Serious Candidates

Assume that, as is often the case, no one has any idea who the next nominee for President of the United States will be, of the party not currently in power. In particular, at the moment no person has probability as high as 20% of being chosen.

As the politics and primaries proceed, probabilities change continuously and some candidates will exceed the 20% threshold while others will never do so. Eventually one candidate's probability will rise to 100% while everyone else's drops to 0. Let us say that after the

convention, a candidate is entitled to say he or she was a "serious" candidate if at some point his or her probability of being nominated exceeded 20%.

Can anything be said about the expected number of serious candidates?

Solution: It seems that you'd need to know a lot about the political process, and the current circumstances, to answer this question. For example, doesn't it seem plausible that under some conditions, the first candidate to exceed 20% is likely to continue rising and eventually secure the nomination? Wait, that can't be right; if he/she really had probability only 20% at the critical point, then most likely someone else would end up with the nomination.

In fact, the expected number of "serious" candidates has to be 5. To see this, imagine that you enter the prediction market with the following strategy: Anytime a candidate hits 20% for his or her first time, you bet $1 on that candidate. If the market's probabilities are "correct" and the market is fair, you should win $5 if that candidate is eventually nominated.

In this game, you are bound to win $5; no matter which candidate gets nominated, you will have bet on him or her at some point. So the only variable is how many bets you made, and since the game is presumed to be fair, your expected expenditure must be $5.

Why is the word "correct" in quotes above? The problem is that when everything is done and (say) Smith has won the nomination, one might argue that a good political prognosticator would, or should, have known all along that that would be the outcome. Thus when Smith hit 20% he or she should probably have been rated at 100%, or at least a lot higher than 20%. Without getting into the philosophy of probability, let's just say that a candidate's rating represents some sort of consensus which you, the bettor, have no reason to doubt one way or the other.

Rolling a Six

How many rolls of a die does it take, on average, to get a 6—given that you didn't roll any odd numbers *en route*?

Solution: Obviously, three. Right? It's the same as if the die had only three numbers, 2, 4, and 6, each equally likely to appear. Then the probability of rolling a 6 is 1/3, and as we have seen in the puzzle *Rolling All the Numbers*, in an experiment that has probability of success p, it takes, on average, $1/p$ trials to get a success.

Except that it's not the same at all. Imagine that you arrange to compute your answer to this puzzle via multiple experiments, making use of the law of large numbers. You start rolling your die, counting trials as you go. If you hit a "6" before you get an odd number, you record the number of trials and repeat. (For example, if you roll "4,4,6" you record a 3.) What happens if you do roll an odd number? Then you can't simply throw out that roll and continue counting trials (if you did that, you would eventually converge to 3 as your answer). Instead, you must throw out that whole experiment, recording nothing, and start anew, beginning with roll #1 of your new experiment. That will produce a smaller average than 3, but how much smaller?

An intuitive way to look at it: When you've rolled a 6 without rolling an odd number, there's an inference that you succeeded quickly—since if you took a long time to get your 6, you'd probably have hit an odd number.

To figure out the correct answer, it helps to think about what you're doing in your multi-experiment. Basically, each sub-experiment consists of rolling the die until you get a 6 *or* an odd number. If you counted rolls in all of these, you'd discover their average length to be 3/2, since the probability of rolling a 6, 1, 3, or 5 is 2/3.

In your *gedankenexperiment* you're only counting the sub-experiments that end in a 6, but does that matter? The finishing roll (be it 1, 3, 5, or 6) is independent of all that has gone before, and independence is a reciprocal notion; thus, the number of rolls "doesn't care" what the sub-experiment ended with. It follows that the average length of the subexperiments that ended with a 6 is that same 3/2.

Napkins in a Random Setting

At a conference banquet for a meeting of the Association for Women in Mathematics, the participants find themselves assigned to a big circular table. On the table, between each pair of adjacent settings, is a coffee cup containing a cloth napkin. As each mathematician sits down, she takes a napkin from her left or right; if both napkins are present, she chooses randomly. If the seats are occupied in random order and the number of mathematicians is large, what fraction of them (asymptotically) will end up without a napkin?

Solution: We want to compute the probability that the diner in position 0 (modulo n) is deprived of a napkin. The limit of this quantity, as $n \to \infty$, is the desired limiting fraction of napkinless diners.

We may assume that everyone decides in advance whether to go for her right or left napkin, in case both are available; later, of course, some will have to change their minds or go without.

Say that diners $1, 2, \ldots, i-1$ choose "right" (away from 0) while i chooses "left"; and diners $-1, -2, \ldots, -j+1$ choose "left" (again, away from 0) while diner $-j$ chooses "right." Note that i and j might both be equal to 1, meaning that the diners on both sides of diner 0 are planning to go for the napkin between them and diner 0.

If $k = i+j+1$, then as long as $k \leq n$ (which is the case with high probability), the probability of this configuration is 2^{1-k}.

Observe that diner 0 loses out only when she is last to pick among $-j, \ldots, i$ and *none* of the diners $-j+1, \ldots, -2, -1; 1, 2, \ldots, i-1$ get the napkins they wanted. If $t(x)$ is the time at which diner x makes her grab, then this happens exactly when $t(0)$ is the unique local maximum of t in the range $[-j, -j+1, \ldots, 0, \ldots, i-1, i]$.

If t is plotted on this interval, it looks like a mountain with $(0, t(0))$ on top; more precisely, $t(-j) < t(-j+1) < \cdots < t(-1) < t(0) > t(1) > t(2) > \cdots > t(i)$.

Instead of evaluating the probability of this event for fixed i and j, it is convenient to lump all pairs (i, j) together which satisfy $i+j+1 = k$ for fixed k. Altogether there are $k!$ ways the values $t(-j), \ldots, t(i)$ can be ordered. If T is the set of all k grabbing times and t_{max} is the last of these, then each mountain-ordering is uniquely identified by the nontrivial subset of $T \setminus \{t_{max}\}$ which constitutes the values $\{t(1), \ldots, t(i)\}$. Thus, the number of orderings that make a valid mountain is $2^{k-1} - 2$.

Finally, the total probability that diner 0 is deprived of her napkin is

$$\sum_{k=3}^{\infty} \frac{2^{1-k} \cdot 2^{k-1} - 2}{k!} = \left(2 - \sqrt{e}\right)^2 \approx 0.12339675. \ \heartsuit$$

Roulette for Parking Money

You're in Las Vegas with only \$2 and in desperate need of \$5 to feed a parking meter. You run in through the nearest door and find yourself at a roulette table. You can bet any whole dollar amount on any allowable set of numbers. What roulette-betting strategy will maximize your probability of walking out with \$5?

Solution: There are two major considerations in trying to maximize your probability of boosting your fortune to \$5. One is to avoid waste: That is, try to hit \$5 exactly. Overshooting gains nothing and since your expectation is limited by your initial \$2, minus what the "house" is expecting to take from you, you don't want to spend it on overshooting your goal.

The second is speed. As we saw above in *Roulette for the Unwary*, it costs you about a nickel in expectation every time you bet \$1 playing roulette, no matter what you bet on.

Of course, you must do *some* betting. Ideal would be if there were a roulette bet that paid off 5:2, so you could just put your \$2 down once and win what you need or lose everything in one shot.

Unfortunately there is no such bet in roulette, so it looks like you are doomed to either risk overshooting, or risk having to make many bets. Example: You could bet \$1 on the numbers 1 through 8. If you win you're home, otherwise you could bet your remaining \$1 on red and if it wins, start over. That gives you probability p of success where

$$p = \frac{8}{38} + \frac{30}{38} \cdot \frac{18}{38} \cdot p.$$

That gives $p = \frac{8}{38}/(1 - \frac{30}{38} \cdot \frac{18}{38}) = 0.33628318584$, about 1/3.

But you can do better by making correlation work for you. How? By making simultaneous bets of \$1 each of the right kind. Specifically: Bet one of your dollars on the numbers 1 through 12, and *simultaneously*, the other on 1 through 18. Then, if you do get a number between 1 and 12, you hit your \$5 on the nose. If you hit a high number, you lose it all (but at least you've lost it all quickly). If you happen to hit 13, 14, 15, 16, 17, or 18, you again have \$2 and can repeat your double bet.

The probability p of success for this scheme is given by

$$p = \frac{12}{38} + \frac{6}{38} \cdot p,$$

which yields $p = \frac{12}{38}/(1 - \frac{6}{38}) = 3/8 = 0.375$, quite a bit better. In fact this scheme can be shown to be optimal.

Next is a puzzle that you might find familiar. Posed and solved by Georges-Louis Leclerc, Comte de Buffon in 1733, it is said to be the very first solved problem in geometric probability.

But to solve it we will use almost no geometry, relying instead on linearity of expectation.

Buffon's Needle

A needle one inch in length is tossed onto a large mat marked with parallel lines one inch apart. What is the probability that the needle lands across a line?

Solution: Let C be a "noodle": any smooth plane curve—self-intersections permitted—of length, say, ℓ. Toss the noodle randomly onto the mat, and let $\mathbf{X}_C$ be the number of crossings you get, that is, points where C crosses a line. We claim that the average number of crossings, $\mathbb{E}\mathbf{X}_C$, is proportional to ℓ, irrespective of the choice of C! The figure below shows the results of five tosses of a particular noodle, each accompanied by its number of crossings.

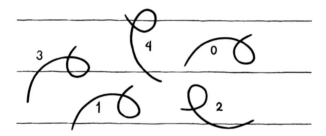

To verify the claim, imagine that C is composed of many short line segments, each of length ε. Throwing just one line segment L of length $\varepsilon < 1$ onto the mat produces a crossing with some probability p, but never more than one crossing, thus $\mathbb{E}\mathbf{X}_L = p$. At the moment, we don't know what p is.

But, by linearity of expectation, we know that if C is composed of n such segments, then $\mathbb{E}\mathbf{X}_C = np$. It follows, by letting ε shrink, that $\mathbb{E}\mathbf{X}_C = \alpha\ell$ for some constant α, since a smooth C is approximable by a string of short line segments. If we can determine what α is, we can infer that $\mathbb{E}\mathbf{X}_N = \alpha$, since Buffon's needle N has length 1.

To find α we only have to pick C cleverly, and this is how: Let C be a circle of diameter 1! Then no matter how C lands on the mat, it produces exactly 2 crossings. Thus $\mathbb{E}\mathbf{X}_C = 2 = \alpha \cdot \pi$, and we solve to get $\alpha = 2/\pi$. We deduce that $\mathbb{E}\mathbf{X}_N = 2/\pi$, and since (with probability 1) the needle either produces one crossing or none, the probability that Buffon's needle crosses a line is that same $2/\pi$. ♡

This wonderful proof was published by T. F. Ramaley in 1969, in a paper entitled "Buffon's Noodle Problem."

Here are a couple of more modern geometry problems, that seem not to have any probability in them.

Covering the Stains

Just as a big event is about to begin, the queen's caterer notices, to his horror, that there are ten tiny gravy stains on the tablecloth. All he has the time to do is to cover the stains with non-overlapping plates. He has plenty of plates, each a unit disk. Can he do it, no matter how the stains are distributed?

Solution: The naked mathematical question being asked here is: Can any set of 10 points in the plane be covered by disjoint unit disks? Sure, if the points are close together, you can cover them with one disk; if they're spread widely apart, you can cover each with its own disk. It's the *medium* distances that could cause problems.

It's a good idea, as advertised in Chapter 7, to think about replacing 10 by other numbers. It's pretty obvious that any one or two points can be covered, and easy to show that any three can be covered.

On the other hand, the answer is certainly "no" if 10 is replaced by 10,000. The reason is that you can't "tile" the plane with unit disks; in other words, you can't arrange disjoint unit disks so as to cover the whole plane (as you could, say, with unit squares). If you put down 10,000 points in a 100×100 grid pattern covering a square somewhat bigger than a unit disk, it won't be possible to cover every grid point with disjoint unit disks.

So it's a question of quantity—is 10 more like 3, or more like 10,000? Here's a thought—how much of the plane *can* be covered by disjoint unit disks? You would probably guess that the best way to pack unit disks is in a hexagonal array, as shown in the figure below. In fact, it is provable that this is the best packing.

How good is it—in other words, if you choose a large region of the plane (say, a $1,000,000 \times 1,000,000$) square) and pack it hexagonally with disks, what fraction of the region will be covered? In the limit, that fraction must be the fraction of the area of one of those hexagonal cells that is occupied by an inscribed unit disk. Of course, the disk,

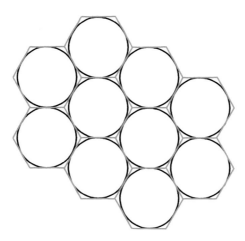

having radius 1, has area π. The hexagon is made up of six equilateral triangles of altitude 1 and area $1/\sqrt{3}$, thus total area $6/\sqrt{3} = \sqrt{12}$. It follows that the fraction of area covered by the disk is $\pi/\sqrt{12} \sim 0.9069$.

Hmm. The disks cover more than 90% of the plane, that is, more than 9 points out of 10. Can we make use of that? Trouble is, the gravy stains are not random points, they are (in the worst case) arranged as if by an adversary.

But we can shift the hexagonal disk-packing randomly. To do this, just fix some hexagonal cell and pick a point in it uniformly at random; then shift (without rotating) the whole picture by moving the center of the cell to the selected point. Now we can say that for any *fixed* point of the plane, the probability that it is covered by some disk of the randomly shifted configuration is 0.9096.

It follows by linearity of expectation that the *expected* number of our 10 gravy stains covered by disks of the randomly shifted hexagonal packing is 9.096 stains, that is, on average more than 9 of the 10 points are covered. But then some configurations must cover all 10 points, and the problem is solved. $\heartsuit$

This technique, in which we prove that something can be done by showing that if we do it randomly it succeeds with probability > 0, is often known by the fancy name "*the probabilistic method*" and associated with Paul Erdős and Alfréd Rényi. We'll see another example in the theorem at the end of this chapter.

The next puzzle provides another example where introduction of randomness is useful.

Colors and Distances

In the town of Hoegaarden, Belgium, exactly half the houses are occupied by Flemish families, the rest by French-speaking Walloons. Can it be that the town is so well mixed that the sum of the distances between pairs of houses of like ethnicity exceeds the sum of the distances between pairs of houses of different ethnicity?

Solution: No. Represent each house in Hoegaarden as a distinct point in the plane, and fix a large disk that contains all the points. Now consider a random line that intersects the disk. (To get such a line, choose a radius of the disk uniformly at random, then take the line perpendicular to the radius that crosses it at a uniformly random point along the radius.)

Your random line will cut the house-points into two sets (one of which could be empty); we claim it separates at least as many house-pairs of different ethnicity as pairs of the same ethnicity. Why? Suppose there are n houses of each ethnicity, and that f Flemish houses and w Walloon houses lie on one side of the line. Then the number of heterogeneous crossings is $(n-w)f + (n-f)w$ while the number of homogeneous crossings is $(n-w)w + (n-f)f$; the former minus the latter is $f^2 + w^2 - 2wf = (f-w)^2 \geq 0$.

Now we make use of the critical fact that for any two particular points, the probability that they are separated by our random line is proportional to the distance between the points. Adding expectations, we deduce that the total distance between heterogeneous (resp., homogeneous) pairs of points is proportional to the expected number of separated heterogeneous (resp., homogeneous) pairs. Since every line cuts at least as many heterogeneous as homogeneous pairs, we conclude that the former distances add up to at least as much as the latter.

Note that for distinct points the inequality is strict, because as long as there is one point which is not simultaneously of both ethnicities, there is positive probability that a line will result in $f \neq w$ causing an imbalance in favor of heterogeneous pairs.

The next puzzle also has a geometric component, but here it is less surprising that we use linearity of expectation.

200

Painting the Fence

Each of n industrious people chooses a random point on a circular fence, and begins painting toward her farthest neighbor until she encounters a painted section. On average, how much of the fence gets painted? How about if instead, each person paints toward her *nearest* neighbor?

Solution: For $n = 2$, the answers are easily determined to be 3/4 and 1/4, respectively. For larger n, we consider any interval I between painters and compare it to its two adjacent intervals. In Case (A), where each painter paints toward her farthest neighbor, I will go unpainted exactly when it is the shortest of the three; in Case (B), when it is the longest. Each of these events happens with probability 1/3.

It seems like we're going to need to know the *expected* length of the shortest (resp., longest) of three intervals chosen by cutting a given interval J at two random points. You can get this noting that choosing two random points from the unit interval is equivalent to choosing a uniformly random point in an equilateral triangle; the barycentric coordinates of the point give your interval lengths. Then, if you identify the regions of the triangle which result in the middle third being shortest (resp. longest) of the three intervals and compute conditional expectations, you get 1/9 and 11/18, respectively.

That calculation was for the unit interval, so if the length of I together with its two neighbor intervals is x, then the expected length of I given that it is the shortest is $x/9$ and given that it is the longest, $11x/18$.

On average, the total length of I and its two neighbors is $3/n$ since adding up this figure for all I counts every interval three times. But in both versions, I is unpainted only a third of the time (in expectation). It follows that in Case (A) the expected amount of unpainted fence is $\frac{n}{3} \cdot \frac{3/n}{9} = 1/9$, and in Case (B), $\frac{n}{3} \cdot \frac{11 \cdot 3/n}{18} = 11/18$.

For the next puzzle, it is useful to have an alternative formula for expectation. Suppose $\mathbf{X}$ is a "counting" random variable, that is, a random variable that takes values in the non-negative integers. Then

$$\mathbb{E}\mathbf{X} = \sum_{i=0}^{\infty} \mathbb{P}(\mathbf{X} > i).$$

To see that this formula is equivalent to the definition at the top of the chapter, we just need to verify that for each j, the probability that

$X = j$ is counted j times in the sum. But that's easy: It's counted once for each i between 0 and $j-1$, so j times as required.

Filling the Cup

You go to the grocery store needing one cup of rice. When you push the button on the machine, it dispenses a uniformly random amount of rice between nothing and one cup. On average, how many times do you have to push the button to get your cup?

Solution: We know that at least two pushes are necessary, and in fact two will suffice just half the time; and sometimes more than three will be required. So the answer must be more than $2\frac{1}{2}$. Would you believe 2.718281828459045?

Let X_1, X_2 etc., be the amount dispensed, and let Y_i be the fractional part of $X_1 + \cdots + X_i$. Then the Y_i's, like the X_i's, are independent and uniformly random in the unit interval.

The first decrease in the values of the Y_i's signals the point when the amount of rice exceeds 1 cup. Thus, the question at hand is equivalent to: What is the expected value of Z, the length of the longest increasing initial sequence of the Y_i's?

The probability that the first j values are increasing is $1/j!$, since the increasing order is just one of $j!$ permutations. Here's where we use the special formula above for the expected value of a counting random variable:

$$\mathbb{E}Z = \sum_{j=0}^{\infty} \mathbb{P}(Z > j) = \sum_{j=0}^{\infty} 1/j!$$

which is just exactly e, Euler's number (also known as Napier's constant, but, naturally, discovered by yet someone else—Jacob Bernoulli).

It's time for our theorem, which, as promised, will be proved using the probabilistic method.

Arguably the most famous theorem in combinatorics, the following remarkable fact was proved in 1930 by Frank Plumpton Ramsey, scion of a famous family of British intellectuals (and brother of Arthur Michael Ramsey, Archbishop of Canterbury). In its simplest finite form, Ramsey's Theorem states that for any positive integer k there is a number n such that if you color every unordered pair of numbers from the set $\{1, \ldots, n\}$ either red or green, then there must be a set S of k numbers which is "homogeneous" in the sense that every pair from S is the same color.

The least n for which this is true is called the "kth Ramsey number" and denoted $R(k, k)$. For example, $R(4, 4)$ turns out to be 18; that means that for any two-coloring of the pairs of numbers between 1 and 18, there's a homogeneous set of size 4; but this does not hold for all colorings of pairs from the set $\{1, \ldots, 17\}$. But $R(5, 5)$ is not known! Paul Erdős, who wrote more papers than any other mathematician in history and was fascinated by Ramsey's theorem, enjoyed giving the following advice. If a powerful alien force demands, on penalty of Earth's annihilation, that we tell them the value of $R(5, 5)$, we should put all the computing power on the planet to work and perhaps succeed in computing that number.

But if they ask for $R(6, 6)$, we should attack them before they attack us.

It is not hard to prove—and, indeed, we will prove it in Chapter 15—that $R(k, k)$ is at most $\binom{2r-2}{r-1}$, which in turn is less than 2^{2k}. What we're going to do is use linearity of expectation to get an exponential *lower* bound for $R(k, k)$.

Theorem. *For all $k > 2$, $R(k, k) > 2^{k/2}$.*

The idea of the proof, published by Erdős in 1947, is to show that a *random* coloring of the pairs of numbers from 1 to $n = 2^{k/2}$ will, with positive probability, have no monochromatic set of size k. Thus, such colorings exist (although the proof does not exhibit one).

By a random coloring, we mean this: For each pair $\{i, j\}$ of numbers between 1 and n, we flip a fair coin to decide whether to color it red or green. If we pick a fixed set S of size k, it contains $\binom{k}{2} = k(k-1)/2$ pairs and thus will be homogeneous with probability $2/2^{k(k-1)/2} = 2^{(k+1)/2 - k^2/2}$.

The number of candidates for the set S, that is, the number of subsets of $\{1, \ldots, n\}$ of size k, is $\binom{n}{k} < (2^{k/2})^k/k! = 2^{k^2/2}/k!$, thus by linearity of expectation, the *expected* number of homogeneous sets is less than

$$2^{(k+1)/2 - k^2/2} \cdot 2^{k^2/2}/k! = \frac{2^{(k+1)/2}}{k!}$$

which is equal to 1 when $k = 2$ and declines as k increases.

So the expected number of homogeous sets is less than 1, and we claim that means the probability that there are no homogeneous sets is greater than zero. One way to see this is to use the formula for expectation of a counting random variable from *Filling the Cup* above. The expected value of $\mathbf{X}$, if $\mathbf{X}$ is the number of homogeneous sets, equals $\Pr(\mathbf{X} > 0)$ plus more non-negative terms; so if $\mathbb{E}\mathbf{X}$ is less than 1, so is $\Pr(\mathbf{X} > 0)$. ♡

The theorem tells us that $R(5,5)$ is at least $\lceil 2^{5/2} \rceil = 6$ and our later induction proof implies an upper bound of $\binom{8}{4} = 70$. In fact $R(5,5)$ is, at the time of this writing, known to be between 43 and 48 inclusive.

15. Brilliant Induction

Induction (from the verb *induce*, not induct!) is one of the most elegant and effective tools in mathematics. In its simplest form, you want to prove something for all positive integers n; you do it by proving the statement for $n = 1$ and then showing that for all $n > 1$, if it's true for $n-1$, it's true for n.

More generally, when proving your statement for n, you may assume it's already true for *all* positive integers $m < n$. That assumption is known as the *induction hypothesis*, or "IHOP" for short (see *Notes & Sources*).

I also like the following formulation, which is even more general. You want to show that some statement is true in all instances satisfying certain conditions. You do this by assuming that there is some instance in which it is false, and focusing on such an instance which is in some sense *minimal*. Then you show how this bad instance can be turned into another bad instance that is even smaller, contradicting your assumption that the instance you began with was minimal. You conclude that there are no bad instances, therefore your statement is always true.

So in a sense, induction is a special case of *reductio ad absurdum*—in other words, showing that a contrary assumption leads to a contradiction.

Is this all a bit too abstract for you? That's what the puzzles are for.

Uniform Unit Distances

Can it be that for any positive integer n, there's a configuration of points on the plane with the property that every point is at distance 1 from exactly n other points?

Solution: For $n = 1$, the endpoints of a unit-length line segment will do. For $n = 2$, we can translate the segment by a unit distance, and add the result to the original segment, to get four points in a diamond. In what direction? Almost any direction will do (as long as it's not at an angle to the original segment that is multiple of $60°$). So if we make it a *random* direction, it'll work with probability 1.

We now proceed in that manner by induction. Assume the set S_n of points (containing 2^n points, but we don't care) has the property that every point in it is at unit distance from exactly n others. Now translate S_n by a unit distance in a random direction to form the set S'_n, and let S_{n+1} be the union of S'_n and S_n. Each point of S_{n+1} is now at unit distance from n others plus the translated copy of itself, for a total of $n+1$. (See figure below for the construction in the $n=2$ case.) Only finitely many translation angles could cause an "accidental" unit distance, so with probability one this works. ♡

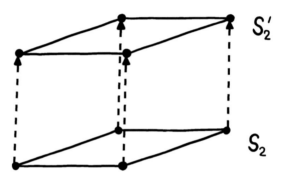

Swapping Executives

The executives of Women in Action, Inc., are seated at a long table facing the stockholders. Unfortunately, according to the meeting organizer's chart, every one is in the wrong seat. The organizer can persuade two executives to switch seats, but only if they are adjacent and neither one is already in her correct seat.

Can the organizer organize the seat-switching so as to get everyone in her correct seat?

Solution: Yes. Let the positions (and the executives that belong in them) be numbered 1 to n from left to right. By induction on n, it suffices to get executives m through n into their correct seats (for some $m \le n$) provided the rest are still all incorrectly seated.

To do that we move executive n to the right by successive swaps until she encounters an executive, say the ith, who's sitting in position $i+1$ and therefore would be swapped into her correct seat if we proceeded. It may also be that executive $i+1$ is in seat $i+2$, executive $i+2$ in seat $i+3$, etc. forming what we will call a "blockade" containing at least executive i.

If the blockade goes all the way to the end of the row, we go ahead and perform all the swaps, ending with executives i through n all in their correct seats while the rest remain to be sorted.

If the blockade terminates before the end, let executive j be the executive immediately to the right of the blockade. Instead of swapping n with i, we swap j toward the left until j swaps with n; at no point in this process will j be in her correct seat because all the seats she is swapped into belong to members of the blockade.

Now we redirect our attention to n, who has moved one step to the right, and continue swapping her to the right until the next encounter with some executive that starts a shifted block, handling that situation as before.

Eventually executive n reaches the right-hand end and the inductive proof is complete. ♡

How many swaps might be needed for this? While executive n is being dealt with, no executive i can be in a blockade twice, because when she is in a blockade she is swapped to the right (from seat $i+1$ to seat $i+2$) and then passed by executive n. We conclude that it takes no more than $2(n-1)$ steps to handle executive n, thus at most $2(1 + 2 + \cdots + n-1) = n(n-1)$ swaps to get everyone into her correct seat.

Odd Light Flips

Suppose you are presented with a collection of light bulbs controlled by switches. Each switch flips the state of some subset of the bulbs, that is, turns on all the ones in the subset that were off, and turns off those that were on. You are told that for any nonempty set of bulbs, there is a switch that flips an odd number of bulbs *from that subset* (and perhaps other bulbs as well).

Show that, no matter what the initial state of the bulbs, you can use these switches to turn off all the bulbs.

Solution: The proof is by induction on number n of bulbs. Note that the order of switch-flipping does not matter; all that counts is whether each switch is flipped an odd or even number of times. The statement is trivially true for $n = 1$, so suppose $n \geq 2$ and we are given a configuration of n bulbs and a set of switches for them. By the IHOP, for every bulb i there is a set S_i of switches that can turn off all the bulbs except possibly bulb i. If for any i the set S_i does turn off bulb i we are done, so let's assume no i has this property.

Choose a pair of distinct bulbs, i and j, and perform S_i followed by S_j. The result is that except for i and j, every bulb that was initially off stays off, while every bulb that was initially on was flipped off and then on again—thus, remains in the state it was in before. The exceptions are bulbs i and j, which are both flipped.

It follows that if bulbs i and j are both on, we can use $S_i \cup S_j$ to turn them both off without changing the state of any other bulb. So if an even number of bulbs are on, we can turn them off pairwise and thus turn all the bulbs off.

What if an odd number of bulbs are on? Then we use the given condition, the relevant set of bulbs being *all of them*, to change the number of lit bulbs to an even number. ♡

There's something a bit mysterious about this proof—didn't we use the given condition only in one special case, where the set of bulbs that we're trying to flip an odd number from is the whole set of bulbs? Maybe that's all we needed!? No, that would fail even for two bulbs, when there's just one switch and it flips one bulb; then there'd be no way to turn the other bulb off.

The issue here is that to execute the induction, we needed a condition that "persists downward"—in this case, one that continues to hold when we remove a bulb. Otherwise, we could not apply the IHOP after removing bulb i.

In fact we really do need the condition for all subsets. If there's a set S of bulbs we can't flip an odd number from, and we begin with just one bulb from S on, we're stuck.

Truly Even Split

Can you partition the integers from 1 to 16 into two sets of equal sizes so that each set has the same sum, the same sum of squares, and the same sum of cubes?

Solution: There is indeed such a partition: One set is $\{1, 4, 6, 7, 10, 11, 13, 16\}$, the other $\{2, 3, 5, 8, 9, 12, 14, 15\}$.

To see this, you might think to yourself: Hmm, 16 is a power of 2; is it possible that this is an example of a more general statement? Can I partition 1 through 8 into two equal-sized sets with the same sum and sum of squares, for example? How about partitioning 1 through 4 into two equal-sized sets with the same sum? The latter is certainly easy:

$\{1,4\}$ versus $\{2,3\}$. Then of course $\{5,8\}$ versus $\{6,7\}$ also works for the numbers from 5 to 8, and if you put these together cross-wise, you get $\{1,4,6,7\}$ versus $\{2,3,5,8\}$ which works perforce for sums and now also seems to work for sums of squares.

In general, you can prove by induction that the integers from 1 to 2^k can be partitioned into sets X and Y so that each part has the same sum of jth powers, where j runs from 0 to $k-1$; equivalently, such that for any polynomial P of degree less than k, the number $P(X)$, which we define as $\sum\{P(x) : x \in X\}$, is equal to $P(Y)$.

To move up to 2^{k+1}, take $X' = X \cup (Y+2^k)$ (where you get $Y+2^k$ by adding 2^k to each element of Y) and $Y' = Y \cup (X+2^k)$. We need to prove that for any polynomial P of degree at most k,

$$P(X) + P(Y+2^k) = P(Y) + P(X+2^k).$$

If P has degree less than k, then by induction $P(X) = P(Y)$ and we also get $P(X+2^k) = P(Y+2^k)$ since each term is merely another polynomial (Q, say) in x_i or y_i. Thus X' and Y' certainly agree for polynomials of degree less than k, but what if P has degree k?

But we're OK here too, because the kth power terms on both sides of the displayed equation are the same as the kth power terms of $P(X) + P(Y)$. ♡

Non-Repeating String

Is there a finite string of letters from the Latin alphabet with the property that there is no pair of adjacent identical substrings, but the addition of any letter to either end would create one?

Solution: Yes, by induction on the number n of letters in the alphabet (then taking $n = 26$). Let $L_1, \ldots, L_n$ be the letters; for $n = 1$ the one-letter string $S_1 = \langle L_1 \rangle$ does the trick. We can take $S_2 = \langle L_1 L_2 L_1 \rangle$, $S_3 = \langle L_1 L_2 L_1 L_3 L_1 L_2 L_1 \rangle$, and in general $S_n = S_{n-1} L_n S_{n-1}$.

Proof that this works: Assume S_{n-1} works for the alphabet $\{L_1, \ldots, L_{n-1}\}$; we want to show $S_n = S_{n-1} L_n S_{n-1}$ works for the alphabet $\{L_1, \ldots, L_n\}$. First we note that S_n can't have a repeated substring, because such a substring could not contain the once-appearing letter L_n, and would thus have to occur entirely inside one of the S_{n-1}'s, contradicting our IHOP.

Now suppose the letter L_k, $1 \le k \le n$, is added to the right-hand end of S_n; we want to show that this *does* create a repeated substring.

Certainly it does if $k = n$, as our whole string is now $S_{n-1}L_n$ followed by itself. Otherwise, our IHOP tells us that there is already a repeated substring inside $S_{n-1}L_k$. The argument when L_k is added to the left-hand end of S_n is essentially the same. ♡

For the Latin alphabet, the length of our string is $2^{26} - 1 = 67{,}108{,}863$ letters.

Baby Frog

To give a baby frog jumping practice, her four grandparents station themselves at the corners of a large square field. When a grandparent croaks, the baby leaps halfway to it. In the field is a small round clearing. Can the grandparents get the baby to that clearing, no matter where in the field she starts?

Solution: Yes. Cover the field with a $2^n \times 2^n$ square grid; we will show by induction on n that from any starting point, the baby frog can be guided to any grid cell we want. Suppose the target cell is in the SW quadrant. Double that smaller grid in both directions and position it so that it covers the whole field as a $2^{n-1} \times 2^{n-1}$ grid. Our IHOP tells us that the baby frog can be croaked into the doubled target cell (which is the union of four cells of the original grid).

Now one croak from the SW corner grandparent will get the baby to the original target cell.

That concludes our induction proof, and it remains only to take n large enough so that some cell of the $2^n \times 2^n$ grid lies entirely inside the clearing that the baby frog is meant to visit.

Guarding the Gallery

A certain museum room is shaped like a highly irregular, non-convex, 11-sided polygon. How many guard-posts are needed in the room, in the worst case, to ensure that every part of the room can be seen from at least one guard-post?

Solution: Three guard-posts are sufficient (and sometimes necessary, as for the 11-gon illustrated below).

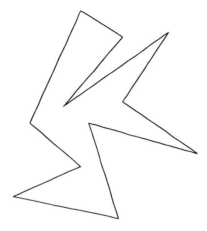

In general, in any n-gon, $\lfloor n/3 \rfloor$ (that is, the greatest integer in $n/3$)
guard-posts suffice, and this number is the best bound possible. The
next figure shows, by extrapolation, that $\lfloor n/3 \rfloor$ posts may be necessary; the following proof shows that they suffice, and in fact can be
placed in corners.

The first move is to *triangulate* the polygon. This can done by drawing non-crossing *diagonals* until no more will fit; a diagonal is a line
segment between two vertices of the polygon whose interior lies entirely in the interior of the polygon.

Next we show by induction on n that the polygon's vertices can be
colored with three colors so that every triangle's vertices get all three
colors. Choose any diagonal D in the triangulation and cut through it
"lengthwise" to form two polygons, each with fewer than n vertices,
and each having D as an edge. Color both using the IHOP and permute the colors of one so that the colors at the endpoints of D match,
then put the two small polygons back together to get a coloring of the
original polygon (as is done in the figure below).

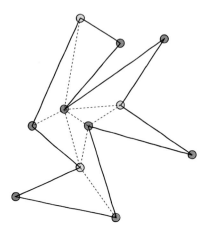

The least-frequently used color is found on at most $\lfloor n/3 \rfloor$ vertices, and stationing guards there covers all the triangles and thus the whole room. ♡

Path Through the Cells

Cells in a certain cellular telephone network are assigned frequencies in such a way that no two adjacent cells use the same frequency. Show that if the number of frequencies used is minimal (subject to this condition), then it's possible to design a path that moves from cell to adjacent cell and hits each frequency exactly once—in ascending order of frequencies!

Solution: This is really a problem in abstract graph theory; an assignment of frequencies, or "colors," to vertices in a graph in such a way that no two adjacent nodes get the same color is called a *proper coloring*. The statement to be proved is that in any graph properly colored with the minimum number of colors, there is a path that hits every color exactly once; and moreover, if the colors are ordered (e.g., they correspond to the numbers from 1 to k) we can choose the path so as to hit the colors in ascending order. The graph G in our case is the one whose vertices are cells, two vertices constituting an edge if the corresponding cells are adjacent (thus susceptible, presumably, to frequency interference).

The idea is to re-color G in a "greedy" fashion, trying to use high-numbered colors. First, re-color any vertex colored $k-1$ by color k, if it isn't adjacent to a vertex of color k. Then recolor any vertex of color $k-2$ by color $k-1$, if it isn't adjacent to a vertex of color $k-1$. Continue in this way to get a new coloring which, by our minimality assumption, still uses all k colors.

Now start the path at any vertex v_1 whose new color (and therefore whose old color as well) is 1; we know there's an adjacent vertex v_2 of new color 2, since otherwise the color of v_1 would have been incremented. Note v_2's old color is also 2; it cannot have been recolored from color 1 since it is adjacent to v_1. Similarly, there's a vertex v_3 adjacent to v_2 of color 3, which could not have been recolored from color 2 on account of v_2. Proceeding in this manner we get a path $v_1, v_2, \ldots$ that hits every color in both the new coloring and the old in increasing order.

This puzzle is one of many examples where asking for a proof of a weaker statement (e.g., just that there is a path that hits every frequency exactly once) can make the puzzle harder.

Profit and Loss

At a recent stockholders' meeting of Widget Industrials Inc., the Chief Financial Officer presented a chart of the month-by-month profits (or losses) since the last meeting. "Note," said she, "that we made a profit over every consecutive eight-month period."

"Maybe so," complained a shareholder, "but I see we *lost* money over every consecutive *five*-month period!"

What's the maximum number of months that could have passed since the last meeting?

Solution: What's needed, of course, is a maximum-length sequence of numbers such that every substring of length 8 adds up to more than 0, but every substring of length 5 adds up to less than 0. The string must certainly be finite, in fact less than 40 in length, else you could express the sum of the first 40 entries both as the (positive) sum of 5 substrings of length 8 and the (negative) sum of 8 substrings of length 5.

Let's tackle the problem more generally and let $f(x, y)$ be the length of the longest string such that every x-substring has positive sum and every y-substring negative; we may suppose $x > y$. If x is a *multiple* of y, then $f(x, y) = x - 1$ and we must accept vacuous truth with respect to the x-substrings.

What if $y = 2$ and x is odd? Then you can have a string of length x itself, with entries that alternate between, say, $x-1$ and $-x$. But you can't have $x+1$ numbers, because in each x-substring the odd entries must be positive (since you can cover it with 2-substrings leaving out any odd entry). But there are two x-substrings and together they imply that the middle two numbers are both positive, a contradiction.

Applying this reasoning more generally suggests that $f(x, y) \leq x + y - 2$ when x and y are relatively prime, that is, they have no common divisor other than 1. We can prove this by induction as follows. Suppose to the contrary that we have a string of length $x+y-1$ which satisfies the given conditions. Write $x = ay + b$ where $0 < b < y$, and look at the last $y + b - 1$ numbers of the sequence. Observe that any consecutive b of them can be expressed as an x-substring of the full string, with a y-substrings removed; therefore, it has positive sum. On the other hand, any $(y - b)$-substring of the last $y + b - 1$ can be expressed as $a+1$ y-strings with an x-string removed, hence has negative sum. It follows that $f(b, y - b) \geq y + b - 1$, but this contradicts our induction assumption because b and $y - b$ are relatively prime.

To show that $f(x, y)$ is actually equal to $x+y-2$ when x and y are relatively prime, we construct a string which has the required properties and more: It takes only two distinct values, and it is periodic with periods *both* x and y. Call the two values u and v, and imagine at first that we assign them arbitrarily as the first y entries of our string.

Then these assignments are repeated until the end of the string, making the string perforce periodic in y. To be periodic in x as well, we only need to ensure that the last $y - 2$ entries match up with the first $y - 2$, which entails satisfying $y - 2$ equalities among the original y choices we made. Since there are not enough equalities to force all the choices to be the same, we can ensure that there is at least one u and one v.

Let us do this, for example, with $x = 8$ and $y = 5$. Call the first five string entries $c_1, \ldots, c_5$, so the string itself will be $c_1 c_2 c_3 c_4 c_5 c_1 c_2 c_3 c_4 c_5 c_1$. To be periodic with period 8, we must have $c_4 = c_1$, $c_5 = c_2$, and $c_1 = c_3$. This allows us to have $c_1 = c_3 = c_4 = u$, for example, and $c_2 = c_5 = v$; the whole sequence is thus $uvuuvuvuuvu$.

Getting back to general x and y, we note that a string which is periodic in x automatically has the property that every x-substring has the same sum; because, as you shift the substring one step at a time, the entry picked up at one end is the same as the entry dropped at the other end. Of course the same applies to y-substrings if the string is periodic in y.

Let S_x be the x-substring sum and S_y similarly; we claim $S_x/x \neq S_y/y$. The reason is that if there are, say, p copies of u in each x-substring and q copies of v in each y-substring, then $S_x/x = S_y/y$ would imply $y(pu + (x-p)v) = x(qu + (y-q)v)$ which reduces to $yp = xq$. Since x and y are relatively prime, this cannot happen for $0 < p < x$ and $0 < y < q$.

It follows that we can adjust u and v so that S_x is positive and S_y is negative. In the case above, for example, each 8-substring contains 5 copies of u and 3 of v, while each 5-substring contains 3 copies of u and 2 of v. If we take $u = 5$ and $v = -8$, we get $S_x = 1$ and $S_y = -1$. The final sequence, solving the original problem, is then $5, -8, 5, 5, -8, 5, -8, 5, 5, -8, 5$. ♡

Uniformity at the Bakery

A baker's dozen (thirteen) bagels have the property that any 12 of them can be split into 2 piles of 6 each, which balance perfectly on the scale. Suppose each bagel weighs an integer number of grams. Must all the bagels have the same weight?

Solution: Yes. Suppose otherwise, and let $b_1, \ldots, b_{13}$ be a set of weights, not all the same, with the stated property. We can assume (by shifting all the weights down) that some bagel weighs 0 grams. Let 2^k be the greatest power of 2 that divides all the weights evenly, and suppose our counterexample was chosen to minimize k. There can't be a bagel of odd weight, because to be balanced the remaining bagels must have an even total weight; but leaving out the 0-gram bagel must also leave even weight behind, an impossibility. But if all the weights are even we can divide them all by 2, reducing k and contradicting our IHOP. ♡

It is worth noting that the requirement that the weighings have six bagels on a side is necessary. Otherwise, for example, a baker's dozen consisting of 7 bagels of 50 grams each and 6 bagels of 70 grams each would have the stated property. The place where the proof breaks down is where we shift all the weights down; that relies on there being the same number of items on each side during the weighings.

Summing Fractions

Gail asks Henry to think of a number n between 10 and 100, but not to tell her what it is. She now tells Henry to find all (unordered) pairs of numbers j, k that are relatively prime and less than n, but add up to more than n. He now adds all the fractions $1/jk$.

Whew! Finally, Gail tells Henry what his sum is. How does she do it?

Solution: For $n = 2$, the only eligible pair is $\{1, 2\}$ so the sum is $1/2$. For $n = 3$, we have pairs $\{1, 3\}$ and $\{2, 3\}$ for a total of $1/3 + 1/6 = 1/2$. For $n = 4$, the pairs are $\{1, 4\}$, $\{2, 3\}$, and $\{3, 4\}$ giving $1/4 + 1/6 + 1/12 = 1/2$. Hmm. Is it possible that the sum is always $1/2$?

Let's try to prove it by induction, starting from $n = 2$. Moving from n to $n+1$, we gain $1/jn$ for each j with $\gcd(j, n) = 1$, and lose $1/jk$ for each j and k with $\gcd(j, k) = 1$ and $j + k = n$. But if $j + k = n$, then any pair of j, k, and n being relatively prime implies that all three pairs are. Thus, each pair j, k that sum to n with $\gcd(j, k) = 1$ signifies a loss of $1/jk$ but a gain of $1/jn + 1/kn = 1/jk$, neatly canceling. ♡

Tiling with L's

Can you tile the positive quadrant of the plane grid with unrotated triominoes each of which is either shaped like a letter L, or a backwards L?

Solution: The answer is perhaps a bit surprising: You can indeed do it, but in only one way, and that way is not periodic.

If you try to do it by hand, starting at the origin with a full row along the X-axis and proceeding upward row by row, you will find that a little forethought (thinking one row ahead) is all you need.

That first row must start (at the left) with a forward L, which we'll call an "el." To its right we'll need to put a backward L, which we'll call a "le," so that the space above these two triominoes will be fillable. Then you'll need another el, another le, another el, and so forth, alternating out to infinity.

The next row is already half-covered; to cover the rest of those cells, you'll need to start with a le. Each successive choice of el or le must leave an even-length gap on row 3, otherwise you'll be stumped later. The good news is that the choice of el or le always changes the length of the previous gap by one, thus you can always make that gap even.

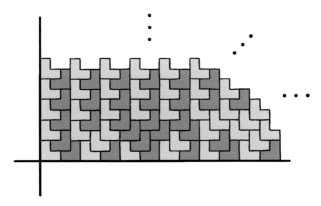

This sets us up nicely for a proof by induction that the construction is possible. What should the induction hypothesis be? We'd like to say that all rows are covered up to and including row k (say), and the next row, row $k+1$, is covered up to position m (which could be zero). After position m, row $k+1$ has even-length gaps out to infinity.

What else? Well, we also need to say that row $k+2$ has even-length gaps out to position m, and is empty after that; and no higher row is covered at all.

That's a lot of hypotheses, but we need them all and once we have them, the argument is easy. We insert a new triomino in such a way that its bottom covers cells $m+1$ and $m+2$ in row $k+1$, choosing el or le so as to make the new gap, above in row $k+2$, even. That will also have the effect of shortening the next gap on row $k+1$ by two, which kills the gap or leaves it of even length, either of which is OK.

Everything seems to be fine, but hold on a minute—what, exactly, are we inducing *on*? It can't be the number of triominoes placed, because that's already infinite after the first row goes down. What we're really doing is a *series* of inductions, one for each row. Each induction ends with a tiling in which rows 1 through k are completely covered, row $k+1$ partly covered, no higher row covered at all, and all gaps on row $k+1$ of even length. This is just an "$m = 0$" case of our induction hypothesis.

Uniqueness of our tiling follows by assuming we have any valid tiling, choosing any k and m, and looking at that part of the tiling corresponding to the k, m case of our induction hypothesis. It must satisfy the hypothesis in order to be extendible, and since our construction leaves no choice at any point, the tiling must be identical to ours.

Traveling Salesmen

Suppose that between every pair of major cities in Russia, there's a fixed one-way air fare for going from either city to the other. Traveling salesman Alexei Frugal begins in St. Petersburg and tours the cities, always choosing the cheapest flight to a city not yet visited (he does not need to return to St. Petersburg). Salesman Boris Lavish also needs to visit every city, but he starts in Kaliningrad, and his policy is to choose the most *expensive* flight to an unvisited city at each step.

It looks obvious that Lavish's tour costs at least as much as Frugal's, but can you prove it?

Solution: One way is to show that for any k, the kth cheapest flight (call it f) taken by Lavish is at least as costly as the kth cheapest flight taken by Frugal. This seems like a stronger statement than what was requested, but it really isn't; if there were a counterexample, we could adjust the flight costs, without changing their order, in such a way that Lavish paid less than Frugal.

For convenience, imagine that Lavish ends up visiting the cities in west-to-east order. Let F be the set of Lavish's k cheapest flights, X the departure cities for these flights, and Y the arrival cities. Note that X and Y may overlap.

Call a flight "cheap" if its cost is no more than f's; we want to show that Frugal takes at least k cheap flights. Note that every flight eastward out of a city in X is cheap, since otherwise it would have been taken by Lavish instead of the cheap flight in F that he actually took.

Call a city "good" if Frugal leaves it on a cheap flight, "bad" otherwise. If all the cities in X are good, we are done; Frugal's departures from those cities constitute k cheap flights. Otherwise, let x be the westernmost bad city in X; then when Frugal gets to x, he has already visited every city to the east of x, else Frugal could have departed x cheaply. But then every city to the east of x, when visited by Frugal, had its cheap flight to x available to leave on, so all are good. In particular, all cities in Y east of x are good, as well as all cities in X west of x; that is k good cities in all. ♡

Lame Rook

A *lame rook* moves like an ordinary rook in chess—straight up, down, left, or right—but only one square at a time. Suppose that the lame rook begins at some square and tours the 8×8 chessboard, visiting each square once and returning to the starting square on the 64th move. Show that the number of horizontal moves of the tour, and the number of vertical moves of the tour, are *not* equal!

Solution: You should be asking yourself: Is this true for $n \times n$ boards as well? Some experimentation will remind you that there is no such tour if n is odd, because (for one thing) such a board has more squares of one color than another but a tour that returns to its starting position must visit the same number of squares of each color.

Moreover, for $n = 2$ and $n = 6$, it is easy to construct tours where the numbers of horizontal and vertical steps are equal. So you might guess that the puzzle works on $n \times n$ boards when $n \equiv 0 \mod 4$, and moreover that the issue could be that the number of (say) horizontal moves must be equivalent to 2 mod 4 and therefore can't be 64/2. All this is correct, but how to prove it?

Assume the rook is a point in the middle of a square, and let P be the polygon thus traced. Note that the vertices of P lie on the *dual* grid of the chessboard, whose vertices lie at the centers of the chessboard squares.

The polygon P outlines a tree T (pink in the figure below) on the grid lines of the chessboard.

We will show by induction that the following holds when P is the outline of *any* tree on the chessboard grid: If there are n_0 dual-grid vertices on P's perimeter that lie on even columns, and n_1 on odd columns, and h is the sum of the lengths of the horizontal borders of P, then $n \equiv n_1 - n_0 + 2 \mod 4$.

The base case is easy: If the tree is a single vertex, we have $n_1 = n_0 = 2$ and $h = 2$. Now let u be any leaf of T; if it's attached vertically, cutting it off decrements both n_1 and n_0 by 1 and does not change h. If u is attached horizontally, it changes $n_1 - n_0$ by two while decrementing h by two.

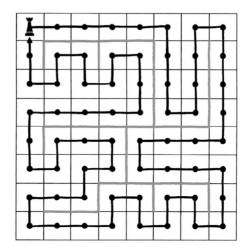

Since the polygon P generated by the lame rook has $n_1 = n_0$, the number of horizontal moves is 2 mod 4 and thus cannot be 32. ♡

For our theorem we return to Ramsey theory, using induction to get an upper bound for Ramsey numbers (and incidentally to show that they exist).

Given any two positive integers s and t, Ramsey's Theorem claims that for sufficiently large n, for every red–green coloring of the unordered pairs of numbers in $\{1, 2, \ldots, n\}$ there is either a set S of size s all of whose pairs are colored red, or a set T of size t all of whose pairs are colored green. The least n for which this holds is, by definition, the Ramsey number $R(s, t)$.

Theorem. *For any s and t, $R(s, t)$ exists and is no more than $\binom{s+t-2}{s-1}$.*

We note first that since there are no pairs inside a set of size 1, $R(1, k) = R(k, 1) = 1$ for any k. Further, to get (say) a red set (i.e., a set all of whose pairs are red) of size 2 we only need one red edge. It follows that $R(2, k) = R(k, 2) = k$, because (in the case of $R(2, k)$) if there's any red pair its members give us a red set of size 2, else all the pairs are green and we can use all the numbers to get our green set of size k.

These observations conform to the theorem. Now we will proceed by induction on the *sum* $s+t$.

Fix s and t and suppose $R(s', t') \leq \binom{s'+t'-2}{s'-1}$ for all s', t' with $s'+t' < s+t$. Let U be the set $\{1, \ldots, n\}$ where $n = \binom{s+t-2}{s-1}$, and suppose some adversary has colored all the unordered pairs in U red or green. We want to find in U either a red subset S of size s or a green subset T of size t. That would show that the conclusion of Ramsey's theorem holds when $n = \binom{s+t-2}{s-1}$, therefore that $R(s,t)$ is at most this number.

Recall that (from Pascal's triangle) $n = j+k$, where $j = \binom{s+t-3}{s-2}$ and $k = \binom{s+t-3}{s-1}$. Note that j is our "n" when $s' = s-1$ and $t' = t$, while k is our "n" when $s' = s$ and $t' = t-1$.

Let's focus our attention on the pairs containing the number 1. There are $n-1$ of those, and we claim that *either* at least j of these pairs are red, or at least k of them are green. Why? Because otherwise there are at most $j-1$ red pairs and at most $k-1$ green pairs, and that adds up to only $j+k-2 = n-2$. (Yes, you can think of it as an application of the pigeonhole principle.)

Suppose the former, that is, there are j red pairs of the form $\{1, i\}$, and let V be the set of numbers i that appear in them (not including 1 itself). Since $(s-1) + t < s+t$, and $j = \binom{(s-1)+t-2}{(s-1)-1}$, we know from our IHOP that there's either a red set $S' \subset V$ of size $s-1$ or a green set $T' \subset V$ of size t. If it's the latter, we are happy; and if the former, we can add the number 1 to our set—since all of the pairs involving 1 are red—and get a red set $S = S' \cup \{1\}$ of size s.

The other case is similar: If there are k green pairs of the form $\{1, i\}$, let W be the set of numbers i that appear in them; since $s + (t-1) < s + t$, and $k = \binom{s+(t-1)-2}{s-1}$, we know from our IHOP that there's either a red set $S' \subset V$ of size s or a green set $T' \subset V$ of size $t-1$. If it's the former, we have our S; if the latter, we can add the number 1 to T—since all of the pairs involving 1 are green—and get a green set $T = T' \cup \{1\}$ of size t. Done! $\heartsuit$

As it happens, $R(3,3)$ is equal to $\binom{3+3-2}{3-1} = \binom{4}{2} = 6$, but $R(4,4) = 18$ which is strictly less than our bound of $\binom{4+4-2}{4-1} = \binom{6}{3} = 20$. Putting our theorem together with the one proved in the previous chapter, we know that $R(k,k)$ lies somewhere between $2^{k/2}$ and 2^{2k}. Most combinatorialists would guess that there's some number α between $\frac{1}{2}$ and 2 such that $R(k,k)$ behaves like $2^{\alpha k}$ as k grows—more precisely, that the logarithm base 2 of the kth root of $R(k,k)$ tends to some α as k goes to infinity. Finding α would have earned you \$5000 from "Uncle Paul" Erdős, and he'd have paid happily.

16. Journey into Space

We all live in three spacial dimensions, and are attuned to thinking three-dimensionally even though we see, draw and paint, for the most part, only in two. Thus, geometrical problems in three dimensions, while potentially much harder than problems in the plane, can appeal to our intuition in pleasant and even useful ways.

Some of the puzzles that follow are set naturally in three dimensions; others can be placed there to good effect.

Easy Cake Division

Can you cut a cubical cake, iced on top and on the sides, into three pieces each of which contains the same amount of cake and the same amount of icing?

Solution: You can get any number k of pieces with the same amount of cake and the same amount of icing in the following neat way: Looking at the cake from above, divide the perimeter of the square top of the cake into k equal parts and make straight, vertical cuts to the middle of the square from each boundary point on the perimeter.

That works because each piece, looked at from the top, is the union of triangles with the same altitude (namely, half the side of the cake) and the same total base (namely, $1/k$ times the perimeter), as in the figure below.

Speaking of cubes, here's a question about painting them.

Painting the Cubes

Can you paint 1000 unit cubes with 10 different colors in such a way that for any of the 10 colors, the cubelets can be arranged into a $10 \times 10 \times 10$ cube with only that color showing?

Solution: Note that the numbers work out perfectly: 6000 faces need to be colored, and the number of faces showing on the big cube is $6 \times 100 = 600$. Thus we'll need to paint exactly 600 faces with each color, and when we put the big cube together, we can't afford to bury even a single face of the designated color inside.

It turns out that despite these restrictions, there are many ways to color the unit cubes that will do the trick. But finding one by hammer and tongs is a pain in the neck.

Instead, number the colors 0 through 9. Apply paint to all the cubes in the infinite 3-dimensional cubic grid, as follows: Paint both sides of all the faces on the plane $x = 0$ with color 0, all faces on $x = 1$ with color 1, etc., rotating back to color 0 for the plane $x = 10$, and so forth. Do the same for the y-planes and z-planes.

This scheme colors every face of every unit cube with a color corresponding to the rightmost digit of that face's u-coordinate, where u is x, y or z according to which axis is perpendicular to the face. In particular, the scheme colors unit cubes in 10^3 different ways. But now, if you want your big cube to display color i on its outside, cut out your big cube from space via planes $x = i$, $x = i+10$, $y = i$, $y = i+10$, $z = i$, and $z = i+10$. This will use each type of unit cube once, and you are done! ♡

Of course, you can replace 10 by any positive integer n (that is, you can color n^3 unit cubes with which you can then make an $n \times n \times n$ cube with any given color the only one showing). The above argument still works, and in fact versions of it work in other dimensions, as well.

Just don't think too hard about what it means to "paint" a 3-dimensional face of a 4-dimensional hypercube.

Curves on Potatoes

Given two potatoes, can you draw a closed curve on the surface of each so that the two curves are identical as curves in three-dimensional space?

Solution: Yes, you can: intersect the potatoes!

Well, maybe a few more words of explanation would not go amiss.

224

Think of the potatoes as holograms and bring them together so that they overlap in space. Then the intersection of their surfaces will describe a curve (possibly more than one curve) that lives on both surfaces, and solves the puzzle.

Painting the Polyhedron

Let P be a polyhedron with red and green faces such that every red face is surrounded by green ones, but the total red area exceeds the total green area. Prove that you can't inscribe a sphere in P.

Solution: Assume the sphere is inscribed, and triangulate the faces of P using the points where the sphere is tangent. Then the triangles on either side of any edge of P are congruent, thus have the same area; at most, one of every such pair of triangles is red. It follows that the red area is at most equal to the green area, contradicting our assumptions. ♡

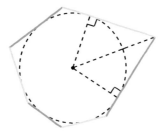

The illustration shows a two-dimensional version, where sides and vertices of a polygon take the place of faces and edges of P.

Boarding the Manhole

An open manhole 4 meters in diameter has to be covered by boards of total width w. The boards are each more than 4 meters in length, so if $w \geq 4$ it's obvious that you can cover the manhole by laying the boards side by side (see figure below). If w is only 3.9, say, you still have plenty of wood and you are allowed to overlap the boards if you want. Can you still cover the manhole?

Solution: Intuitively speaking, even though you have plenty of wood, it seems that you have to waste a lot of it either by covering

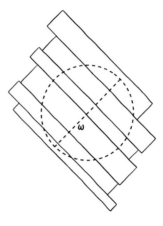

the center of the manhole many times, or laying boards too close to the edge to be of much use. Putting it another way, it's relatively expensive to cover parts of the manhole far from the center. Can you quantify that statement in a useful way?

It turns out you can, using a famous fact known sometimes as Archimedes' hat box theorem. Archimedes used his "method of exhaustion" (these days, we would use calculus) to show that if a sphere of radius r is intersected by two parallel planes a distance d apart, as in the figure below, the surface area of the sphere between the planes is $2\pi r d$. In particular, it doesn't depend on where the planes cut the sphere, but only on how far apart the planes are. (The "hat box" presumably alludes to a cylinder of radius r perpendicular to the planes; the cylinder's area between the planes is that same $2\pi r d$.)

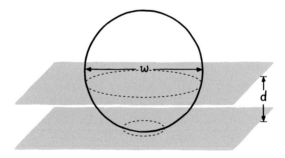

Where do we get our planes from? Let's replace the manhole by a sphere, also of diameter w, whose equator is the circumference of the manhole. Each board is replaced by a vertical *slab* consisting of two parallel vertical planes whose distance d apart is the width of the board. Now the key observation: Arranging the boards to cover the manhole is equivalent to arranging the slabs to cover the sphere.

But by Archimedes' theorem, the slabs cannot cover more than $2\pi \frac{w}{2} d = \pi w d$ of the surface of the sphere, which has area πw^2. So if $d < w$, you're stuck. $\heartsuit$

Now we are set up perfectly to conquer the next puzzle.

Slabs in 3-Space

A "slab" is the region between two parallel planes in three-dimensional space. Prove that you cannot cover all of 3-space with a set of slabs the sum of whose thicknesses is finite.

Solution: In a way the conclusion seems obvious; if you can't cover space when the slabs are parallel and disjoint, *surely* you can't cover when they waste space by overlapping. But infinity is a tricky concept. The slabs all have infinite volume, so what's to prevent them from covering anything they want?

Actually, we know from the previous puzzle that they have trouble covering a large ball. If the total thickness of the slabs is T, then, by Archimedes' theorem, they can't cover the surface of a ball of diameter greater than T, thus they can't cover the ball itself.

Thus they certainly can't cover all of 3-space. But it's odd that to show the latter, we seem to need to reduce the problem to just a finite piece of space.

Bugs on Four Lines

You are given four lines in a plane in general position (no two parallel, no three intersecting in a common point). On each line a ghost bug crawls at some constant velocity (possibly different for each bug). Being ghosts, if two bugs happen to cross paths they just continue crawling through each other uninterrupted.

Suppose that five of the possible six meetings actually happen. Prove that the sixth does as well.

Solution: As the title of this chapter suggests, lifting the puzzle into space provides an elegant solution. How? By means of a time axis. Suppose every pair of bugs meets except bug 3 with bug 4. Construct a time axis perpendicular to the plane of the bugs, and let g_i be the

graph (this time, in the sense of graphing a function) of the ith bug in space. Since each bug travels at constant speed, each such graph is a straight line; its projection onto the plane of the bugs is the line that bug travels on. If (and only if) two bugs meet, their graphs intersect.

The lines g_1, g_2, and g_3 are coplanar since all three pairs intersect, and the same applies to g_1, g_2, and g_4. Hence all four graphs are coplanar. Now g_3 and g_4 are certainly not parallel, since their projections onto the original plane intersect, thus they intersect on the new common plane. So bugs 3 and 4 meet as well.

Circular Shadows II

Show that if the projections of a solid body onto two planes are perfect disks, then the two projections have the same radius.

Solution: The conclusion seems eminently reasonable, yet it's not completely obvious how to prove it.

An easy way to make your intuition rigorous is to select a plane which is simultaneously perpendicular to the two projection planes, and move parallel copies of it toward the body from each side. They hit the body at the opposite edges of each projection, so that the distance between the parallel planes at that moment is the common diameter of the two projected circles. ♡

Box in a Box

Suppose the cost of shipping a rectangular box is given by the sum of its length, width, and height. Might it be possible to save money by fitting your box into a cheaper box?

Solution: The issue here is that if packed diagonally, the inside box could have some dimensions that exceed the greatest dimension of the outside box. For example, a narrow needle-like box of length almost $\sqrt{3}$ could be packed inside a unit cube. This particular example wouldn't provide a way to cheat, because in this case the total cost of the outside cube would be 3, much more than the 1.7 or so you might have to pay for the inside box. But how do we know that some better example won't turn up?

Here's a delightful argument that lets ε go to infinity(!) in order to show that you can't cheat the system.

Suppose your box is $a \times b \times c$, and let R stand for the region of space occupied by your box including its interior. Suppose R can be packed into an $a' \times b' \times c'$ box R' with $a + b + c > a' + b' + c'$. Let $\varepsilon > 0$ and consider the region R_ε consisting of all points in space that are within distance ε of R. This R_ε region will be a rounded convex shape containing your box R. The volume of R_ε is equal to your box's volume plus $2\varepsilon(ab + bc + ac)$s for the volume added to the sides, plus $\pi\varepsilon^2(a + b + c)$ for the volume added to the edges, plus $\frac{4}{3}\pi\varepsilon^3$ for the volume (consisting of the eight octants of a ball) added to the corners. If you do the same to the outside box R' you get a region R'_ε which of course must contain R_ε. But that's impossible! In the expression for $\text{vol}(R'_\varepsilon) - \text{vol}(R_\varepsilon)$, the ε^3 terms cancel, so if you take ε to be large, the dominant term is the ε^2 term, namely $\pi\varepsilon^2((a' + b' + c') - (a + b + c))$, which is negative. ♡

Angles in Space

Prove that among any set of more than 2^n points in $\mathbb{R}^n$, there are three that determine an obtuse angle.

Solution: It makes sense that the 2^n corners of a hypercube represent the most points you can have in n-space without an obtuse angle. But how to prove it?

Let $x_1, \ldots, x_k$ be distinct points (vectors) in $\mathbb{R}^n$, and let P be their convex closure. We may assume P has volume 1 by reducing the dimension of the space to the dimension of P, then scaling appropriately; we may also assume x_1 is the origin (i.e., the 0 vector). If there are no obtuse angles among the points, then we claim that for each $i > 1$, the interior of the translate $P + x_i$ is disjoint from the interior of P; this is because the plane through x_i perpendicular to the vector x_i separates the two polytopes.

Furthermore, the interiors of $P + x_i$ and $P + x_j$, for $i \neq j$, are disjoint as well; this time via the separating plane running through $x_i + x_j$ perpendicular to the edge $x_j - x_i$ of P. We conclude that the volume of the union of the $P + x_i$, for $1 \leq i \leq k$, is k.

However, all these polytopes lie inside the doubled polytope $2P = P + P$, whose volume is 2^n. Hence, $k \leq 2^n$ as claimed!

Curve and Three Shadows

Is there a simple closed curve in 3-space, all three of whose projections onto the coordinate planes are trees? This means that the shadows of the curve, from the three coordinate directions, may not contain any loops.

Solution: No one knew the answer to this question until such a curve was actually found, by John Terrell Rickard of Advanced Telecommunication Modules Ltd. in Cambridge, UK. His beautiful, symmetrical solution is shown below with its three shadows. I have no idea how Rickard found it.

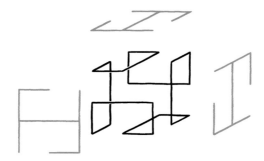

Later, Donald Knuth, author of the multi-volume classic *The Art of Computer Programming*, programmed a computer to see if there were any other solutions that, like Rickard's, could be carved from a $3 \times 3 \times 3$ region of a cubical grid. In the time it took to push his ENTER button (according to Knuth), the computer found that there was just one other solution, shown below. Knuth's solution, in contrast to Rickard's, has no symmetry at all.

If you found either one of these curves, more power to you!

Our theorem is perhaps the world's best-known example of proving a theorem about plane geometry by moving into 3-dimensional space. But there's an ironic twist.

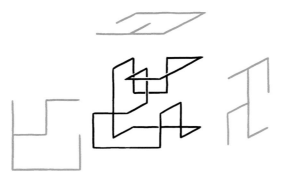

Given any two circles in the plane of different radii, neither contained in the other, we can construct two lines that are tangent to the circles such that for each line, the circles are on the same side of the line. (See picture below). These lines will intersect at some point on the plane, known as the *Monge point* of the two circles.

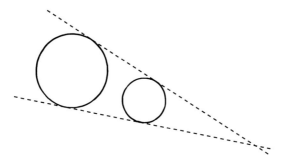

Theorem. *Suppose three circles are given, all of different radii, none contained in another. Then the three Monge points determined by the three pairs of circles all lie on a line.*

This is what the picture looks like (we have taken the circles to be disjoint, but in fact they're allowed to overlap).

The theorem is attributed to Gaspard Monge, 1746–1818, a distinguished French mathematician and engineer. On Wikipedia (as I write this) you will find a proof which is elegant, famous, and wrong! Here's how it goes:

Replace each circle by a sphere whose equator is that circle, so now you have three spheres (of different diameters) intersecting the original plane at their equators. Pick two of them: Instead of two

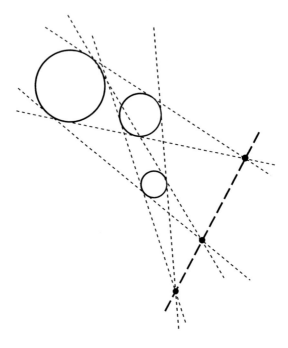

tangent lines you now have a whole tangent *cone* whose apex is the Monge point of the circles.

Now, take a plane tangent to the three spheres, thus also tangent to the three cones. The three Monge points will all lie on that plane, as well as on the original plane. But the two planes intersect in a line, so the Monge points are all on that line, QED!

It's not easy to spot the hole in this proof. The difficulty is that there may not *be* a plane tangent to the three spheres; for example, if two of the circles are large and the third is between them and smaller.

The following proof, attributed by cut-the-knot.org to Nathan Bowler of Trinity College, Cambridge, works by erecting *cones* instead of spheres on top of the circles. Call them C_1, C_2, and C_3, and let them all be "right" cones—that is, they support $90°$ angles at their apices. (Actually, we only need all the cones to have the same angle.) Each pair of cones determines two (outside) tangent planes, say P_1 and Q_1 (for cones C_2 and C_3), P_2 and Q_2 (for cones C_1 and C_3), and finally P_3 and Q_3 (for cones C_1 and C_2).

Each pair of planes P_i, Q_i intersect in a line L_i which passes through the apex of both tangent cones, as well as through the point where the corresponding circle tangents meet. Thus, in particular, L_1

and L_2 both meet at the apex of C_3, L_1, and L_3 at the apex of C_2, and L_2 and L_3 at the apex of C_1. Hence the three lines of intersection are coplanar (all lie on the plane determined by the three apices); the intersection of that plane with the original plane of the circles is a line through the three foci, and this time we are really done. ♡

There are other ways to prove Monge's theorem and as you might suspect, one of them is to transform the plane in a circle-preserving way so as to escape annoying initial configurations that screw up the spheres proof. But it's curious, and a bit sobering, that the unfixed sphere argument lives on as the "standard" proof of Monge's theorem.

17. Nimbers and the Hamming Code

Probably the most famous equation in mathematics is not $E = mc^2$ or $e^{i\pi} = -1$ but $1 + 1 = 2$. But if, like many computer scientists, you are fond of binary arithmetic, you might prefer $1 + 1 = 10$; if you are an algebraist or logician who works with numbers modulo 2, you might prefer $1 + 1 = 0$.

It turns out to be surprisingly useful, in certain circumstances, to work in a world where any number added to itself is zero. Such is the world of *nimbers*.

Nimbers are a lot like numbers, and in fact you can think of them as non-negative integers, written in binary. But their rules for addition and subtraction are different from the usual ones. In particular, you add nimbers *without carrying*.

Here's another way to say the same thing: To compute the ith bit (from the right) of the sum of a bunch of nimbers, you only need to know the ith bit of each nimber. If an odd number of these are 1's, the ith bit of the sum will be a 1; otherwise it will be a 0.

Let's try this. To avoid confusion, we'll write the *decimal* expression of a nimber with an overline, so, for example, $\overline{7} = 111$ and $\overline{10} = 1010$. To distinguish nimber addition from ordinary addition of binary numbers, we'll use the symbol $\oplus$ instead of $+$.

Thus, $\overline{7} \oplus \overline{10} = 111 \oplus 1010 = 1101 = \overline{13}$. Of course, $\overline{1} \oplus \overline{1} = 1 \oplus 1 = 0 = \overline{0}$ and in fact, as we said, $n \oplus n = 0$ for any n.

Nimber arithmetic, once you get used to it, is really rather nice. It obeys the usual laws of arithmetic—it's commutative, associative, and satisfies $x \oplus 0 = x$ for any x. It has additive inverses, as we saw: Any nimber is the additive inverse of itself! Thus, we have no use for minus signs in nimber arithmetic, and subtraction is the same as addition. If we have two nimbers m and n and we want to know which nimber to add to m to get n, why, it's the nimber $m \oplus n$, since $m \oplus (m \oplus n) = (m \oplus m) \oplus n = 0 \oplus n = n$.

If you've studied linear algebra, you may recognize nimbers as vectors over a two-element field. When you add vectors in any vector space, the coordinates are added independently; that's what is happening with our nimbers, since we don't carry. One difference is that for nimbers we don't bother to specify the dimension (number of coordinates) in advance. We can always add more zeroes to the left of a binary number if we need more coordinates (bits), without changing the number's value or its nimber properties.

Whence the name? As we shall see, nimbers are numbers that arise in the game of Nim, so the term "nimber" was irresistible to the himself irresistible mathematician John H. Conway. Understandably, "nimber" has largely replaced an old term, "Grundy number."

Consider:

Life Is a Bowl of Cherries

In front of you and your friend Amit are 4 bowls of cherries, containing, respectively, 5, 6, 7, and 8 cherries. You and Amit will alternately pick a bowl and take one or more cherries from it. If you go first, and you want to be sure to get the last cherry, how many cherries should you take—and from which bowl?

Solution: In the traditional game of Nim the bowls of cherries are replaced by piles of stones or (my preference) stacks of chips, and the problem is how best to play from *any* starting position.

We denote positions in Nim by listing the stack sizes from largest to smallest; for example, in the cherries puzzle, the given position is 8|7|6|5. The rules of Nim specify that from a given position you can reach any position with the same numbers except that one number has been either lowered or (if all the chips from that stack were removed) deleted. If, for example, you remove two cherries from the bowl of seven, you reach the new position 8|6|5|5.

The goal is to be the player to achieve the empty position. This can be done immediately if there is only one stack; just take all of it.

If there are just two stacks, and they are of different sizes, you (as first player) have the following winning strategy: Take chips from the big stack to reduce it to exactly the size of the smaller one, and repeat.

If there are more than two stacks, things start to get complicated. Nimbers make it simple—they enable you to win whenever you should, and often (in practice) when you shouldn't!

Here's how it works. We define the *nimsum* of a position to be the nimber-sum of the stack sizes. For example, the nimsum of the initial position in the cherries puzzle is

$$\overline{8} \oplus \overline{7} \oplus \overline{6} \oplus \overline{5} = 1000 \oplus 111 \oplus 110 \oplus 101 = 1100 = \overline{12}.$$

The fact is that you (as first player) can force a win from any position whose nimsum is not zero. How do you do it? Easy: Change it to a position whose nimsum *is* zero. Your opponent will have to change the position to one whose nimsum is again non-zero, and then you repeat. Eventually you get to the empty position, whose nimsum is of course zero, and you win!

There are some things to verify here. First of all, how do you know that your opponent can't change a zero-nimsum position to another zero-nimsum position? Well, your opponent (like you) must remove one or more chips from just one stack, so he reduces one of the stack-sizes by some positive number k. That will have the effect of nimber-subtracting k from the previous position nimsum, but remember, nimber subtraction is the same as nimber addition. So the nimsum of the new position will be $0 \oplus k = k \neq 0$.

A bit more subtle is showing that from any position whose nimsum is not zero, you can get to one whose nimsum *is* zero. Here's how you can do it. Suppose the current nimsum is s, and suppose the leftmost 1 in the binary representation of s is in the ith bit from the right (that is, the bit representing 2^{i-1}). Then there must be at least one stack whose size (s_j, say) also has a 1 as its ith bit—otherwise, the ith bit of s would have been zero. We claim we can remove some chips from that stack to change its size to $s_j \oplus s$. Doing so will change the position's nimsum to $s \oplus s = 0$. The only thing we need to check is that $s_j \oplus s$ is a smaller number than s_j (since we're not allowed to *add* chips to a stack). But that is easy because nimber-adding s to s_j will change the ith bit from 1 to 0, and won't affect any bits of s_j that are to the left of the ith. It follows that $s \oplus s_j$ is smaller than s_j, so reducing the s_j-stack to size $s \oplus s_j$ is a legal move.

Let's try that with the cherries. We saw that the nimsum of the given position was $\overline{12} = 1100$, whose leftmost 1 is in the fourth position from the right. Here there happens to be only one stack whose size (in binary) has a 1 in the fourth position from the right: the stack of size 8. Nimber-adding 1100 to 1000 gives $100 = \overline{4}$ so we reduce the number of cherries in the bowl of 8 to 4.

The resulting position, 7|6|5|4, has nimsum $111 \oplus 110 \oplus 101 \oplus 100 = 0$ and we have Amit where we want him.

In some positions there are several stacks whose sizes, in binary, have a 1 in the same position as the leftmost 1 in the position's nimsum; in such cases, there will be one winning move in each of those stacks. For our initial cherries position, there is only one such stack, thus only one winning move. After any other first move, it is Amit who can force a win.

Even within the specialized field of combinatorial games (where two players alternate and the first one who is unable to move loses), nimbers play a much bigger role than just their part in Nim. It turns out that any position in an "impartial" combinatorial game—that is, one in which both players have the same move choices from any given position—can be represented by a nimber! This is the content of the celebrated Sprague-Grundy Theorem, about which the interested reader can read in *Winning Ways* or any other book on combinatorial games.

In fact, here's another impartial game to which we can apply nimbers—but perhaps not in the way you would think.

Life Isn't a Bowl of Cherries?

In front of you and your friend Amit are 4 bowls of cherries, containing, respectively, 5, 6, 7, and 8 cherries. You and Amit will alternately pick a bowl and take one or more cherries from it. If you go first, and you want to be sure Amit gets the last cherry, how many cherries should you take—and from which bowl?

Solution: So, in effect, in this second puzzle you want your friend to win. (In such a situation you are said to be playing the *misère* version of the game.) It looks like since your objective is the opposite of the previous puzzle's, you should make any move *but* the one that reduces the number of cherries in the bowl of eight to four.

That's fine if Amit cooperates, that is, if he himself wants the last cherry, but it will not do if you want to *insist* that Amit get the last cherry. Surprisingly, your unique winning first move for the second puzzle is the same as the one for the first!

The philosophy behind this counterintuitive solution is that playing for zero-nimber positions allows you to control the game, until the crucial moment when you decide whom you want to get the last cherry. That moment comes when you first encounter a position containing *exactly one bowl with multiple cherries*, that is, in general Nim terms, a position in which every remaining stack except one contains only a single chip.

Note that such a position never has a zero nimber, because the big stack's size, in binary, has at least one 1 in it that can't be cancelled. Thus, when such a position arises, as it inevitably must, it will be you facing it, not your opponent. Now you need consider only two moves: Removing all of the big stack, and removing all but one chip from it. Either way there will only be "singleton" stacks left for your opponent. If there are an even number of them, you will get the last chip; if there are an odd number of them, he will. Since one of the above two moves will leave an even number of stacks and the other an odd number, you control who gets the last chip.

Let's do a game of cherries to see an example of how this works. In the following pictures, we zigzag through time with your move (hollow cherries are those to be removed) on the left and Amit's on the right. The nimsum for each position (prior to the indicated move) and the nimbers for each stack are indicated in binary.

At the last position shown, you will take all the cherries in the 4-cherry bowl if you want to end up with the last cherry; if you want Amit to get it, you'll take just three of them.

Whim-Nim

You and a friend, bored with Nim and Nim Misère, decide to play a variation in which at any point, either player may declare "Nim" or "Misère" instead of removing chips. This happens at most once in a game, and then of course the game proceeds normally according to that variation of Nim. (Taking the whole single remaining stack in an undeclared game *loses*, as your opponent can then declare "Nim" as his last move.)

What's the correct strategy for this game, which its inventor, the late John Horton Conway, called "Whim"?

Solution: Owing to our previous analyses, the game is effectively over once someone declares. So we concentrate on how to play the game pre-declaration.

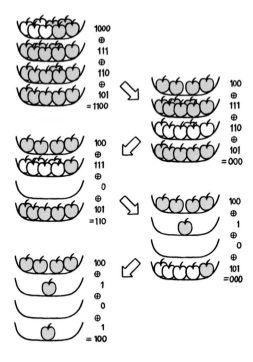

It's easy to see that if you're in a 0-position you win by simply declaring "Nim." Thus, you must not create a 0-position for your opponent. Moreover, you don't want to be handing your opponent nothing but singleton stacks, because then he will declare appropriately ("Nim" if the number of singletons is even, else "Misère") and win. On the other hand, you'd love to give your opponent a single stack of size 2. For that matter, you can give him any odd number of stacks of size 2, together with any even number of stacks of size 1 or 3. Then eventually he'll be the one to reduce to all singletons, and you will declare and win.

John Conway had a nice way to look at this: Imagine an additional stack, the "whim stack," of size 2; then play the normal Nim strategy of handing your opponent 0-positions. If the only way for you to achieve a 0-position is to remove the whim stack, you declare appropriately instead.

The only trouble is: What if achieving a 0-position requires removing just 1 chip from the whim stack? That won't happen if there's a stack of size 2 or another stack of size 3, and you're all set if all the stacks are singletons, so you can be in trouble only if there's one or more "tall" stack of size at least 4.

Conway's solution to that problem is to imagine that the whim stack has size only 1, when there's a tall stack in play. We need to check that the transition causes no problem, but that turns out to be easy. Your opponent will never be the one to kill the only tall stack because, being in a 0-position with the whim stack (thus a 1-position without), he can't be looking at just one tall stack. If it is you who are facing one tall stack, and leaving your opponent with a 0-position requires reducing that stack to size 0, 1, 2, or 3, then you can instead reduce it to 3, 2, 1, or 0 respectively and have a 0-position with whim stack of size 2.

In summary: If it is your move, and there's no tall stack, you can win as long as the position's Nim value, counting an imaginary stack of size 2, is not 0; if there is a tall stack, then the Nim value, counting an imaginary stack of size only 1, must not be 0. ♡

Let's put nimbers to work on hats.

Option Hats

One hundred prisoners are told that at midnight, in the dark, each will be fitted with a red or black hat according to a fair coinflip. The prisoners will be arranged in a circle and the lights turned on, enabling each prisoner to see every other prisoner's hat color. Once the lights are on, the prisoners will have no opportunity to signal to one another or to communicate in any way.

Each prisoner will then be taken aside and given the option of trying to guess whether his own hat is red or black, *but he may choose to pass*. The prisoners will all be freed if (1) at least one prisoner chooses to guess his hat color, and (2) every prisoner who chooses to guess guesses correctly.

As usual, the prisoners have a chance to devise a strategy before the game begins. Can they achieve a winning probability greater than 50%?

Solution: The prisoners have virtually no chance of winning if they all guess, so it looks reasonable to pick one prisoner (say, Joe) to guess and have everyone else decline; then, at least, the prisoners are all freed if Joe gets it right. In fact, it almost looks like there's a proof that they can't do better: Every individual guess has only a 50% probability of being right, so what's the advantage of *ever* having more than one person guess?

Suppose we fix some strategy S that tells each prisoner whether to guess black ("B"), guess red ("R"), or decline to guess ("D"), depending on the 99 hats he sees. Imagine that we write down a huge matrix indicating, for each of the 2^{100} possible hat configurations, what each prisoner does according to S. If each guess (B or R) is boldfaced if it is correct, then the matrix will have exactly the same number of boldfaced guesses as lightfaced; because, if prisoner i's guess is (say) a boldface R in a particular configuration, it will be a lightface R for the configuration that is exactly the same except that prisoner i's hat is changed to black.

In other words, no matter what the strategy, over all possible configurations half the guesses are correct. That suggests an idea: What if we crowd as many wrong guesses as possible into a few configurations, and try to put just one *correct* guess into each remaining configuration?

Let's do a little bit of arithmetic. Fixing a strategy S, let w be the number of winning configurations (that is, those in which all prisoners who guess are correct). Let x be the average number of (correct) guesses in winning configurations, and y the average number of wrong guesses in losing configurations. Then $wx \le (2^{100}-w)y$ with equality only when all guesses in the losing configurations are wrong. To maximize w, we would like to have every prisoner guess and all be wrong, in losing configurations; in winning configurations, to have just one prisoner guess and be correct. If we could achieve that for n prisoners, we would get a winning probability of $n/(n+1)$. We call this number the *count bound*. Achieving the count bound would be very good news indeed for our 100 prisoners, as they would all be freed with probability $100/101$—better than 99%.

But let's not get ahead of ourselves. For one thing, we could never get $100/101$ probability of winning for our 100 prisoners, because that would require that the number of losing configurations be exactly $1/101$ times the total number of configurations, and 2^{100} is not divisible by 101.

But $n+1$ does divide evenly into 2^n for $n = 3$; let's see if we can get probability $3/4$ of winning for three prisoners. We need two of the $2^3 = 8$ hat configurations to be losers; the logical ones to try are all red, and all black. If we instruct each prisoner to guess "red" when he sees two black hats and "black" when he sees two red, that will have the desired effect: They'll all guess wrong in the losing configurations.

Let's see: If we ask them to decline to guess otherwise, we get just what we want: In any configuration with both colors present, the prisoner with the odd hat will guess correctly while the other two decline!

Here's the matrix for this strategy:

	Configuration:			Action:		
Prisoner#:	01	10	11	01	10	11
	R	R	R	B	B	B
	R	R	B	D	D	**B**
	R	B	R	D	**B**	D
	R	B	B	**R**	D	D
	B	R	R	**B**	D	D
	B	R	B	D	**R**	D
	B	B	R	D	D	**R**
	B	B	B	R	R	R

Notice that we have numbered the prisoners in binary—actually, we have *nimbered* them. If we define the nimsum of a configuration to be the nimber-sum of the nimbers of the red-hatted prisoners, we find that the two losing configurations, RRR and BBB, are the ones with nimsum 0. Every other configuration has a non-zero nimsum which corresponds to some prisoner—namely, the one who guesses (and is right) when that configuration occurs.

Thought of this way, the above strategy generalizes to any number of prisoners that is one less than a power of two—that is, whenever $n = 2^k - 1$ for some positive integer k. The prisoners are nimbered from 1 to k, that is, from $000 \ldots 001$ to $111 \ldots 111$. The losing configurations are those for which the nimsum of the red-hatted prisoners is 0. The strategy is as follows: Each prisoner assumes that the configuration is not a losing one, and if that assumption tells him what his own hat color is, he guesses accordingly. Otherwise he declines.

What that amounts to is this. Prisoner i computes the nimber-sum m_i of the nimbers of the red-hatted prisoners he sees; we call m_i prisoner i's "personal" nimsum. If $m_i = 0$, prisoner i reasons that his hat must be red (else the nimsum of the whole configuration is 0, which every prisoner is assuming is not the case). If $m_i = i$, prisoner i's own nimber, then his hat must be black (else the nimsum of the whole configuration would be $i \oplus i = 0$). If m_i is anything other than 0 or i, prisoner i knows the configuration is not a losing one. That is great news, but it doesn't tell prisoner i what his hat color is, so he declines to guess.

This works because: (1) If the configuration has nimsum 0, every prisoner will guess, and they will all be wrong; (2) If the configuration's nimsum is $s \neq 0$, then prisoner s will guess correctly while every other prisoner declines. Thus the prisoners will achieve exactly the count bound for their success probability, namely $n/(n+1) = (2^k - 1)/2^k$.

We have exploited the following property of the set $\mathcal{L}$ of configurations with zero nimsum: Every configuration *not* in $\mathcal{L}$ can be changed to one in $\mathcal{L}$ by flipping the color of someone's hat. (Here, there is just one such hat: the one belonging to the prisoner whose nimber is the configuration's nimsum.) Interpreted as a set of binary strings of length n (red=1, black=0), $\mathcal{L}$ is said to be an *error-correcting code*, and the above bit-changing property makes this particular code (called the *Hamming code* of length n) a "perfect 1-error correcting code." The Hamming code and other error-correcting codes are objects of enormous practical as well as theoretical interest—we'll identify some more serious applications later.

What can the prisoners do if their number n is not one less than a power of two? Well, they can find the biggest k such that $2^k - 1 \leq n$, and have prisoners 1 through $2^k - 1$ run the above protocol among themselves, while the others decline to guess no matter what they see. This will not generally be optimal but it won't be far off; the probability of failure will always be less than $2/(n+1)$, thus small (for large n) and never as much as twice the failure probability associated with the count bound. Our group of 100 prisoners can have 63 prisoners run the Hamming code protocol among themselves while the remaining 37 stand by, declining to guess. That will give the prisoners a pretty impressive $63/64 = 98.4375\%$ probability of winning.

To get the precise optimal winning probability for values of n that aren't one less than a power of 2, the prisoners would need to find the smallest set $\mathcal{L}$ of configurations with the property that any configuration is at most one color-flip away from a configuration in $\mathcal{L}$. We know from the count bound that such a set, sometimes called a "radius-1 covering code," must be of size at least $2^n/(n+1)$ rounded up to an integer.

Let's see this reasoning in action for a couple of small numbers. For $n = 4$, the count bound says that we need $|\mathcal{L}| \geq 4$, and we already get winning probability $1 - 4/16 = 3/4$ by having three prisoners employ the Hamming code, so that solution is optimal. For $n = 5$, though, the count bound allows a radius-1 covering code of size $\lceil 32/5 \rceil = 6$. In fact, 6 is impossible but there's a radius-1 covering code of size 7:

$$\{00000, 00111, 01111, 10111, 11001, 11010, 11100\}.$$

With this the prisoners can get winning probability $1 - 7/64 = 57/64$, beating the Hamming code. Notice here that if, for example, the actual configuration is 11111 (all hats red), *two* prisoners—numbers 1 and 2, counting from the left—will guess (correctly) that their hats are red, the first to avoid the codeword 01111, the second to avoid 10111.

It turns out that for large n, radius-1 covering codes that are (relatively) close in size to $2^n/(n+1)$ can always be found. They are much harder to describe than Hamming codes, but if the number of prisoners is large and awkward (e.g., equal to $2^k - 2$ for some k), hard work will enable the prisoners to cut down their losing probability by a factor of almost 2.

Chessboard Guess

Troilus is engaged to marry Cressida but threatened with deportation, and Immigration is questioning the legitimacy of the proposed marriage. To test their connection, Troilus will be brought into a room containing a chessboard, one of whose squares is designated as special. On every square will be a coin, either heads-up or tails-up. Troilus gets to turn over one coin, after which he will be ejected from the room and Cressida brought in.

Cressida, after examining the chessboard, must guess the designated square. If she gets it wrong, Troilus is deported.

Can Troilus and Cressida save their marriage?

Solution: Troilus must somehow impart 6 bits worth of information (since the chessboard has $64 = 2^6$ squares) with his one turned coin. He does have 64 different things he can do, so this is not *a priori* impossible, but he and Cressida are going to need to be very efficient.

The natural approach—if you know about nimbers—is to assign a nimber of length 6 to each square of the chessboard. When Cressida comes in, she sums the nimbers of the squares whose coin shows heads; that gives a nimber and she now guesses that its associated square is the designated special square.

So Troilus needs to do is make sure the sum points to the right square, but that's easy. He computes the sum of the nimbers of the heads-squares of the board as presented to him, and compares it with the nimber of the designated special square. If their sum (i.e., difference!) is the nimber b, he turns the coin on square b. That will add (i.e., subtract) b from the sum and leave the sum pointing to the special square.

Majority Hats

One hundred prisoners are told that at midnight, in the dark, each will be fitted with a red or black hat according to a fair coinflip. The prisoners will be arranged in a circle and the lights turned on, enabling each prisoner to see every other prisoner's hat color. Once the lights are on, the prisoners will have no opportunity to signal to one another or to communicate in any way.

Each prisoner is then taken aside and *must* try to guess his own hat color. The prisoners will all be freed if a majority (here, at least 51) get it right.

As before, the prisoners have a chance to devise a strategy before the game begins. Can they achieve a winning probability greater than 50%? Would you believe 90%? How about 95%?

Solution: Too bad that a tie is insufficient to free the prisoners: They could force a tie (see "Half-Right Hats" in Chapter 2) by having 50 of them guess "black" when they see an even number of red hats and "red" otherwise, while the other 50 do the reverse. That guarantees that exactly 50 prisoners will guess correctly. Unfortunately, the way the problem is phrased, that guarantees that the prisoners lose. But maybe we can use this idea some other way.

Before we begin our analysis, though, it's worth checking one possibility that may already have occurred to you. What if each prisoner simply guesses that his hat color is the same as the majority of those he sees? This works as long as the number of red-hatted prisoners is not exactly 50; in the latter case, the plan comes spectacularly a cropper, with every prisoner guessing incorrectly. But the likelihood of exactly 50 red hats is not very high: There are 100 choose 50 ways this can happen, out of 2^{100} ways to pick the hat colors, so the probability of exactly 50 red hats is $\binom{100}{50}/2^{100} = 0.07958923738...$. Thus, this "preponderance" strategy frees the prisoners with probability better than 92%; not bad!

The fact that when it goes wrong, the preponderance scheme produces 100 wrong guess, is a feature, not a bug. What we have is a problem very similar to the one whose solution recalls the so-called St. Petersburg paradox. You are in Las Vegas, and determined to go home with more money than you came with; how can you maximize your probability of doing so, even though you expectation from gambling is always negative? The answer was discussed in Chapter 14 in connection with the puzzle *Roulette for the Unwary*. Set aside $2^k - 1$ dollars that you can afford to lose, with k as large an integer as possible; then make successive even-money (and as fair as possible) bets of sizes \$1, \$2, \$4, etc. until you win a bet. At that point you quit, \$1 ahead.

You end up *probably* ahead \$1 but possibly broke, and that's the hallmark of a good scheme: You either just barely accomplish your goal, or miss it by a mile.

The same applies to *Majority Hats*. Since every guess is a 50% proposition, the ideal scheme would have just two possible outcomes: Win barely (meaning, in this case, have exactly 51 correct guesses) or lose spectacularly (every guess wrong). If you could do that for 100 prisoners, their probability p of success would satisfy $51p = 49p + 100(1-p)$, since for any scheme, the expected number of right guesses has to equal the expected number of wrong guesses. That works out to $p = 100/102 > 98\%$, very impressive—if it were possible.

In general, with n prisoners, this calculation shows that when n is even their success probability cannot exceed $n/(n+2)$; when n is odd, and the prisoners need only one more right guess than wrong, they can hope for a success probability as high as $n/(n+1)$. We will show that this "count bound" is achievable when n is one or two less than a power of 2.

Suppose, for example, that there are $31 = 2^5 - 1$ prisoners, and let's assign them nimbers from 1 to 11111 (the binary representation of 31). We divide the prisoners into five groups according to the position of the leftmost 1 in their nimbers; thus Group 1 consists of just prisoner 1, Group 2 contains prisoners 2 and 3, and in general each group has twice as many prisoners as the previous group.

Prisoner 1 guesses "black." If he's right—and, remember, everyone else can see if he's going to be right or not—every other group behaves as if it's playing *Half-Right Hats*. The agreement could be, for example: All prisoners with even nimbers guess as if they know the total number of red hats *in their group* is even, while all those with odd nimbers assume the opposite. Then groups 2 through 5 will each be exactly half right, and the prisoners end up with a 16-to-15 majority correct.

If prisoner 1's hat is red, Group 2 plays the *Unanimous Hats* strategy—they *all* guess as if they knew the number of red hats in their group were even. If they're right (probability 1/2), then each remaining group splits itself as before and again the prisoners win by one.

If Group 2 is doomed to failure as well as Group 1—which, again, everyone else can see—then there are three wrong guesses coming, but they are rescued by Group 3 which now attempts to be all correct. This pattern continues and always produces a 16-15 majority unless every one of the five groups contains an odd number of red hats, which happens with probability only 1/32. So the 31 prisoners achieve success probability $1 - 1/32 = 96.875\%$. We know this is unbeatable because it achieves the ideal: The prisoners either enjoy a 1-guess victory, or everyone is wrong.

We can achieve this ideal (though probability of success will be only $1 - 1/16$) with 30 prisoners as well; here there's no group of size 1, and the group of size two plays the *Unanimous Hats* strategy to try to achieve the 2-guess advantage.

What about 100 prisoners? The count bound is never *exactly* achievable unless $n+1$ or $n+2$ is a power of 2, because there are 2^n hat configurations and thus any protocol's probability of winning must be a multiple of $1/2^n$. One hundred is not a good number, but we could let prisoners 1 through 62 use our "St. Petersburg protocol," while the remaining 38 prisoners just split themselves using the *Half-Right Hats* strategy, so as not to get in the way of the others. That frees the prisoners with probability $1 - 1/32 = 96.875\%$, pretty good. But it seems as if this strategy wastes 38 prisoners.

Let's go back to small numbers again. The smallest number that's *not* one or two less than a power of 2 is 4. Can 4 prisoners win with probability better than 1/2? Yes: Have prisoner 1 guess black. If he's wearing a black hat, the other three prisoners can run their St. Petersburg protocol to get 2-out-of-3 right with probability 3/4. If prisoner 1 has a red hat, the other three play the *Unanimous Hats* strategy and thus all three are right with probability 1/2. Altogether the 4 prisoners achieve their 2-guess majority with probability $\frac{1}{2} \cdot \frac{3}{4} + \frac{1}{2} \cdot \frac{1}{2} = 5/8$.

With this observation we can split our 100 prisoners into groups of sizes 4, 6, 12, 24, and 48, with just 6 left over. The group of 4 plays as above, succeeding with probability 5/8. As usual, if they are going to succeed (which everyone else can see), all other prisoners play the *Half-Right Hats* strategy. If they are going to fail, the next group (size 6) plays *Unanimous Hats*, and if they fail the group of 12 plays *Unanimous Hats*, etc. The final group of 6 always plays *Half-Right Hats*.

Result? The prisoners win unless the first group of 4 fails and the next four groups all suffer an odd number of red hats. This strategy thus frees the prisoners with probability $1 - \frac{3}{8} \cdot \frac{1}{2} \cdot \frac{1}{2} \cdot \frac{1}{2} \cdot \frac{1}{2} = 1 - 3/128 = 97.65625\%$.

Can we tweak this any further? Yes, by recursively computing the best strategies of this kind for achieving k correct guesses out of n, for all n and k up to 100. This "dynamic programming" approach can reach a success probability of 1129480068741774213/ 1152921504606846976, about 97.96677954%; very close indeed to the count bound of 100/102 = 98.039215686%.

That last solution didn't end up making much use of nimbers, but the next puzzle brings us back to the Hamming code.

Fifteen Bits and a Spy

A spy's only chance to communicate with her control lies in the daily broadcast of 15 zeros and ones by a local radio station. She does not know how the bits are chosen, but each day she has the opportunity to *alter* any one bit, changing it from a 0 to a 1 or vice-versa.

How much daily information can she communicate?

Solution: Since there are 16 things the spy can do (change any bit or none), she can *in principle* communicate as much as four bits of information each day to her control. But how?

The answer is easy once you have nimbers in your arsenal. The spy and her control assign the 4-bit nimber corresponding to the number

k to the kth bit of the broadcast, and their "message" is defined to be the sum of the nimbers of the ones in the broadcast—the "nimsum" of the broadcast.

The claim is that the spy can send any of the 16 possible nimsums at will, thus achieving a full four bits of communication. Suppose she wishes to send the nimsum n but the sum of the nimbers corresponding to the ones in the station's intended broadcast is $m \neq n$. Then she flips the $n \oplus m$th bit. The resulting nimsum is $m \oplus (n \oplus m) = n$. Of course, of the broadcast's nimsum already happens to be n, she leaves it alone. ♡

We will finish this chapter with a brief introduction to the *raison d'etre* of error-correcting codes: sending messages over an imperfect channel.

Almost every means of communication suffers from some probability of error, whether it be voice, telegraph, radio, TV, typed letters, hand-written words, photographs, or bit streams over the internet. It's pretty amazing that you can send a billion bits to your friend in an email attachment, and not a single one is accidentally flipped!

This comes about through the use of error-correcting codes. If bits had to be sent so slowly that every one was guaranteed to arrive intact with probability $1 - 1/10^9$, we could forget about streaming movies. No, the individual electical, radio and light pulses that zip around the internet are imperfect and are mis-sent or mis-read with alarming frequency. What saves us is that information is coded in such a way that errors can be detected and corrected.

The simplest error-detection method is one you may already know about, and is typical for, say, credit card numbers. By choosing the last "check" digit correctly, your credit card company can ensure that the sum of the digits in your credit card number is (say) a multiple of 10. Then, if you mis-enter *one* digit of your credit card during a purchase, the digit-sum will be thrown off, and a warning will appear.

Error-*correction* is trickier but you've seen it above in the person of the Hamming code. Simplest case: You want to send a string of bits and be confident that they all arrive intact, despite some small probability that any given bit will be corrupted. Send each bit three times! Only in the unlikely event that two bits in the same group of three are corrupted will your message get unfixably mangled.

In fact, this is the Hamming code of length 3 that we used when three prisoners were faced with either of our hat problems. The codewords were 000 (all hats black) and 111 (all hats red).

Of course, tripling the lengths of our messages is rather a heavy price to pay. If we are confident that the likelihood of more than one error in a given group of $2^k - 1$ bits is small, we can use the Hamming code of length $n = 2^k - 1$, as follows.

Suppose the message to be sent is (or could be represented by) a long string of N bits. Break up the message into "blocks" of size $m = n - k$. Before transmission, each block is replaced by a codeword of length n; at the receiving end, these are "decoded," and the original blocks recovered correctly, provided the codeword was received with at most one flipped bit.

We can state this as a theorem:

Theorem. *At the cost of adding k bits to the transmission, a block of bits of length $n-k$, where $n = 2^k - 1$, can be sent in such a way that as long as at most one bit is corrupted, the original block can be faithfully recovered at the receiving end.*

Proof. We will do a little more than is necessary for the proof, in order to show that all this can be done with very little computational effort by transmitter and receiver.

First, we nimber the bits in the n-bit codeword, using k-bit non-zero nimbers from $000 \ldots 001$ to $111 \ldots 111$. There are many ways to do this, but it's especially convenient to assign nimbers $2^0 = 000 \ldots 001$, $2^1 = 000 \ldots 010$, $2^2 = 000 \ldots 100, \ldots, 2^{n-1} = 100 \ldots 000$ to the rightmost k positions, reading from right to left.

The same nimbering is used for a block of length $n-k$, regarded as the leftmost $n-k$ bits of a codeword. For example, if $k = 3$, the bits of our 7-bit codewords, and the corresponding 4-bit blocks, could be nimbered as in the following figure:

The *nimsum* of a block or codeword is the nimber sum of the positions in which a "1" is found. The nimsum of the codeword in the figure is $\bar{0} = 000$; indeed, we define our codewords to be the exactly the n-bit strings with zero nimsum. The nimsum of the block in the figure happens to be $\bar{3} = 011$.

The easiest way to transform a block into a codeword is simply to append k "check bits" to the right-hand end of the block, in such a way that the nimsum of the resulting n-bit string is zero. That is particularly easy to do with the nimbering we recommend, because the check bits will then be exactly the nimsum of the block!

The reason for this pleasant coincidence is that by choosing the "basis" nimbering of the check bits, we ensure that the nimsum of the check bits is the string of check bits itself. Thus, if the nimsum of the block is s, when we add the check bits the new nimsum will be $s \oplus s = 0$.

In this way the "encoding" part of our Hamming code is made as easy as possible; what about the "decoding" part? No problem. As receiver of an n-bit string, we compute its nimsum; if we get 0, we assume it was transmitted correctly and accept its leftmost $n-k$ bits as the intended message block.

If we get nimsum $s \neq 0$, we assume that one, and only one, error has been made. If so that error must be in the bit nimbered s; so we flip that bit, and only now take the leftmost $n-k$ bits as the intended message block. (The transmission error could have occurred in a check bit, in which case the extracted message will not change.) $\quad\square$

We have seen that the property we needed from codewords in the hat puzzles—that each non-codeword is one bit-flip away from a unique codeword—is used to correct transmission errors. The Hamming code is one of many codes, some of which detect or correct many errors, and have lots of other neat properties. (Example: Reed-Muller codes, Reed-Solomon codes, and Golay codes were used to transmit photos of Mars and Jupiter from various Mariner and Voyager spacecraft.) Error-correcting codes have played a major role in the theory of computing, as well, leading (among other things) to miraculous strings called "probabilistically checkable proofs" and thence to huge advances in our understanding of how well computers can do approximate calculations. And error-correcting codes will need to play a major role if quantum computing is to become practical.

18. Unlimited Potentials

Puzzles often involve *processes*—systems that evolve in time randomly, deterministically, or under your control. Popular questions include: Can the system reach a certain state? Does it *always* reach a certain state? If so, how long does it take?

The key to such a puzzle may be to identify some parameter of the current state of the system that measures progress toward your goal. We call such a parameter a *potential*.

For example, suppose you can show that the potential never increases, but starts below the potential of the desired end state. Then you have proved that the end state is unreachable.

On the other hand, perhaps the end state has potential zero, the beginning state has positive potential $p > 0$, and every step of the process reduces the potential by at least some fixed amount $\varepsilon > 0$. Then you *must* reach the end state, and moreover must do so in time at most p/ε.

If the loss per step in potential is not bounded, it could be that the potential never reaches zero and may not even approach zero. But if the system has only finitely many states, you don't need to worry about this possibility.

Can you find an appropriate potential for the following puzzle?

Signs in an Array

Suppose that you are given an $m \times n$ array of real numbers and permitted, at any time, to flip the signs of all the numbers in any row or column. Can you always arrange matters so that all the row sums and column sums are non-negative?

Solution: We note first that there are only finitely many states here: Each number in the array has at most two states (positive or negative) so there can't be more than 2^{mn} possible states of the array.

Actually, there's a better bound: If a line (a row or column) is flipped twice, it's as if it had never been flipped. Thus the state of each line is either "flipped an odd number of times" or "flipped an even number of times," and it follows that the number of states of the array can't exceed 2^{m+n}.

Back to the problem at hand. The obvious algorithm is to find a line with negative sum and flip it; does this eventually work? After all, flipping a row (say) with negative sum could result in changing several column sums from positive to negative. Thus, the number of lines with negative sum would not work as a potential.

Instead, let's take the sum s of all the entries in the array! That number goes up by $2c$ if you flip a line whose sum was $-c$. Since there are only finitely many possible states, the potential cannot keep going up, and you must reach a point where no line sum is negative.

Fair warning: From here on, it's going to be trickier to spot the right potential.

Righting the Pancakes

An underchef of the great and persnickety Chef Bouillon has made a stack of pancakes, but, alas, some of them are upside-down—that is, according to the great Chef, they don't have their best side up. The underchef wants to fix the problem as follows: He finds a contiguous substack of pancakes (of size at least one) with the property that both the top pancake and the bottom pancake of the substack are upside-down. Then he removes the substack, flips it as a block and slips it back into the big stack in the same place.

Prove that this procedure, no matter which upside-down pancakes are chosen, will eventually result in all the pancakes being right side up.

Solution: The idea is that even though a flip might result in turning more pancakes upside-down, it rights some pancake that's relatively high in the stack. Let's penalize the underchef 2^k points for an upside-down pancake k pancakes from the bottom, and keep track of the total of the penalties at any given time.

Every flip decreases the total penalty, because the penalty of the target pancake exceeds the sum of all possible penalties below it. Thus the penalty total must reach 0, at which time the task is finished.

An equivalent way of thinking about this potential function is to code a stack of pancakes by a binary number whose left-most digit is 1 if the top pancake is upside-down, and 0 otherwise. The next-to-top pancake determines the next digit, and so forth. Flipping always decreases the code until the code reaches 0.

Breaking a Chocolate Bar

You have a rectangular chocolate bar scored in a 6×4 grid of squares, and you wish to break up the bar into its constituent squares. At each step, you may pick up one piece and break it along any of its marked vertical or horizontal lines.

For example, you can break three times to form 4 rows of 6 squares each, then break each row five times into its constituent squares, accomplishing the desired task in $3 + 4 \times 5 = 23$ breaks.

Can you do better?

Solution: This delightful problem has embarrassed some brilliant professional mathematicians, as well as many amateurs, all looking for geometrical insight.

But geometry is a red herring: All you need for a potential function is the number of pieces! You start with one piece, increase the number of pieces by one every time you break, and end with 24 pieces. So there are 23 breaks no matter how you do it.

Red Points and Blue

Given n red points and n blue points on the plane, no three on a line, prove that there is a matching between them so that line segments from each red point to its corresponding blue point do not cross.

Solution: Suppose you pair the red and blue points up arbitrarily and draw the connecting line segments. If the line segment connecting r_1 and b_1 intersects the segment connecting r_2 and b_2, then these segments are the diagonals of a quadrilateral and by the triangle inequality, the sum of the lengths of the *non-crossing* segments from r_1 to b_2 and from r_2 to b_1 is less then sum of the lengths of the diagonals.

The trouble is, if you re-pair the points as suggested above, you may have created new intersections between these segments and others. So "number of intersections" is not the right potential; how about "total length of line segments?"

That works beautifully! Just take any matching that minimizes the sum of the lengths of its line segments, and by the above argument, it contains no crossings.

Bacteria on the Plane

Suppose the world begins with a single bacterium at the origin of the infinite plane grid. When it divides its two successors move one vertex north and one vertex east, so that there are now two bacteria, one at (0,1) and one at (1,0). Bacteria continue to divide, each time with one successor moving north and one east, provided both of those points are unoccupied.

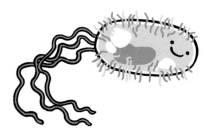

Show that no matter how long this process continues, there's always a bacterium inside the circle of radius 3 about the origin.

Solution: It's worth trying this to see how things get clogged up near the origin, preventing that area from clearing. How can we demonstrate this using a potential function?

Since each bacterium divides into two, it's natural to give each child half the potential of its parent. We can do this by penalizing its potential by a factor of 2 for each additional edge in its shortest path to the origin.

This can be done by assigning value 2^{-x-y} to a bacterium at (x,y), so that the total assigned value begins at 1 and never changes.

The total potential of all bacteria on the half line $y = 0$ (i.e., the positive X-axis) cannot exceed $1 + 1/2 + 1/4 + \cdots = 2$, nor can the potential of the points on the half-line $y = 1$ exceed $1/2 + 1/4 + 1/8 + \cdots = 1$; continuing in this manner, we see that even if the whole north-east quadrant were full of bacteria, the total potential would only add up to 4.

But the potentials of bacteria at the nine non-negative integer points inside the circle $x^2 + y^2 = 9$ already add up to $49/16$, more than 3. Thus the potentials of all the bacteria not inside this circle add up to less than 1, and it follows that there must be bacteria left inside the circle no matter how long the process continues.

Pegs on the Half-Plane

Each grid point on the XY plane on or below the X-axis is occupied by a peg. At any time, a peg can be made to jump over a neighbor peg (horizontally, vertically, or diagonally adjacent) and onto the next grid point in line, provided that point was unoccupied. The jumped peg is then removed.

Can you get a peg arbitrarily far above the X-axis?

Solution: The difficulty is that as pegs rise higher, grid points beneath them are denuded. What is needed is a parameter P which is rewarded by highly placed pegs, but compensatingly punished for holes left behind. A natural choice would be a sum over all pegs of some function of the peg's position. Since there are infinitely many pegs, we must be careful to ensure that the sum converges.

We could, for example, assign value r^y to a peg on $(0, y)$, where r is some real number greater than 1, so that the values of the pegs on the lower Y-axis sum to the finite number $\sum_{y=-\infty}^{0} r^y = r/(r-1)$. Values on adjacent columns will have to be reduced, though, to keep the sum over the whole plane finite; if we cut by a factor of r for each step away from the Y-axis, we get a weight of $r^{y-|x|}$ for the peg at

(x, y), and a total weight of

$$\frac{r}{r-1} + \frac{1}{r-1} + \frac{1}{r-1} + \frac{1}{r(r-1)} + \frac{1}{r(r-1)} + \cdots = \frac{r^2 + r}{(r-1)^2} < \infty$$

for the initial position.

If a jump is executed, then at best (when the jump is diagonally upward and toward the Y-axis), the gain to P is vr^4 and the loss $v + vr^2$, where v is the previous value of the jumping peg. As long as r is at most the square root of the "golden ratio" $\theta = (1+\sqrt{5})/2 \approx 1.618$, which satisfies $\theta^2 = \theta + 1$, this gain can never be positive.

If we go ahead and assign $r = \sqrt{\theta}$, then the initial value of P works out to about 39.0576; but the value of a peg at the point $(0, 16)$ is $\theta^8 \approx 46.9788$ *by itself*. Since we cannot increase P, it follows that we cannot get a peg to the point $(0, 16)$. But if we could get a peg to *any* point on or above the line $y = 16$, then we could get one to $(0, 16)$ by stopping when some peg reaches a point $(x, 16)$, then redoing the whole algorithm shifted left or right by $|x|$. ♡

Pegs in a Square

Suppose we begin with n^2 pegs on a plane grid, one peg occupying each vertex of an n-vertex by n-vertex square. Pegs jump only horizontally or vertically, by passing over a neighboring peg and into an unoccupied vertex; the jumped peg is then removed. The goal is to reduce the n^2 pegs to only 1.

Prove that if n is a multiple of 3, it can't be done!

Solution: Color the points (x, y) of the grid red if neither x nor y is a multiple of 3, otherwise white. This leaves a regular pattern of 2×2 red squares (as in the figure).

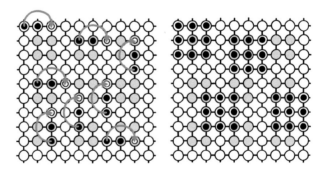

If two pegs are (orthogonally) adjacent on the grid, both on red points or both on white, the peg remaining after the jump will be on white. If one is on red and the other on white, however, the peg remaining after the jump will be on red. It follows that if you start with any configuration having an even number of pegs on red squares, then this property will persist forever regardless of what jumps are made.

It is easy to check that a 3×3 square of pegs, no matter where it is placed on the plane grid, hits an even number of red points. Since an $n \times n$ square with n a multiple of 3 is composed of such squares, it too will always hit an even number of red points. Suppose, however, that it were possible to reduce such a square to a single peg. Then we could shift the original square so that the surviving peg ended up on a red point, and this contradiction concludes the proof.

It is routine, but not particularly easy or enlightening, to show that if n is *not* a multiple of 3, you *can* reduce an $n \times n$ square to a single peg.

First-Grade Division

On the first day of class Miss Feldman divides her first-grade class into k working groups. On the second day, she picks the working groups a different way, this time ending up with $k+1$ of them.

Show that there are at least two kids who are in smaller groups on the second day than they were on the first day.

Solution: Most people find this simple-seeming combinatorial statement frustratingly difficult to prove, unless they hit upon the "magic" potential. Think of each working group as a task force for a job that requires a total of one unit of effort; then, assume that each child in a working group of size s contributes an amount $1/s$ worth of effort.

The sum of these "effort" fractions is evidently k on the first day, $k+1$ on the second. A given child's contribution to these numbers cannot increase by as much as 1 from one day to the next (since his contribution is greater than 0 on day 1, and at most 1 on day 2). So at least two kids must be contributing more to the total effort on their second day—meaning that both have moved to smaller groups.

Infected Checkerboard

An infection spreads among the squares of an $n \times n$ checkerboard in the following manner: If a square has two or more infected neighbors, then it becomes infected itself. (Neighbors are orthogonal only, so each square has at most four neighbors.)

For example, suppose that we begin with all n squares on the main diagonal infected. Then the infection will spread to neighboring diagonals and eventually to the whole board.

Prove that you *cannot* infect the whole board if you begin with fewer than n infected squares.

Solution: This puzzle was presented to me (by NYU's Joel Spencer) as having a "one-word solution." That's an exaggeration, perhaps, but not a huge one.

At first it might seem that to infect the whole board, you need to start with a sick square in every row (and in every column). That would imply the conclusion, but it's not true. For example, sick squares in alternating positions in the leftmost column and bottom row can infect the whole board.

What we really need is a potential function, and the one that works like a charm—and the one word Spencer had in mind—is "perimeter."

The perimeter of the infected region is just the total length of its boundary. We may as well assign each grid edge length 1, so the perimeter could as well be defined as the number of edges that have a sick square on one side and either a well square or nothing on the other side.

Here is the key observation: When a square becomes infected (by two or more neighbors), the perimeter cannot increase! Indeed, it is easy to check that when just two neighbors administer the dose, the perimeter remains the same; if the well square had three or four sick neighbors, the perimeter actually goes down.

If the whole board gets infected, the final perimeter is $4n$; since the perimeter never increased, the initial perimeter must have been at least $4n$. But to start with perimeter $4n$ there must have at least n sick squares.

Impressionable Thinkers

The citizens of Floptown meet each week to talk about town politics, and in particular whether or not to support the building of a new shopping mall downtown. During the meetings each citizen talks to his friends—of whom there are always an odd number, for some reason—and the next day, changes (if necessary) his opinion regarding the mall so as to conform to the opinion of the majority of his friends.

Prove that eventually, the opinions held every *other* week will be the same.

Solution: To prove that the opinions eventually either become fixed or cycle every other week, think of each acquaintanceship between citizens as a pair of arrows, one in each direction. Let us say that an arrow is currently "bad" if the opinion of the citizen at its tail differs from the *next week's* opinion of the citizen at its head.

Consider the arrows pointing out from citizen Clyde at week $t-1$, during which (say) Clyde is pro-shopping mall. Suppose that m of these are bad. If Clyde is still (or again) pro on week $t+1$, then the number n of bad arrows pointing *toward* Clyde at week t will be exactly m.

If, however, Clyde is anti-shopping mall on week $t+1$, n will be strictly less than m since it must have been that the majority of his friends were anti on week t. Therefore a majority of the arrows out of Clyde were bad on week $t-1$ and now only a minority of the arrows into Clyde on week t are bad.

The same observations hold, of course, if Clyde is anti on week $t-1$.

But, here's the thing: *Every* arrow is out of *someone* on week $t-1$, and into someone on week t. Thus, the total number of bad arrows cannot rise between weeks $t-1$ and t and, in fact, will go strictly down unless every citizen had the same opinion on week $t-1$ as on week $t+1$.

But, of course, the total number of bad arrows on a given week cannot go down forever and must eventually reach some number k from which it never descends. At that point, every citizen will either stick with his opinion forever or flop back and forth every week.

Frames on a Chessboard

You have an ordinary 8×8 chessboard with red and black squares. A genie gives you two "magic frames," one 2×2 and one 3×3. When you place one of these frames neatly on the chessboard, the 4 or 9 squares they enclose instantly flip their colors.

Can you reach all 2^{64} possible color configurations?

Solution: There are $2^{49} \times 2^{36}$ ways to select frame locations, enough in theory to get all 2^{64} color configurations. But of course there may be many ways to achieve a given configuration, and maybe we can find a potential that shows some configurations are unobtainable.

In fact, you might recall that the puzzle *Odd Light Flips*, of Chapter 15, offered a useful criterion: the ability to change the color (or bulb status) of an odd number of items in any set. Is there a troublesome set of chessboard squares?

A bit of experimenting might lead you to the following set. Label squares as "special" if they are in row 3 or 6 or file c or f but not both (see figure below), Then every 2×2 or 3×3 frame covers an even number of special squares.

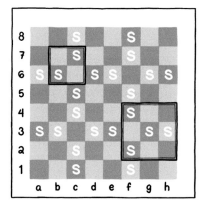

If the board starts with an even number of red special squares (as the usual chessboard does), you cannot reach any configuration where the number of red special squares is odd, and vice-versa. Either way, you cannot reach all color configurations.

Bugs on a Polyhedron

Associated with each face of a solid convex polyhedron is a bug which crawls along the perimeter of the face, at varying speed, but only in the clockwise direction. Prove that no schedule will permit all the bugs to circumnavigate their faces and return to their initial positions without incurring a collision.

Solution: Let us observe first that we may assume no bug begins on a vertex (by advancing or retarding bugs slightly). We may also assume that the bugs move one at a time, crossing a vertex each time.

At any time, we may draw an imaginary arrow from the center of each face F, through F's bug, to the center of the face on the other side of the bug. If we start at any face and follow these arrows, we must eventually hit some face a second time, completing a cycle of arrows on the polyhedron.

This cycle divides the surface of the polyhedron into two portions; let us define the "inside" of the cycle to be that portion surrounded clockwise by the cycle. Let P be the number of vertices of the polyhedron inside the cycle.

Initially, P could be anything from 0 to all of the polyhedron's vertices; the extremes occur if there are two bugs on the same edge, causing a cycle of length 2. In the $P = 0$ case, the two bugs are facing each other, and doomed to collide.

When a bug on the cycle moves to its next edge, the arrow through it rotates clockwise. The vertex through which it passed, previously on the inside of the cycle, is now outside; other vertices may also have passed from inside to outside the cycle, but there is no way for a vertex to move *inside*. To see this, note that the new arrow now points inside the cycle. The chain of arrows emanating from its head has no way to escape the cycle so must hit the tail of some cycle arrow, creating a new cycle with smaller interior. In particular, P has now dropped by at least 1.

Since we can never restore P to its starting value, there is nothing to do but hope that the bugs are carrying collision insurance. ♡

Bugs on a Line

Each positive integer on the number line is equipped with a green, yellow, or red light. A bug is dropped on "1" and obeys the following rules at all times: If it sees a green light, it turns the light yellow and moves one step to the right; if it sees a yellow light, it turns the light red and moves one step to the right; if it sees a red light, it turns the light green and moves one step to the *left*.

Eventually, the bug will fall off the line to the left, or run out to infinity on the right. A second bug is then dropped on "1," again following the traffic lights starting from the state the last bug left them in; then, a third bug makes the trip.

Prove that if the second bug falls off to the left, the third will march off to infinity on the right.

Solution: We first need to convince ourselves that the bug will *either* fall of to the left, or go to infinity on the right; it cannot wander forever. To do so, it would have to visit some numbers infinitely often; let n be the least of those numbers, but now observe that every third visit to n will find it red and thus will incur a visit to $n - 1$, contradicting the assumption that $n - 1$ was visited only finitely often.

With that out of the way, it will be useful to think of a green light as the digit 0, red as 1, and yellow, perversely, as the "digit" $\frac{1}{2}$. The configuration of lights can then be thought of as a number between 0 and 1 written out in binary,

$$x = .x_1 x_2 x_3 \ldots,$$

where, numerically,

$$x = x_1 \cdot \left(\frac{1}{2}\right)^1 + x_2 \cdot \left(\frac{1}{2}\right)^2 + \cdots .$$

Think of the bug at i as an additional "1" in the ith position, defining

$$y = x + \left(\frac{1}{2}\right)^i .$$

The point of this exercise is that y is an *invariant*, that is, it does not change as the bug moves. When the bug moves to the right from point i, the digit upon which it sat goes up in value by $\frac{1}{2}$; therefore, x increases by $\left(\frac{1}{2}\right)^{i+1}$, but the bug's own value diminishes by the same amount. If the bug moves to the left from i, it gains in value by $\left(\frac{1}{2}\right)^i$, but x decreases by a whole digit in the ith place to compensate.

The exception is when the bug falls off to the left, in which case both x and the bug's own value drop by $\frac{1}{2}$, for a loss of 1 overall. When the next bug is added, y goes up by $\frac{1}{2}$. To put it another way, the value of x goes up by $\frac{1}{2}$ if a bug is introduced and disappears to the right; and drops by $\frac{1}{2}$ if a bug is introduced and falls off to the left.

Of course, x must always lie in the unit interval. If its initial value lies strictly between 0 and $\frac{1}{2}$, the bugs must alternate right, left, right, left; if between $\frac{1}{2}$ and 1, the alternation will be left, right, left, right.

The remaining cases can be checked by hand. If $x = 1$ initially (all points red) the first bug turns point 1 green and drops off to the left; the second wiggles off to infinity leaving all points red again, so the alternation is left, right, left, right. If $x = 0$ initially (all points green), the bugs will begin right, right (as the points change to all yellow, then all red), and then left, right, left, right as before.

The $x = \frac{1}{2}$ case is the most interesting because there are several ways to represent $\frac{1}{2}$ in our modified binary system: x can be all $\frac{1}{2}$'s, or it can start with any finite number (including 0) of $\frac{1}{2}$'s, followed either by $0111\ldots$ or $1000\ldots$. In the first case, the leadoff bug turns all the yellows to red as it zooms off to the right; thus, we get a right, left, right, left alternation. The second case is similar, the first bug wiggling off to the right, but again leaving all points red behind it. In the third case, the bug changes the yellows to red as it marches out, but when it reaches the red point, it reverses and heads left, turning reds to green on its way to dropping off the left end. Thereafter, we are in the $x = 0$ case, so the final pattern is left, right, right, left, right, left, right.

Checking back all the cases, we see that indeed, whenever the second bug went left, the third went right. ♡

Flipping the Pentagon

The vertices of a pentagon are labeled with integers, the sum of which is positive. At any time, you may change the sign of a negative label, but then the new value is subtracted from both neighbors' values so as to maintain the same sum.

Prove that, inevitably, no matter which negative labels are flipped, the process will terminate after finitely many flips, with all values non-negative.

Solution: Even with experimentation (highly recommended), it's a little hard to see what can be used as a potential. The number of negative entries doesn't necessarily go down, nor does the size of the largest negative entry.

Similarly, neither the sum of the absolute values of the differences between adjacent numbers, nor the sum of the squares of the differences, seems to go reliably down (or up).

But it turns out, if you take the sum of the squares of the differences between *non*-adjacent numbers, it works!

Suppose the numbers are, reading around the pentagon, a, b, c, d and e. Then the sum in question is $s = (a-c)^2 + (b-d)^2 + (c-e)^2 + (d-a)^2 + (e-b)^2$. Suppose b is negative and we flip it, replacing b by $-b$, a by $a+b$, and c by $c+b$. Then the new s is equal to the old s plus $2b(a+b+c+d+e)$, and since b is negative and $a+b+c+d+e$ is positive, s goes down. And since s is an integer, it goes down by at least 1.

But s is non-negative (being a sum of squares) so it can't go below zero. That means we must reach a point where there's no negative entry left to flip, and we are done.

But there's an even better potential—better because it works for all polygons, not just the pentagon. It's the sum, over all sets of consecutive vertices, of the absolute value of the sum of the numbers in the set.

An even more remarkable solution was later found by Princeton computer scientist Bernard Chazelle. Construct a doubly-infinite sequence of numbers whose successive differences are the pentagon (or polygon) values, reading clockwise around the figure. If the pentagon's labels began as 1, 2, −2, −3, 3, the sequence might be

$$\ldots, -1, 0, 2, 0, -3, 0, 1, 3, 1, -2, 1, 2, 4, 2, -1, 2, 3, 5, \ldots .$$

Notice that the sequence climbs gradually (since the sum of the numbers around the polygon is positive) but not steadily (since some of the numbers around the polygon are negative).

Now the key observation: Flipping a vertex has the effect of transposing pairs of entries of the above sequence that were in the wrong order—in other words, the flipping process turns into a sorting process for our infinite sequence!

For example, if we flip the −2 on our pentagon to get 1, 0, 2, −5, 3, the sequence changes to

$$\ldots, -1, 0, 0, 2, -3, 0, 1, 1, 3, -2, 1, 2, 2, 4, -1, 2, 3, 3, \ldots .$$

It turns out to be easy to find a potential for the sorting progress that declines by exactly one per turn, and to conclude that not only does the process always terminate, but it terminates in the same number of steps and in the same final configuration, no matter which labels you flip!

Picking the Athletic Committee

The Athletic Committee is a popular service option among the faculty of Quincunx University, because while you are on it, you get free tickets to the university's sports events. In an effort to keep the committee from becoming cliquish, the university specifies that no one with three or more friends on the committee may serve on the committee—but, in compensation, if you're not on the committee but have three or more friends on it, you can get free tickets to any athletic event of your choice.

To keep everyone happy, it is therefore desirable to construct the committee in such a way that even though no one on it has three or more friends on it, everyone *not* on the committee *does* have three or more friends on it.

Can this always be arranged?

Solution: What's the dumbest way to try to find a good committee? How about this: Start with an arbitrary set of faculty members, as a prospective Athletic Committee. Oops, Fred is on the committee and already has three friends on the committee? Throw Fred out. Mona is not on the committee, but has fewer than three friends on it? Put Mona on. Continue fixing in this haphazard manner.

Now, why in the world would you expect this to work? Clearly, the above actions could make things worse; for example, throwing Fred off the committee might create many more Monas; maybe we should have thrown off one of Fred's on-committee friends instead. So there doesn't seem to be anything to prevent cycling back to the same bad committee. Moreover, even if you don't cycle back, there are exponentially many possible committees and you can't afford to consider every one. Suppose there are 100 faculty members in all; then the number of possible committees is $2^{100} > 10^{30}$ which, even if you spent only a nanosecond considering each committee, would take a thousand times longer than all the time that has passed since the Big Bang.

But if you try it, you will find that after shockingly few corrections, you end up with a valid committee. And this happens in situations where there is only one valid committee, as well as where there are many.

How can this be? Sounds like there must be some potential function at work here, something that is improving each time you throw someone off or add someone to the current prospective committee. Let's see: When you throw someone off, you destroy at least three on-committee friendships; when you put someone on, you add at most two. Let $F(t)$ be the number of friendships on the committee *minus $2\frac{1}{2}$ times the number of people on the committee* at time t. Then when Fred is thrown off, $F(t)$ goes down by at least $\frac{1}{2}$. When Mona is put on, $F(t)$ *again* goes down by at least $\frac{1}{2}$. But $F(0)$ can't be more than $(100 \times 99)/2 - 250 = 245$ and $F(t)$ can never dip below -250, so there can't be more than $2 \times (245 - (-250)) = 990$ steps total. (A computer scientist would say that the number of steps in the process is at worst quadratic in the number of faculty members.)

In practice, the number of steps is so small that if there are 100 faculty members and you start with (say) the empty committee, you will reach a solution easily by hand. Of course, you'll need access to the friendship graph, so you might need to do some advance polling. It might be interesting to see who claims friendship with whom that isn't reciprocated.

For the last puzzle, we use the mother of all potentials—potential energy!

Bulgarian Solitaire

Fifty-five chips are organized into some number of stacks, of arbitrary heights, on a table. At each tick of a clock, one chip is removed from each stack and those collected chips are used to create a new stack.

What eventually happens?

Solution: It's worth actually trying this—for example, with a deck of cards (which often comes with two jokers and an order card, making 55 cards in all.) If all you have is 52 cards, remove seven and try the experiment with 45 cards.

You will discover, if you have sufficient patience, that you eventually reach the "staircase" configuration where the stack heights are 10, 9, 8, 7, 6, 5, 4, 3, 2, and 1; and it's easy to see that this configuration never changes. Proving that the staircase is always reached is a typical job for a potential function, but what potential is minimized by the staircase?

It helps to think of each chip as a square block, and each stack as a column of square blocks. We can arrange the columns side-by-side and descending in height, as in the figure below.

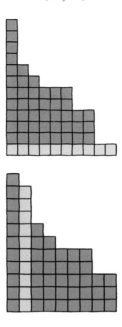

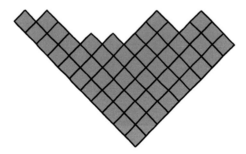

Then one "operation" can be thought of as stripping the *bottom* blocks from all the columns, turning them 90°, then adding the resulting new column to the picture in the proper position to keep descending order of heights, as in the next figure.

The configuration we're aiming for still does not seem to minimize anything natural—until we rotate the whole picture 45° to counterclockwise, as shown in the next figure. The staircase now represents the "lowest" way to pack our 55 square blocks into a V-shaped cradle.

By "lowest" we mean, or at least should mean, the configuration with least potential energy under gravity. Let the stacks be numbered in descending order of height order, and the blocks in each stack numbered from bottom to top; then let s_{ij} be the ith block in the jth column. We can size the V-frame and blocks so that the height of the center of square s_{ij} is $i+j$, and then the potential energy of a configuration will be proportional to the sum over all the blocks of the quantities $i+j$.

Now, what happens when we perform an operation? The blocks along the right-hand arm of the V move to the left-hand arm; their potential energy does not change. Nor does the potential of the other blocks, since each of those just moves one position to the right. But wait, there's one more step: We might have to rearrange blocks to get the (original) columns in descending order. That's equivalent to letting gravity push blocks downward (diagonally to the left), as in the next figure, reducing the potential energy of the configuration.

If we're already in the staircase configuration, we are at minimum potential energy and rearrangement of the stacks is never again required. Suppose we're not in the staircase configuration; what then?

The staircase configuration is possible, to begin with, only because 55 is a "triangular number," equal to $1 + 2 + 3 + \cdots + k$ for some k; these are just the numbers that can be expressed as $\binom{k+1}{2}$ for some positive integer k. In the staircase configuration, there's a block in every position (i, j) with $2 \leq i+j \leq k+1$, and no block any higher. It follows that if the number of blocks is triangular but we're not in the triangular configuration, there's a "hole" somewhere at height $k+1$ and a block somewhere at height $k+2$.

As long as no rearrangement is needed, the hole and the block each move one square to the right at every operation, regularly hitting the right arm of the V and cycling back to the left. But the hole's cycles are of length k and the block's of length $k+1$, so eventually the block will find itself directly above and to the right of the hole. At that point the block will fall into the hole, decreasing the potential energy of the configuration.

We have shown, therefore, that if the number of blocks is 55 or any other triangular number, and the current configuration is not a staircase, that its potential energy will eventually go down. There are only finitely many possible configurations, thus only finitely many values for this potential. It follows that eventually we must reach the minimum-potential triangular configuration.

Our theorem is a relatively easy application of potentials to graph theory. An *unfriendly partition* of a graph $G = \langle V, E \rangle$ is a partition of the vertex set V into two sets with the property that every vertex in V has at least as many neighbors in the *other* part as it has in its own part.

Theorem. *Every finite graph has an unfriendly partition.*

Proof. Start with any partition $V = X \cup Y$, and suppose it is not unfriendly; then there is some vertex v, say in X, that has more neighbors in X than in Y. Move v to Y. What happens? Unfortunately it may not be the case that the number of "friendly" vertices decreases, because it could be that other vertices are negatively affected by moving v. But the number of edges that cross the partition *increases*, because the only affected edges are those emanating from v and the majority of those didn't cross before but now do.

That makes things easy: The number of crossing edges cannot keep increasing indefinitely, so we must eventually come to an unfriendly partition. Equivalently: Start with a partition that maximizes the number of crossing edges, and you're done! ♡

It's worth noting that this argument fails for infinite graphs, even though it seems like when a vertex has infinitely many neighbors, it's easier to ensure that there are as many on the opposite side (now considering infinite cardinalities) as there are at home. In fact the infinite case is tied up with optional axioms of set theory, and currently it's not even known whether every countably infinite graph has an unfriendly partition.

19. Hammer and Tongs

Often, to solve a puzzle (or prove a theorem), you need to try things, see where problems arise, then fix them. We call this the "hammer-and-tongs" approach, and sometimes it works wonders.

Phone Call

A phone call in the continental United States is made from a west coast state to an east coast state, and it's the same time of day at both ends of the call. How is this possible?

Solution: Generally speaking, the west coast of the United States is on Pacific Time, 3 hours later than Eastern Time. But you may know that Florida, an east coast state, has a panhandle that extends quite far west; in fact Pensacola, FL, is on Central Time. So that picks up one of the three hours we need to make up.

If you check you will find that there is a small part of Oregon that's on Mountain Time, including, for example, the town of Ontario. That picks up our second hour.

To get the third, we time the call so that Daylight Saving has just clicked off in Pensacola. An instant after 1:59:59 a.m. in Pensacola, typically on the first Sunday in November, the clock jumps back to 1 a.m. while it's still just the "first" 1 am in Ontario, OR. So you've got an hour to make that phone call.

As I write this I note that a revolt against Daylight Savings is brewing among the states. So before you attempt to earn drinks at the bar with this tidbit of a puzzle, you might want to check that it still works!

Crossing the River

In eighth century Europe, it was considered unseemly for a man to be in the presence of a married woman, even briefly, unless her husband was there as well. This posed problems for three married couples who wished to cross a river, the only means being a rowboat that could carry at most two people. Can they get to the other side without violating their social norms? If so, what's the minimum number of crossings needed?

Solution: This is just a question of trying it out. Call the women 1, 2, and 3, the corresponding husbands A, B, and C. The ideal plan would require nine crossings, with the boat carrying two people across and one back with each round trip. If 1 and A cross, it must be A that returns with the boat (else 1 would find herself on the near shore with B and C but no husband). Now 2 and 3 must cross with one woman returning; two men can then join their wives on the far side. Now, however, a married couple must come back together in the boat, costing us a couple of crossings and making the nine-crossing ideal impossible.

Now the remaining men cross the river for good and the women finish the job. The final plan has eleven crossings, the minimum possible, and is illustrated below.

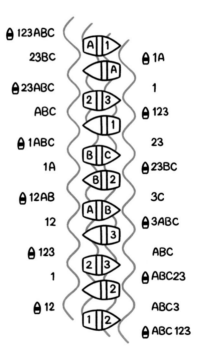

Sprinklers in a Field

Sprinklers in a large field are located at the vertices of a square grid. Each point of land is supposed to be watered by exactly the three closest sprinklers. What shape is covered by each sprinkler?

Solution: We first must ask ourselves: What *are* the three sprinklers closest to a given point in the field? Suppose the point is in the grid square bounded by sprinklers at a, b, c, and d, labeled clockwise. Divide the square into four congruent subsquares, and call them A, B, C, and D respectively, according to whether the corner shared with the big square is a, b, c, or d. It's not hard to see that for any point in A the closest three sprinklers are at a, b, and d, and similarly for points in the other small squares.

It follows that the sprinkler at a has to reach all the points in A, B, and D, plus the points in corresponding subsquares of the other three grid squares incident to a. This amounts to 12 little subsquares in the shape of a fat Greek cross.

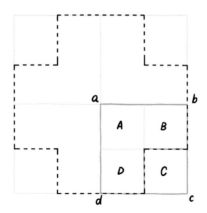

Fair Play

How can you get a 50-50 decision by flipping a bent coin?

Solution: Flip the bent coin *twice* hoping to get a head and a tail; if the head comes first, call the result HEADS; if the tail comes first, call it TAILS. If the result is two heads or two tails, repeat the experiment.

This solution is often attributed to the late, great mathematician and computing pioneer John von Neumann, and called "von

Neumann's trick." It relies on the fact that even if the coin is bent, successive flips are (or at least should be) independent events. Of course, it also relies on it being at least *possible* for the bent coin to land on either side!

If you want to minimize the number of flips to get your decision, the above scheme can be improved upon. For example, if you get HH for the first pair of flips and TT for the second, you can quit and call the result HEADS (TT followed by HH would then be called TAILS).

Finding the Missing Number

All but one of the numbers from 1 to 100 are read to you, one every ten seconds, but in no particular order. You have a good mind, but only a normal memory, and no means of recording information during the process. How can you ensure that you can determine afterward which number was not called out?

Solution: Easy—you keep track of the sum of the numbers being called out, adding each one in turn to your accumulated total. The sum of *all* numbers from 1 to 100 is 100 times the average number ($50\frac{1}{2}$), namely 5,050; that minus your final sum will be the missing number.

There is, by the way, no need to keep the hundreds digit or thousands digit during the process. Addition modulo 100 is good enough. At the end you subtract the result from 50 or 150 to get an answer in the correct range.

Dealing with streams of data, when handicapped by limited computing and memory resources, is a major subject of study in modern computer science. Here's another streaming puzzle.

Identifying the Majority

A long list of names is read out, some names many times. Your object is to end up with a name that is guaranteed to be the name which was a called a majority of the time, if there is such a name.

However, you have only one counter, plus the ability to keep just one name at a time in your mind. Can you do it?

Solution: It makes sense to use the counter to keep track of how well the name currently in memory is doing. Then, whenever that name is heard again, you can increment the counter; when some other name is heard, decrement it. If the counter reaches 0, where it began, you replace whatever name was in memory by the currently-heard name and then increment.

You could, of course, finish with a name in mind which occurred only once (e.g., if the list were "Alice, Bob, Alice, Bob, Alice, Bob, Charlie"). However, if a name occurs more than half the time, it's guaranteed to be the one in your memory at the end. The reason is that when this name is in memory, the counter is more often incremented than decremented. ♡

Poorly-Placed Dominoes

What's the smallest number of dominoes one can place on a chessboard (each covering two adjacent squares) so that no more fit?

Solution: The goal is to place dominoes so that the number h of "holes" (uncovered squares of the chessboard) is as large as possible, without having two adjacent holes. The number of dominoes used would then be $d = (64 - h)/2$.

It seems natural to arrange dominoes in diagonal lines, as in the first figure below; worst fit is achieved by switching from horizontal in half the board to vertical in the other half, then filling in adjacent holes as necessary. But alternating, gapped horizontal lines work better provided you switch to vertical (or vice-versa), as in the second figure, when the edge of the board causes problems. This gives 20 holes, thus 22 dominoes.

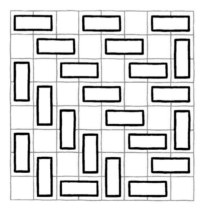

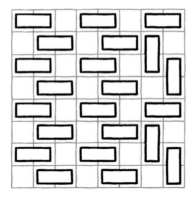

Can we do any better? Note that except for holes on the top row, every hole has a domino directly above it—and no domino can be directly above two holes. Moreover, dominoes that touch the bottom row (of which there must be at least 3) aren't helping.

In the best case, therefore, where there are four holes on the top row and three dominoes touching the bottom, $d \geq h-1$ and therefore $(64 - h)/2 \geq h - 1$, $33 \geq 3h/2$, $h \leq 22$.

But wait, if we really did have four holes on the top row and only three dominoes touching the bottom, then we could up-end the above argument to show that $h \leq 23$. But this is impossible since the number of holes must be even. Conclusion: we can't beat our 22-hole solution.

Dominoes appear in many intriguing puzzles.

Unbreakable Domino Cover

A 6×5 rectangle can be covered with 2×1 dominoes, as in the figure below, in such a way that no line between dominoes cuts all the way across the board. Can you cover a 6×6 square that way?

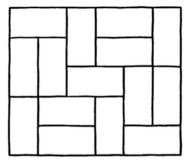

Solution: No. You need two dominoes, not just one, to cross each interior line—with just one, there's be an odd number of squares on each side of that line, but that number is a multiple of 6. Since there are 10 interior lines, you'll need 20 crossing dominoes, but there's room for only $6^2/2 = 18$ dominoes on the board. The 6×5 rectangle is in fact the only one with both dimensions under 7 that can be "unbreakably" covered.

Filling a Bucket

Before you are 12 two-gallon buckets and a 1-gallon scoop. At each turn, you may fill the scoop with water and distribute the water any way you like among the buckets.

However, each time you do this your opponent will empty two buckets of her choice.

You win if you can get one of the big buckets to overflow. Can you force a win? If so, how long will it take you?

Solution: It's safe to assume your opponent will always empty the fullest two buckets. To force your opponent to pour out as little water as possible, it's natural to begin by keeping all the buckets at the same level. How far will this get you?

You'll start by putting $1/12$ of a gallon in each bucket; your opponent will empty two buckets, reducing the total amount of water to $10/12$. You can then add a gallon total to bring each bucket up to $(1 + 10/12)/12 = 11/72$. Continuing in this manner, you make progress as long as the amount in each bucket is less than half a gallon (since then your opponent is pouring off less than you're adding). You can get close to, but never quite reach, half a gallon per bucket by this method. Then what?

Then you're going to need to give up on keeping all the buckets level. Suppose you build up to x gallons per bucket, then give up on the two buckets your opponent just emptied, and fill the rest evenly. Then you can get $x + 1/10$ gallons in those; she empties two of them and you build up to $x + 1/10 + 1/8$ in the remaining eight, and so on.

You end with $x+1/10+1/8+1/6+1/4+1/2 = x+1.141666$ in the last two buckets, not good enough to cause overflow, since $x < 1/2$.

But wait—you don't need two buckets overflowing, only one. So, you start by building up only 11 buckets, with the idea of later reducing to 9, 7, 5, 3, and finally, 1 bucket. That will get you up to $x + 1/9 + 1/7 + 1/5 + 1/3 + 1 \sim x + 1.7873015873$. That's more like it! So it's enough to get $x \geq 0.2127$.

You start with 1/11 of a gallon in each of 11 buckets, abandoning the twelfth forever. You build that up to $(1 + 9 \cdot 1/11)/11 = 20/121 \sim$ 0.1653 gallons in each of the 11 buckets after the second round, and $(1 + 9 \cdot 20/121)/11 \sim 0.2261$ after the third. That's enough to go on to the reduction phase. Altogether it takes you $3 + 5 = 8$ rounds, and with some effort it can be shown that you can't do any better.

Polygon on the Grid

A convex polygon is drawn on the coordinate plane with all its vertices on integer points, but no side parallel to the x- or y-axis. Let h be the sum of the lengths of the horizontal line segments at integer height that intersect the (filled-in) polygon, and v the equivalent for the vertical line segments. Prove that $h = v$.

Solution: Both are equal to the area of the polygon. One way to see that is to let $L_1, \ldots, L_k$ be the integer-height horizontal lines that intersect the interior of the polygon, dividing it into two triangles (at the top and bottom ends) and $k-1$ trapezoids. If the length of the segment of L_i that intersects the polygon is ℓ_i, then the sum of the areas of the triangles and trapezoids is

$$\frac{1}{2}\ell_1 + \frac{1}{2}(\ell_1 + \ell_2) + \frac{1}{2}(\ell_2 + \ell_3) + \cdots + \frac{1}{2}(\ell_{k-1} + \ell_k) + \frac{1}{2}\ell_k,$$

which equals h, and a similar argument holds for v.

One-Bulb Room

Each of n prisoners will be sent alone into a certain room, infinitely often, but in some arbitrary order determined by their jailer. The prisoners have a chance to confer in advance, but once the visits begin, their only means of communication will be via a light in the room which they can turn on or off. Help them design a protocol which will ensure that *some* prisoner will eventually be able to deduce that everyone has visited the room.

Solution: It will, of course, be necessary to assume that no one fools with the room's light between visits by prisoners; but prisoners do not need to know the initial state of the light. The idea is that one prisoner (say, Alice) repeatedly tries to turn the light on, and each of the others turns it off *twice*.

More precisely, Alice always turns on the light if she finds it off, otherwise she leaves it on. The rest of the prisoners turn it off the first two times they find it on, but otherwise leave the light alone.

Alice keeps track of how many times she finds the room dark after her initial visit; after $2n-3$ dark revisits she can conclude that everyone has visited. Why? Every dark revisit signals that one of the other $n-1$ prisoners has visited. If one of them, say Bob, hasn't been in the room, then the light cannot have been turned off more than $2(n-2) = 2n-4$ times. On the other hand, Alice *must* eventually achieve her $2n-3$ dark revisits because eventually the light will have been turned off $2(n-1) = 2n-2$ times and only one of these (caused by a prisoner darkening an initially light room before Alice's first visit) can fail to cause a dark revisit by Alice.

Sums and Differences

Given 25 different positive numbers, can you always choose two of them such that none of the other numbers equals either their sum or their difference?

Solution: Yes. Let the numbers be $x_1 < x_2 < \cdots < x_{25}$, and suppose that every two of them have either their sum or their difference represented among the other numbers. For $i < 25$, $x_{25} + x_i$ can't be on the list, so $x_{25} - x_i$ must be there; it follows that the first 24 numbers are paired with $x_i + x_{n-i} = x_{25}$. Now consider x_{24} together with any of $x_2, \ldots, x_{23}$; these pairs sum to more than $x_{25} = x_{24} + x_1$ and so $x_2, \ldots, x_{23}$ must also be paired, and in particular $x_2 + x_{23} = x_{24}$. But we just had $x_2 + x_{23} = x_{25}$, a contradiction.

Prisoner and Dog

A woman is imprisoned in a large field surrounded by a circular fence. Outside the fence is a vicious guard dog that can run four times as fast as the woman, but is trained to stay near the fence. If the woman can contrive to get to an unguarded point on the fence, she can quickly scale the fence and escape. But can she get to a point on the fence ahead of the dog?

Solution: We may as well take our unit of distance to be the radius of the field. If the prisoner were constrained to a smaller, concentric circle of radius r, where $r < \frac{1}{4}$, she would be able to maneuver herself to the farthest available point from the dog (point P in the figure below); this is because the circumference of the small circle would be less than $\frac{1}{4}$ of the circumference of the field. But if r is close enough to $\frac{1}{4}$, the woman can then make a run for it straight to the fence. Her distance to the fence would only be a hair over $\frac{3}{4}$ of a unit, but the dog has to go half way around the field, a distance of π units. Since $\pi > 3$, this is more than four times further than the woman has to run.

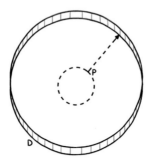

Love in Kleptopia

Jan and Maria have fallen in love (via the internet) and Jan wishes to mail her a ring. Unfortunately, they live in the country of Kleptopia where anything sent through the mail will be stolen unless it is sent in a padlocked box. Jan and Maria each have plenty of padlocks, but none to which the other has a key. How can Jan get the ring safely into Maria's hands?

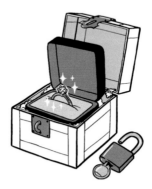

Solution: In one solution, Jan sends Maria a box with the ring in it and one of his padlocks on it. Upon receipt Maria affixes her own padlock to the box and mails it back with both padlocks on it. When

Jan gets it he removes his padlock and sends the box back to Maria, who can now open her own padlock and enjoy the ring. This solution is not just play; the idea is fundamental in Diffie-Hellman key exchange, an historic breakthrough in cryptography.

Depending on one's assumptions, other solutions are possible as well. A nice one was suggested by several persons at the Gathering for Gardner VII, including origami artist Robert Lang: It requires that Jan find a padlock whose key has a large hole, or at least a hole which can be sufficiently enlarged by drilling, so that the key can be hooked onto a second padlock's hasp.

Jan uses this second padlock, with the aforementioned key hooked on its hasp, to lock a small empty box which he then sends to Maria. When enough time has passed for it to get there (perhaps he awaits an email acknowledgment from Maria) he sends the ring in another box, locked by the first padlock. When Maria gets the ring box, she picks up the whole first box and uses the key affixed to it to access her ring.

Badly designed Clock

The hour and minute hands of a certain clock are indistinguishable. How many moments are there in a day when it is not possible to tell from this clock what time it is?

Solution: Let us first note that for the problem to make sense, we must assume that the hands move continuously, and that we are not tasked with deciding whether a time is a.m. or p.m. (We assume there is no "second hand.") Note that we *can* tell what time it is when the hour and minute hands coincide, even though we can't tell which hand is which; this happens 22 times a day, since the minute hand goes around 24 times while the hour hand goes around twice, in the same direction.

This reasoning turns out to be good practice for the proof. Imagine that we add to our clock a third "fast" hand, which starts at 12 midnight and runs exactly 12 times as fast as the minute hand.

Now we claim that whenever the hour hand and the fast hand coincide, the hour and minute hands are in an ambiguous position. Why? Because later, when the minute hand has traveled in all 12 times as far as it had moved since midnight, it will be where the fast hand (and thus also the hour hand) is now, while the hour hand is where the minute hand is now. That's the definition of an ambiguous moment.

Conversely, by the same reasoning, all ambiguous positions occur when the hour hand and fast hand coincide.

So, we need only compute the number of times a day this coincidence of fast hand and hour hand occurs. The fast hand goes around $12^2 \times 2 = 288$ times a day, while the hour goes around just twice, so this happens 286 times.

Of these 286 moments, 22 are times when the hour hand and minute hand (thus all three hands) are coincident, leaving 264 ambiguous moments.

Worms and Water

Lori is having trouble with worms crawling into her bed. To stop them, she places the legs of the bed into pails of water; since the worms can't swim, they can't reach the bed via the floor. But they instead crawl up the walls and across the ceiling, dropping onto her bed from above. Yuck!

How can Lori stop the worms from getting to her bed?

Solution: This curious problem is arguably more about engineering than mathematics.

Lori can indeed keep the worms off her bed by (in addition to putting the bed's legs in water buckets) hanging a large canopy from the ceiling, extending well over the bed. But the canopy must curve underneath itself at its edges, creating a ring-gutter *underneath* that is filled with water. (See the figure below for a cross section of the canopy.)

The under-slung gutter prevents the worms from dropping onto the canopy's edge, crawling under the canopy's surface to some point over the bed, and dropping onto Lori.

If the worms have no high-up way to get in her bedroom, Lori can more easily accomplish this same task by encircling the room itself with a water-filled gutter.

Generating the Rationals

You are given a set S of numbers that contains 0 and 1, and contains the mean of every finite nonempty subset of S. Prove that S contains all the rational numbers between 0 and 1.

Solution: First note that S contains all the "dyadic" rationals, that is, rationals of form $p/2^n$; we can obtain all those with denominator 2^n and odd numerator by averaging two adjacent ones with lower-powered denominators.

Now any general p/q is of course the average of p ones and $q-p$ zeros. We choose n large and replace the zeros by $1/2^n$, $-1/2^n$, $2/2^n$, $-2/2^n$, $3/2^n$, etc. including one 0 if p is odd. Similarly, we replace the ones by $1 - 1/2^n$, $1 + 1/2^n$, $1 - 2/2^n$, and so forth. Of course, some of these numbers lie outside the unit interval, but we can rescale the procedure to fit some dyadic interval containing p/q and lying strictly between 0 and 1. ♡

Funny Dice

You have a date with your friend Katrina to play a game with three dice, as follows. She chooses a die, then you choose one of the other two dice. She rolls her die while you roll yours, and whoever rolls the higher number wins. If you roll the same number, Katrina wins.

Wait, it's not as bad as you think; you get to design the dice! Each will be a regular cube, but you can put any number of pips from 1 to 6 on any face, and the three dice don't have to be the same.

Can you make these dice in such a way that you have the advantage in your game?

Solution: What you'd like to do is design the dice so that no matter which of the three Katrina chooses, you can pick one of the remaining two which will roll a higher number more than half the time. Such a trio is known as a "set of non-transitive dice," and there are many ways to design them.

The key is that if die A's roll has the same average as die B's, that doesn't mean A beats B half of the time. It could be that when A beats B, it wins by a lot (e.g., 6 versus 1) but when it loses, it does so by only a little bit (e.g., 5 versus 4). Then it would follow that B would win more often.

To take an extreme example, put a 6 and a five 3's on die A (averaging 3.5), and a 1 and five 4's on die B (same average). Then with probability $\frac{5}{6} \times \frac{5}{6} = 25/36 > 1/2$, there'll be a 3 on die A and a 4 on die B. And you've eliminated ties by using different numbers on the two dice.

Let's use the remaining numbers, 2 and 5, on die C—three of each, so that again the average roll is 3.5. It remains only to check that, indeed, die C is a favorite over die B (winning anytime C rolls a 5, and also occasionally 2 to 1, altogether with probability $7/12$), but die A is the expected winner over die C. Since we already know B beats A, your strategy against Katrina is clear.

Sharing a Pizza

Alice and Bob are preparing to share a circular pizza, divided by radial cuts into some arbitrary number of slices of various sizes. They will be using the "polite pizza protocol": Alice picks any slice to start; thereafter, starting with Bob, they alternate taking slices but always from one side or the other of the gap. Thus after the first slice, there are just two choices at each turn until the last slice is taken (by Bob if the number of slices is even, otherwise by Alice).

Is it possible for the pizza to have been cut in such a way that Bob has the advantage—in other words, so that with best play, Bob gets more than half the pizza?

Solution: If the number of slices is even (as with most pizzas), the puzzle *Coins in a Row* from Chapter 7 applies and Alice can always get at least half the pizza. In fact that is so even if Bob gets to start by choosing one radial cut and insisting that Alice's choice of first slice be on one side or the other of that cut. That makes the problem exactly the same as *Coins in a Row*; Alice can just number the slices 1, 2, etc. starting clockwise (say) from the cut, and play so as to take all the even-numbered slices or all the odd, whichever is better for her.

The argument fails if the number of slices is odd. But the odd case sounds even better for Alice since then she ends up with more slices. How can we get a handle on the odd case?

Taking another cue from the Chapter 7, let's limit the slice sizes to 0 or 1. (Can a slice be of size zero? Mathematically, no problem; gastronomically, think of a slice of size ε = one one-hundredth of a pizza, say.) To try to give the advantage to Bob, we somehow need to organize these so that no matter where Alice begins, Bob can use his parity advantage to overcome Alice's one-slice head start.

That takes a fair number of slices, but it turns out that 21 of these $\{0, 1\}$-sized slices is enough to turn the advantage to Bob. With a bit more fooling around, you will find that you can combine some of the slices to get a 15-slice pizza with slice-sizes 0, 1, and 2 of which nothing can stop Bob from acquiring 5/9. (Two schemes work: 010100102002020 and 010100201002020.) Pictured below is one of the Bob-friendly 15-slice schemes, with pepperoni distributed to indicate the slice sizes. No matter how she plays, Alice can never get more than 4 of the 9 pepperoni chunks against a smart, hungry Bob.

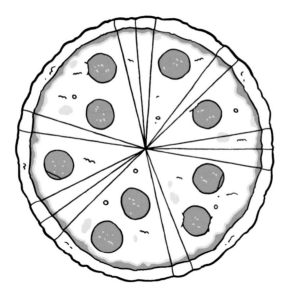

It has been shown that this pizza is best possible for Bob. Summary: If the number of slices is even, or odd but at most 13, Alice can get at least half the pizza; if the number of slices is odd and at least 15, she can guarantee at least 4/9 but no larger fraction.

Names in Boxes

The names of 100 prisoners are placed in 100 wooden boxes, one name to a box, and the boxes are lined up on a table in a room. One by one, the prisoners are led into the room; each may look in at most 50 boxes, but must then leave the room exactly as he found it and is permitted no further communication with the others.

The prisoners have a chance to plot their strategy in advance, and they are going to need it, because unless *every single prisoner finds his own name* all will subsequently be executed.

Find a strategy that gives the prisoners a decent chance of survival.

Solution: To solve it, the prisoners must first agree on a random labeling of the boxes by their own names. (The point of making it random is that it makes it impossible for the warden to place names in boxes in such a way as to foil the protocol described below.) When admitted to the room, each prisoner inspects his own box (that is, the box with which his own name has been associated). He finds a name in that box, probably not his own. He then looks into the box belonging to the name he just found, and then into the box belonging to the name he found in the second box, etc., until he either finds his own name, or has opened 50 boxes.

That's the strategy; now, why on earth should it work? Well, the process which assigns to a box's owner the name found in his box is a permutation of the 100 names, chosen uniformly at random from the set of all such permutations. Each prisoner is following a cycle of the permutation, beginning with his box and (if he doesn't run over the 50-box limit) ending with his name on a piece of paper. If it happens that the permutation *has no cycles of length greater than 50*, this process will work every time and the prisoners will be spared.

In fact, the probability that a uniformly random permutation of the numbers from 1 to $2n$ contains no cycle of length greater than n always exceeds 1 minus the natural logarithm of 2—about 30.6853%.

To see this, let $k > n$ and count the permutations having a cycle C of length exactly k. There are $\binom{2n}{k}$ ways to pick the entries in C, $(k-1)!$ ways to order them, and $(2n-k)!$ ways to permute the rest; the product of these numbers is $(2n)!/k$. Since at most one k-cycle can exist in a given permutation, the probability that there is one is exactly $1/k$.

It follows that the probability that there is no long cycle is

$$1 - \frac{1}{n+1} - \frac{1}{n+2} - \cdots - \frac{1}{2n} = 1 - H_{2n} + H_n,$$

where H_m is the sum of the reciprocals of the first m positive integers, approximately $\ln m$. Thus our probability is about $1 - \ln 2n + \ln n = 1 - \ln 2$, and in fact is always a bit larger. For $n = 50$ we get that the prisoners survive with probability 31.1827821%.

Let's put these new skills to work on a variation.

Life-Saving Transposition

There are just two prisoners this time, Alice and Bob. Alice will be shown a deck of 52 cards spread out in some order, face-up on a table. She will be asked to transpose two cards of her choice. Alice is then dismissed, with no further chance to communicate to Bob. Next, the cards are turned down and Bob is brought into the room. The warden names a card and to stave off execution for both prisoners, Bob must find the card after turning over, sequentially, at most 26 of the cards.

As usual the prisoners have an opportunity to conspire beforehand. This time, they can guarantee success. How?

Solution: This is just a variation on the names-in-boxes puzzle. We may assume the cards are numbered 1 to 52. Bob's algorithm, when asked to search for card i, will be to flip the ith card in the display. If it's value is j, he then flips the jth card; if the value of the jth card is k, he flips the kth card, etc. Thus, he follows a cycle in the permutation that maps the nth card to its value, and if no cycle in this permutation is of length > 26, he will succeed.

Alice can guarantee this condition. If she sees a cycle of length > 26 (of course, there cannot be more than one of these) she bisects it by transposing two antipodal or nearly antipodal cards of the cycle. Otherwise she makes a harmless move, for example, bisects a short cycle or if the ordering happens to be the identity, does any transposition.

We conclude this chapter's puzzles with one that is mind-numbingly easy if you try the right thing.

Self-Referential Number

The first digit of a certain 8-digit integer N is the number of zeroes in the (ordinary, decimal) representation of N. The second digit is the number of ones; the third, the number of twos; the fourth, the number of threes; the fifth, the number of fours; the sixth, the number of fives;

the seventh, the number of sixes; and, finally, the eighth is the total number of *distinct* digits that appear in N. What is N?

Solution: It's pretty tough to work out N by reasoning. But there's an easy way!

Take *any* 8-digit number N_0, and transform it according to the puzzle conditions: that is, make a new number N_1 by letting the first digit of N_1 be the number of 0's in N_0, etc.

If by chance $N_1 = N_0$, you've solved the puzzle. But you'd have to be astonishingly lucky for that to happen.

Not to worry. Repeat the procedure with N_1 to get N_2, then N_3, until you reach a time t when $N_t = N_{t-1}$. Then you're done.

Let's try this. For fun we'll start with 31415926, the first 8 digits of the decimal expansion of π, but you should try it yourself with your favorite 8-digit number.

We obtain:

$$31415926$$

$$02111117$$

$$15100004$$

$$42001104$$

$$32102004$$

$$31211005$$

$$23110105$$

$$23110105$$

It worked! But why should that happen in a reasonable number of steps? Indeed, why should it happen at all? Couldn't we start to cycle at some point, with N_t duplicating some previous N_{t-k} with $k > 1$?

Sadly, I don't know. This method does not work with all problems of this type. But it is surprisingly useful—a great weapon to have in your armory.

We conclude this chapter with a marvelous theorem that's well known to geometers and to folks who do recreational math, but deserves to be better known among serious mathematicians in other areas. Described by Georg Alexander Pick in 1899, it tells you how to compute the area of a lattice polygon just by counting lattice points.

A lattice polygon is a closed figure made from line segments that terminate at vertices of a plane grid, like the one shown in the figure below.

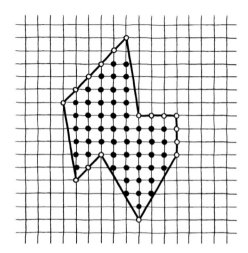

Theorem. *The area of a lattice polygon is exactly $I + B/2 - 1$ times the area of a single square cell of the grid, where I is the number of lattice points strictly inside the polygon, and B the number on its boundary.*

Proof. We will assume that each cell of the grid has unit area, so that the theorem simply says the area of a lattice polygon P is $I + B/2 - 1$.

The formula carries some reasonable intuition: Inside points (filled circles in the figure) are fully counted, and boundary points (empty circles) are arguably half-in and half-out, so should be counted at half value—except that corners are really somewhat less than half-in, so we subtract one to account for them.

The formula certainly works when the polygon is a single grid cell, since there $I = 0$ and $B = 4$.

Unfortunately we can't generally build lattice polygons out of grid cells, but it's worth thinking about how we could build them out of smaller units. Suppose P is the union of two lattice polygons R and S that share one side from each, as in the next figure.

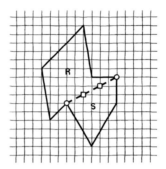

If the formula is correct for R and S, does it follow that it's correct for P? Indeed, that is the case: If there are k lattice points on the common edge, not counting its endpoints, then we have $I_P = I_R + I_S + k$ and $B_P = B_R + B_S - k - 2$, so that the area of P is $I_R + I_S + k + (B_R + B_S - k - 2)/2 - 1 = (I_R + B_R/2 - 1) + (I_S + B_S/2 - 1)$ which is indeed the area of R plus the area of S. So far so good.

Note that this calculation also shows that if the formula is correct for any two of the three polygons P, R, and S, it works for the third—in other words, the formula works with subtraction as well as addition.

That's great—and because we can cut any lattice polygon into lattice triangles, as in the next figure, it's enough to prove the theorem for triangles. This is something of a hammer-and-tongs type operation.

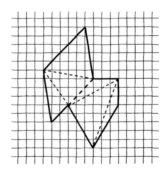

First we note that *right* triangles are easy, if their sides are vertical and horizontal line segments. The reason is that we can copy such a triangle, rotate the copy $180°$, then match its hypotenuse to the original to get an "aligned" lattice rectangle (i.e., one with sides parallel to the axes).

The formula holds for aligned rectangles, because they're made out of unit cells. Since our two triangles have the same values for I and B, Pick's formula assigns them the same area which must therefore be half the area of the aligned rectangle, as it should be.

Finally, we only need observe that *any* triangle can be made into an axis-aligned rectangle by adding (at most) three right triangles of the above sort, as in the final figure. ♡

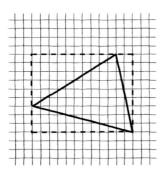

20. Let's Get Physical

Rotating Coin

While you hold a US 25-cent piece firmly to the tabletop with your left thumb, you rotate a second quarter with your right forefinger all the way around the first quarter. Since quarters are ridged, they will interlock like gears and the second will rotate as it moves around the first.

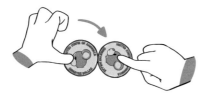

How many times will it rotate?

Solution: A classic. It rotates twice, relative to the table: once relative to the stationary quarter, and once more owing to its revolution around the stationary quarter. Try it!

Pie in the Sky

What fraction of the sky is occupied by a full moon?

Solution: The moon is about half a degree in diameter (actually, between 29.43 arc minutes at apogee and 33.5 minutes at perigee), hence its area in "square degrees" is $\pi r^2 \sim \pi (1/4)^2 = \pi/16$.

How many square degrees is the sky? Its circumference is 360 so its radius in degrees (!) is $180/\pi$ and the sky's area is half the area of the celestial sphere, about $(1/2)4\pi r^2 = 2(180)^2/\pi$ square degrees.

Thus the fraction of sky covered by the moon is about $(\pi/16)/(2(180)^2/\pi) = \pi^2/(32 \cdot 180^2) \sim 1/105{,}050$.

If we repeat this calculation with the more accurate estimate of 31 minutes for the diameter of the moon, we get 1/101,661 of the sky—so it is remarkably accurate to say the moon occupies ten millionths of the sky. The sun, by the way, is about the same size.

Returning Pool Shot

A ball is shot from a corner of a polygonal pool table (not necessarily convex) with right-angle corners; let's say all sides are aligned either exactly east-west or exactly north-south.

The starting corner is a convex corner, that is, has an interior angle of 90°. There are pockets at all corners, so that if the ball hits a corner exactly it falls in. Otherwise, the ball bounces true and without energy loss.

Can the ball ever return to the corner from which it began?

Solution: Suppose it is possible, and imagine such a shot. We may assume the ball starts off going NE; then when it finally returns it is going SW. At each bounce one of the two direction letters changes; for example, the second lap will be either SE (if it bounces off an E-W cushion) or NW (if it bounces off a N-S cushion). It follows that it takes an even number of bounces, thus an odd number of laps.

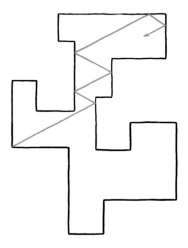

All NE and SW laps are parallel (likewise all NW and SE paths), thus the initial and final laps follow the same line in opposite directions. Since reflection works as well backwards as forwards, the whole path out from the original corner and back is a palindrome— the sequence of bounce points reads the same backwards and forwards. It follows that the two middle bounce points are the same point, which is ridiculous. ♡

Falling Ants

Twenty-four ants are placed randomly on a meter-long rod; each ant is facing east or west with equal probability. At a signal, they proceed to march forward (that is, in whatever direction they are facing) at 1 cm/sec; whenever two ants collide, they reverse directions. How long does it take before you can be certain that all the ants are off the rod?

Solution: The key to this puzzle is that, ants being interchangeable as far as we're concerned, it would make no difference to the process if they passed one another "like ships in the night" instead of bouncing. Then it's clear that each ant is simply walking straight ahead and must fall off within 100 seconds.

You can't afford to wait any shorter amount of time because if there's an ant starting at one end of the rod and facing the other end, *some* ant won't exit the far end until 100 seconds later.

Ants on the Circle

Twenty-four ants are randomly placed on a circular track of length 1 meter; each ant faces randomly clockwise or counterclockwise. At a signal, the ants begin marching at 1 cm/sec; when two ants collide they both reverse directions. What is the probability that after 100 seconds, every ant finds itself exactly where it began?

Solution: This time we have to be a bit more careful about the ants' anonymity; the argument that we can replace bouncing by passing tells us only that the ants' *set of locations* will be exactly the same after 100 seconds, but any particular ant might end up in some other ant's starting spot.

In fact, since the ants cannot pass one another, their final locations will be some rotation of their initial locations. Putting it another way, the whole collection will rotate by some number of ants, and we are in effect being asked to determine the probability that that number will be a multiple of 24.

In fact it could be 24 (clockwise or counterclockwise) only if the ants are all facing the same way, thus each walks once around the track without any collisions. There are 2^{24} ways to choose how the ants face of which only two have this property, so the probability of one of these outcomes is a minuscule $1/2^{23}$.

Much more likely is that the net rotation will be zero. When does that happen? Well, *conservation of angular momentum* tells us that the rate of rotation of the ant collection as a whole is constant. Thus the net rotation will be zero if and only if the initial rate of rotation is zero, meaning that exactly the same number ants start off facing counterclockwise as clockwise. The probability of *that* happening is $\binom{24}{12}/2^{24}$ which is about 16.1180258%. Adding $1/2^{23}$ to that boosts the final answer to about 16.1180377%.

Sphere and Quadrilateral

A quadrilateral in space has all of its edges tangent to a sphere. Prove that the four points of tangency lie on a plane.

Solution: The idea is to weight the vertices of the quadrilateral so that the point where each edge touches the sphere is exactly the center of gravity of the edge. We can do this because each vertex is equidistant from the tangency points of the two edges emanating from it; if we give that vertex weight equal to the reciprocal of that distance, we get the desired weighting.

So what? Well, if we draw a line between opposite points of tangency, the center of gravity of the quadrilateral will have to be on that line. There are two such lines, though, so the center of gravity is on both; thus, the lines intersect. Thus the quadrilateral formed by the points of tangency lies on a plane! ♡

Two Balls and a Wall

On a line are two identical-looking balls and a vertical wall. The balls are perfectly elastic and friction-free; the wall is perfectly rigid; the ground is perfectly level. If the balls are the same mass and the one farther from the wall is rolled toward the closer one, it will knock the close ball toward the wall; that ball will bounce back and hit the first ball, which will then roll (forever) away from the wall. Three bounces altogether.

Now assume the farther ball has mass a million times greater than the closer one. How many bounces now? (You may ignore the effects of angular momentum, relativity, and gravitational attraction.)

Solution: You can expect quite a lot of bouncing now, because the heavy ball will hardly notice that it has been hit the first time, and proceed undaunted toward the wall. But the light ball will bounce off the wall and hit the heavy one many more times, until the heavy ball slows down, reverses direction, and eventually disappears.

In fact the number of collisions will be 3141, the first four digits of the decimal expansion of π.

Coincidence? No indeed. Make the heavy ball a googol (that is, 10^{100}) times heavier than the light one, and the number of bounces will be the first 51 digits of π, namely 314,159,265,358,979,323,846, 264,338,327,950,288,419,716,939,937,510. It doesn't matter how far apart the three objects are, or how hard the heavy ball is pushed.

The expectation is that for any integer $k \geq 0$, the number of bounces when the heavy ball is 10^{2k} times heavier than the light one will be the first $k+1$ digits of π. Almost as remarkable as the statement itself is the fact that it is not a theorem, that is, no one can prove it. But following is an explanation of why we *think* it's true.

The key is to make use of conservation of energy and momentum. Let v be the velocity of the heavy ball and y be the velocity of the light ball. The mass of the heavy ball is $m = 10^{2k}$, and conservation of energy says $mv^2 + y^2$ is a constant. Letting $x = \sqrt{m} \cdot v = 10^k v$ and starting at $x = -1$, $y = 0$ puts the system on the unit circle $x^2 + y^2 = 1$.

Total momentum is $mv + y = 10^k x + y = -10^k$, so $y = -10^k x - 10^k$, putting us initially on a line of slope $s = -10^k$. So when the light ball first hits the wall, the "system point"—that is, the point on the XY-plane that describes the state of the system—moves from $(-1,0)$ south-east along the momentum line to the second point where that line intersects the circle. But upon hitting the wall the light ball's momentum flips, reflecting the new point across the X-axis back to the positive side.

In this way the system point zigzags down and up, moving slightly eastward on the down-zigs, until finally it reaches a point just a bit northwest of (1,0) from which moving southeast along a line of slope s would either miss the circle entirely or hit it above the X-axis. That indicates there will be no more collisions.

The figure below shows our circle, with four system states corresponding to the first two and last two system points.

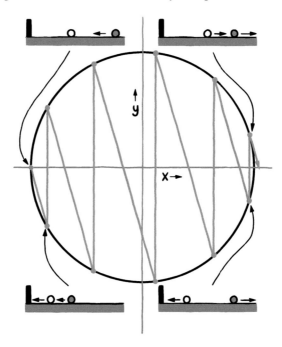

The total number of collisions is the number of zigzags; how many were there? Well, the momentum line's angle to the vertical is $\tan(10^{-k})$ and thus the arc of the unit circle between successive points above the X-axis is $2\tan(10^{-k})$. It follows that the number of collisions is the greatest integer in $2\pi/(2\tan(10^{-k})) = \pi/\tan(10^{-k})$.

Since 10^{-k} is a small angle, the approximation $\tan(10^{-k}) \sim 10^{-k}$ is a very good one (the difference is about the square of the angle, that is, of order only 10^{-2k}). Thus when k is large we expect the number of collisions to be the greatest integer in $\pi/10^{-k} = 10^k\pi$, but that is precisely the first $k+1$ digits of π!

We're a bit lucky that this already works for $k = 1$ (if we change from base 10 to some other base, it sometimes fails for $k = 1$). But as k goes up the approximation above gets better so rapidly that there's only a wisp of a chance that it every fails.

What *would* it take to fail? Roughly speaking, it would miss by 1 if for some k, digits $k+1$ through $2k$ of the decimal expansion of π were all 9's. If, as most mathematicians believe, the digits of π behave like a random sequence, this is ludicrously unlikely. (In 2013 A.J. Yee and S. Kondo computed the first 12.1 trillion digits of π, with no more than a dozen 9's in a row anywhere, about what you'd expect.)

But no one has been able to prove that the digits of π continue to behave randomly, and even if they did the above glitch *could* occur.

So it's not a theorem. But you could bet your life on it.

We conclude with something that really is a theorem.

Let's think of a *polyhedron* as any solid in three-dimensional space with flat faces. It doesn't have to be convex and could even have holes in it.

To each face F, we associate a vector v_F perpendicular to the face, pointing outward, and with length proportional to the area of the face.

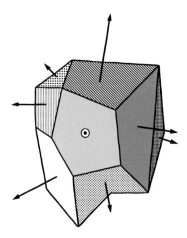

Theorem. *For any polyhedron, the sum over all faces F of the vectors v_F is zero.*

Proof. Pump the polyhedron full of air! The pressure on each face F will be proportional to the area of F, and is exerted outwardly perpendicular to F; in other words, it is represented by v_F. If the sum of these forces were not zero, the polyhedron would move of its own accord.

And we can't have that, can we? ♡

21. Back from the Future

Nothing says you have to try to solve a puzzle by starting with the premise, then working straight toward the answer. Often starting from the end—"retrograde analysis" is the fancy term for it—will make things much easier. (Are you among those that think solving a maze any way other than by drawing a line from "start" to "finish" is tantamount to cheating? That's fine if that makes it more fun for you, but I advise against applying this strategy to every puzzle.)

Consider the following classic.

Portrait

A visitor points to a portrait on the wall and asks who it is. "Brothers and sisters have I none," says the host, "but that man's father is my father's son." Who is pictured?

Solution: Where to start? At the end. Who is "my father's son"? Could be the host, if male. Wait, the host has no brothers and sisters, so it *is* the host.

Substituting, we get "that man's father is me," so the face in the picture is the host's son. Simple!

Another classic, but maybe with a twist to the solution that you haven't seen:

Three-way Duel

Alice, Bob, and Carol arrange a three-way duel. Alice is a poor shot, hitting her target only $\frac{1}{3}$ of the time on average. Bob is better, hitting his target $\frac{2}{3}$ of the time. Carol is a sure shot.

They take turns shooting, first Alice, then Bob, then Carol, then back to Alice, and so on until only one is left. What is Alice's best course of action?

Solution: Let's think about the situation *after* Alice takes her turn. If only Carol survives Alice's turn, Alice is doomed. If only Bob survives, Alice has probability p of survival where $p = \frac{1}{3}(\frac{1}{3} + \frac{2}{3}p)$, giving $p = 1/7$, not very good. If both Bob and Carol survive, things are much better because they will aim at each other, with only one survivor, at whom Alice gets to shoot first. Thus, in that situation, her survival probability is more than $\frac{1}{3}$.

It follows that Alice doesn't want to kill anyone, and her best course of action is to shoot to miss!

But wait. We'e been assuming the others wouldn't do that, but now that we have allowed Alice the option of abstaining, we must surely allow it to the others. They can work out that whenever three duelers are still alive, Alice will never aim to kill. Again applying retrograde analysis, if it gets to Carol's turn with no one dead, should she kill Bob? If Bob tried to shoot her, then yes. But if Bob shot into the air, thereby suggesting a willingness to do so indefinitely, then Carol should do the same—that way, no one's life is at risk and when the ammunition runs out, everyone can go home and do math puzzles. Going back one turn, we deduce that Bob should indeed shoot into the air and the whole duel will be a dud.

Thinking back on it, it's pretty obvious that if the highest priority for all three parties is to stay alive—which is what we've been assuming—then the duel should never have been arranged.

Here's an application of retrograde analysis to experimental design.

Testing Ostrich Eggs

In preparation for an ad campaign, the Flightless Ostrich Farm needs to test its eggs for durability. The world standard for egg-hardness calls for rating an egg according to the highest floor of the Empire State Building from which the egg can be dropped without breaking.

Flightless's official tester, Oskar, realizes that if he takes only one egg along on his trip to New York, he'll need to drop it from (potentially) every one of the building's 102 floors, starting with the first, to determine its rating.

How many drops does he need in the worst case, if he takes *two* eggs?

Solution: Let's look ahead at the position Oskar will be in when he's down to one egg. At that point there will be some minimum

possible egg rating m (namely, the highest floor from which the first egg survived) and some maximum possible egg rating M (one less than the floor from which the first egg went splat). That adds up to $M - m$ drops.

The plan for dropping the first egg will be some increasing sequence of floors, say $f_1, f_2, \ldots, f_k$. The spaces between the f_i's will be decreasing, since (assuming you want to prevent having to make more than some fixed number d of drops) the more drops made with the first egg, the fewer remain for the second.

For example, suppose Oskar's first drop of the first egg is from the 10th floor. If it breaks, he's got 9 floors to try with the 2nd egg, for a total of 10 drops. But then to prevent having to make 11 drops, he'd have to make his second drop of egg number one from the 19th floor, allowing for 8 more drops of the 2nd egg, the 3rd from the 27th floor, 4th from the 34th, 5th from the 40th, 6th from the 45th, 7th from the 49th, 8th from the 52nd, 9th from the 54th, and 10th from the 55th. That's his ten drops, so he'd be OK only if the building had at most 55 floors.

How high must Oskar start, to make it to floor 102? Easy way to figure this out: Start at the other end! Add $1 + 2 + 3 + \cdots$ until you reach or exceed 102, and your last addend, which proves to be 14, is Oskar's starting floor—and also his guaranteed maximum number of drops. Later drops of the first egg, until it breaks, are at floors 27, 39, 50, 60, 69, 77, 84, 90, 95, 99, and 102.

You can get 14 another way by deriving or remembering the formula $1+2+3+\cdots+k = k(k+1)/2$. (To derive this, note that there are k addends and the average size of an addend is $(k+1)/2$.) Then solve the equation $k(k+1)/2 = 102$ and round k up to the next integer.

If there are more than two eggs, you can extend this retrograde analysis in a similar way. For example, for three eggs, you compute the numbers $k(k+1)/2$ (directly or by adding) and then sum *these* to reach 102: $1+3+6+10+15+21+28+36 = 120$ where $35 = 8(8+1)/2$. So Oskar drops the first egg at floor 36, then 64, then 85. If it breaks at (say) 64, he's then playing the two-egg game in a 28-story building (floors 36 through 63 of the Empire State Building) which requires a maximum of 7 more drops. The total maximum number of drops for three eggs is nine.

How about when the number of eggs is unlimited? If you're a computer scientist, then you may already have guessed that *binary search* is the answer: The next drop is made at the midpoint of the current range of ratings. For example, if the building originally had 63 floors (so that the rating could be anything from 0 to 63), the first drop should be from floor 32. If the egg breaks the new range is 0 to 31, else 32 to 63. Since 102 is between $64 = 2^6$ and $128 = 2^7$, Oskar might need as many as 7 drops.

There's a whole class of puzzles where retrograde analysis is the first thing you should try: namely, game puzzles. Suppose you are presented with a two-person, alternating move, deterministic, full-information game. That's a lot of adjectives but they describe an enormous number of popular games, from chess, checkers, go, hex, and reversi to nim and tic-tac-toe.

We'll look at some games of that sort in which no draw is possible. In such a game, a "P" position is one that, if you can create it with your move, enables you to force a win. (Think of "P" as standing for "Previous player wins.") The other positions are "N" positions ("Next player wins") and those are the ones you would like passed to you by your opponent.

It follows that from an N position there is always a move that creates a P position, and from a P position, *all* moves lead to an N position. This completely defines the N and P positions and tells you how to play optimally. But these rules operate backwards in time; to classify positions this way you must start from the end of the game and work backward.

Let's try this on the next puzzle.

Pancake Stacks

At the table are two hungry students, Andrea and Bruce, and two stacks of pancakes, of height m and n. Each student, in turn, must eat from the larger stack a non-zero multiple of the number of pancakes in the smaller stack. Of course, the bottom pancake of each stack is soggy, so the player who first finishes a stack is the loser.

For which pairs (m, n) does Andrea (who plays first) have a winning strategy?

How about if the game's objective is reversed, so that the first player to finish a stack is the winner?

Solution: Let m and n be the two stack sizes. You (if you are playing) are immediately stuck just when $m = n$; this is therefore a P position. If $m > n$ and m is a multiple of n, you are in a winning N position because you can then eat $m-n$ pancakes from the n-stack and reduce your opponent to the above P position. In particular, if the short stack has just one pancake, you are in great shape.

What if m is *close* to a multiple of n? Suppose, for example, that $m = 9$ and $m = 5$. Then you are forced to reduce to $m = 4$, $n = 5$ but your opponent must now give you a 1-stack. If $m = 11$ and $n = 5$ you have a choice but reducing to $m = 6$, $n = 5$ wins for you.

It's beginning to look like you want to make the ratio of the stacks small for your opponent, forcing her to make the ratio big for you. Let's see. Suppose the current ratio $r = m/n$ is strictly between 1 and 2; then the next move is forced and the new ratio is $\frac{1}{1-r}$. These ratios are equal only for $r = \phi = (1+\sqrt{5})/2 \sim 1.618$, the golden mean; since ϕ is irrational, one of the two ratios r and $\frac{1}{1-r}$ must exceed ϕ while the other is smaller than ϕ. Aha! Thus, when you present your opponent with $r < \phi$ she must make it bigger, then you make it smaller, etc., until she's stuck with $r = 1$ and must lose!

We conclude that Andrea wins exactly when the initial ratio of larger to smaller stack exceeds ϕ. In other words, the P positions of the game are exactly those with $m/n < \phi$, where $m \geq n$. To see this, suppose $m > \phi n$, but m is not a multiple of n. Write $m = an + b$, where $0 < b < n$. Then either $n/b < \phi$, in which case Alice eats an, or $n/b > \phi$, in which case she eats only $(a-1)n$. This leaves Bruce with a ratio below ϕ, and faced with a forced move which restores a ratio greater than ϕ.

Eventually, Andrea will reach a point where her ratio m/n is an integer, at which point she can reduce to two equal stacks and stick Bruce with a soggy pancake. But note that she can also, if desired, grab a whole pile for herself, thus winning the alternative game where the eater of a soggy pancake is the victor.

Of course, if Andrea is instead faced with a ratio m/n which is strictly between 1 and ϕ, she is behind the eight-ball and it is Bruce who can force the rest of the play.

We conclude that no matter which form of Pancakes is played, if the stacks are at heights $m > n$, Andrea wins precisely when $m/n > \phi$. Only in the trivial case when the stacks are initially of equal height does it matter what the game's objective is!

Try the above approach on this similar game.

Chinese Nim

On the table are two piles of beans. Alex must either take some beans from one pile or the same number of beans from each pile; then Beth has the same options. They continue alternating until one wins the game by taking the last bean.

What's the correct strategy for this game? For example, if Alex is faced with piles of size 12,000 and 20,000, what should he do? How about 12,000 and 19,000?

Solution: As in classical Nim and many other games, it's easiest to try to characterize the P positions, as there are fewer of those. Once you know the P positions, the correct strategy is automatic. If a player is in an N position, he or she makes a move that puts his or her opponent in a P position; that opponent must then move to another N position.

In Chinese Nim, the empty position (no beans left) is a P position since the previous player has just won the game. Any position with one pile empty, or both piles of the same size, is an N position, since the empty position is reachable in one move. It's not hard to deduce that the simplest non-empty P position is $\{1, 2\}$. After that, you can work out that $\{3, 5\}$, $\{4, 7\}$, and $\{6, 10\}$ are P positions as well. What's the pattern?

Let $\{x_1, y_1\}, \{x_2, y_2\}, \ldots$ be the non-empty P positions, with $x_i < y_i$ and $x_i < x_j$ for $i < j$. Notice you cannot have $x_i = x_j$ for $i \neq j$ because then a player could reduce the larger of y_i and y_j to the smaller, leaving another P position.

Some thought will lead you to conclude that given $\{x_1, y_1\}$ up to $\{x_{n-1}, y_{n-1}\}$, x_n is the least positive number not among $\{x_1, \ldots, x_{n-1}\} \cup \{y_1, \ldots, y_{n-1}\}$, and $y_n = x_n + n$. Notice that this forces y_n to be a higher number than any in the set $\{x_1, \ldots, x_{n-1}\} \cup \{y_1, \ldots, y_{n-1}\}$.

The proof is by induction on n. You have already seen that x_n can't be among the numbers in $\{x_1, \ldots, x_{n-1}\} \cup \{y_1, \ldots, y_{n-1}\}$ and also that there can't be more than one y_n to go with this x_n, so all you need to do is show that this $\{x_n, y_n\}$ really is a P position.

If $\{x_n, y_n\}$ were an N position (for Alex, say), it must be that he can reduce it to $\{x_i, y_i\}$ for some $i < n$; but he cannot get to this position by reducing the smaller pile or by reducing both piles by the same amount, because that would leave the difference of the two piles at n or more. Nor can he get there by reducing the larger pile, because then he would get another y for the same x. Thus $\{x_n, y_n\}$ is indeed a P position.

You now have the means to generate as long a list as you like of P positions. From this Alex's strategy is easy to work out. If he is faced with $\{x_i, y_i\}$ he removes a bean or two and hopes for an error. If he sees $\{x_i, z\}$ for $z > y_i$, he reduces z to y_i. If he sees $\{x_i, z\}$ with $x_i < z < y_i$, the difference $d = z - x_i$ is less than i; he takes from both piles to get down to $\{x_d, y_d\}$ (if $z = y_j$ for some $j < i$ he also has the option of just reducing x_i to x_j). If he sees $\{y_i, z\}$ with $y_i \leq z$ he can reduce z all the way down to x_i, and may have other options as well.

But it might take a while to generate enough P positions to decide what to do with thousands of beans in each pile. Is there a more direct way to characterize the P positions?

Well, you know that for each n, x_n is somewhere between n and $2n$, because it is preceded by all the x_i's for $i < n$ and *some* of the y_i's. It is reasonable to guess that x_n is approximately equal to rn, for some ratio r between 1 and 2. If so, y_n would be approximately $rn + n = (r+1)n$.

If this holds up, it follows that the n x_i's between 1 and x_n are more or less evenly distributed, and therefore a fraction $r/(r+1)$ of them will have their corresponding y_i below x_n. Thus there are about $nr/(r+1)$ y_i's below x_n, together with the n x_i's, adding up to x_n numbers in all; making an equation out of this gives

$$n + n\frac{r}{r+1} = nr,$$

which gives us $r + 1 = r^2$, $r = (1+\sqrt{5})/2$, the familiar "golden ratio" (again!).

If we were *really lucky*, it would turn out that for each n, x_n is exactly $\lfloor rn \rfloor$ (the greatest integer in rn), and $y_n = \lfloor r^2 n \rfloor$. In fact, this is the case, and you can prove it using the theorem at the end of this chapter. The key is that because r and r^2 are irrational numbers that sum to 1, every positive integer can be uniquely represented *either* as $\lfloor rj \rfloor$ for some integer j, or $\lfloor r^2k \rfloor$ for some integer k.

Let's use this to find Alex's move in the example positions. Note that $12000/r$ is a fraction under 7417, and $7417r = 12000.9581\ldots$ so 12000 is an x_i, namely x_{7417}. The corresponding y_{7417} is $\lfloor 7417r \rfloor = 19417$ so if the other pile has 20000 beans, Alex can win by taking $20000 - 19417 = 583$ beans away from it. If there are only 19000 beans in the other pile, Alex can win instead by reducing the piles simultaneously to $\{x_{7000}, y_{7000}\} = \{11326, 18326\}$. Since 1900 happens to be a y_j, namely y_{2674}, Alex can also win by reducing the x-pile to $x_{2674} = \lfloor 2674r \rfloor = 4326$.

Turning the Die

In the game of Turn-Die, a die is rolled and the number that appears is recorded. The first player then turns the die 90 degrees (giving her four options) and the new value is added to the old one. The second player does the same and the two players alternate until a sum of 21 or higher is reached. If 21 was hit exactly, the player who reached it wins; otherwise, the player who exceeded it loses.

Do you want to be first or second?

Solution: This game, which I call Turn-Die, yields nicely to retrograde analysis using P and N positions—but first you have to figure out what, exactly, "positions" are.

In Turn-Die, to know your position you must certainly know the current sum; but you also need to know what numbers are available when you turn the die. A standard die has numbers that sum to 7 on opposite faces; that is, 1 opposite 6, 2 opposite 5, and 3 opposite 4. Thus, for example, if the die is now showing 1 or 6, the options are to turn it to 2, 3, 4, or 5 only.

Thus, what you need to know for strategic purposes is what the current sum is, and whether the die is showing 1 or 6 (case A), 2 or 5 (case B), or 3 or 4 (case C).

To do the retrograde analysis, it helps to make a picture of the positions. Write the numbers from 1 to 21 three times in parallel columns labeled A, B, and C; let's say with the 1's on the bottom, and the 21's on top. You then have a picture of 63 positions (not all reachable), and can assign P's and N's to them from the top down.

Let's see how this process would start (in other words, how the game ends). The 21's are all P positions, since the previous player has now won the game. Position 20A is also a P position, since the 1 is unavailable to turn to; the player facing 20A will have to overshoot and lose. The other 20's, 20B and 20C, are N positions.

Continuing in this way, you eventually determine the state of the six possible opening positions, namely 1A, 2B, 3C, 4C, 5B, and 6A (boxed in the figure below). It turns out that of these only 3C and 4C are P positions, so if you go first, you will with probability $\frac{2}{3}$ start with an N position and be able to force a win.

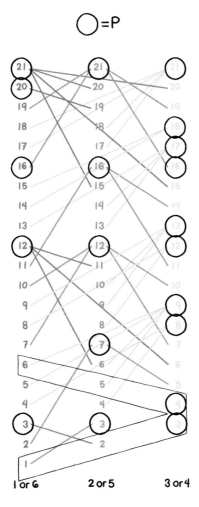

It's perhaps worth noting that in some vague sense, you usually want to be the first player in a combinatorial game that begins in a random position. The reason is that the condition for being an N position (that some move leads to a P) is milder than the condition for being a P position (that *every* move leads to an N position). This is why you usually find more N positions than P. When the game starts in an N position, you want to be first to play.

Game of Desperation

On a piece of paper is a row of n empty boxes. Tristan and Isolde take turns, each writing an "S" or an "O" into a previously blank box. The winner is the one who completes an "SOS" in consecutive boxes. For which n does the second player (Isolde) have a winning strategy?

Solution: The game seems confusing and impenetrable until you realize that the only way you can force a win on your next move is to oblige your opponent to play into the configuration S-blank-blank-S (henceforth to be called a "pit"). Thus, for example, Tristan can win when $n = 7$ by placing an S in the middle, then another at the end farther from Isolde's response, to make a pit. After a move by each player outside the pit, Isolde must play into the pit and lose.

The same applies for any odd n greater than 7, as Tristan can play an S anywhere at least 4 spaces from the ends, then form a pit on one side or the other and wait.

When n is even Tristan has no chance, as there will never be a time when Isolde has only pits to play in; when she moves, there is always an odd number of blank squares to play in. Instead, when n is even and large, Isolde wins by playing an S far from the ends and from Tristan's first move. However, if Tristan begins with an O, Isolde cannot put an S next to it, so she needs extra room.

In the $n = 14$ case, if Tristan writes an O in place 7 (of 1 through 14), Isolde's best response is an S in position 11 (threatening to make a pit with an S in 14). Tristan can counter this with an O in 13 or 14 (or an S in 12), and now Isolde would like to make a pit with S in position 8 but can't, as Tristan would then win with S in 6.

Thus $n = 14$ is a draw; Isolde needs n to be even and at least 16. To wrap up, Tristan wins when n is odd and at least 7, Isolde when n is even and at least 16; all other values of n lead to a draw with best play.

Deterministic Poker

Unhappy with the vagaries of chance, Alice and Bob elect to play a completely deterministic version of draw poker. A deck of cards is spread out face-up on the table. Alice draws five cards, then Bob draws five cards. Alice discards any number of her cards (the discarded cards will remain out of play) and replaces them with a like number of others; then Bob does the same. All actions are taken with the cards face-up in view of the opponent. The player with the better hand wins; since Alice goes first, Bob is declared to be the winner if the final hands are equally strong. Who wins with best play?

Solution: You need to know a little about the ranking of poker hands for this puzzle: Namely, that the best type of hand is the straight flush (five cards in a row of the same suit), and that an ace-high straight flush (known as a "royal flush") beats a king-high straight flush and on down.

That means if Bob is allowed to draw a royal flush, Alice's goose is cooked. For Alice to have a chance, her initial hand must contain a card from each of the four possible royal flushes.

The best card of each suit, for that purpose, is the 10, since it stops all straight flushes which are 10-high or better. Indeed, a moment's thought will convince you that any hand of Alice's containing the four 10s will win. Bob cannot now hope to get a straight flush better than 9-high. To stop Alice getting a royal flush, he must draw at least one high card from each suit, leaving room for only one card below a 10. Alice can now turn in four cards and make herself a 10-high straight flush in a suit other than the suit of Bob's low card, and Bob is helpless.

Alice has other winning hands as well—see if you can find them all!

The next puzzle is a difficult one, asking you to solve a bluffing game (even the simplest of bluffing games need some machinery to analyze; imagine what poker might require!). For this we do need the concept of *equilibrium*—a pair of strategies, one for each player, with the property that neither player can improve her results by changing strategy, if the other player doesn't change hers. Example: In Rock-Paper-Scissors, the unique equilibrium is achieved when both players choose each of the three options with equal probability.

For many games, and all bluffing games, equilibrium strategies are (like the ones for Rock-Paper-Scissors) *randomized*. To find these strategies (which, by the way, are often but not always unique) you can take advantage of the following fact.

Suppose the equilibrium strategy for Player I calls for her to choose among several options, say $A_1, \ldots, A_k$, each with some positive probability. Then each option must give her the same expectation against Player II's strategy. Why? because if the expectation of, say, option A_3 were among the highest and better than that of some other options, she could improve her results by choosing A_3 instead of randomizing; and this contradicts the definition of equilibrium.

Bluffing with Reals

Consider the following simple bluffing game. Louise and Jeremy ante $1 each and each is given a secret random real number between 0 and 1. Louise may decide to pass in which case the $2 pot goes to the player with the higher number. However, if she chooses, Louise may add another dollar to the pot. Jeremy may then "call" by adding another dollar himself, in which case the pot, now with $4 in it, goes again to the player with the higher number. Or Jeremy may fold, ceding the pot, with his $1 in it, to Louise.

Surely Louise has the advantage in this game, or at least equality, since she can break even by always passing. How much is the game worth to her? What are the players' equilibrium strategies?

Solution: For each value x that she could get, Louise has to decide whether to pass or raise; and for each value y that Jeremy could get, he has to decide whether to call or fold if Louise raises. Thus, there are in principle infinitely many strategies for each—a pretty big infinity, too (technically, two to the power of the continuum).

So to find equilibrium strategies, we'll need to restrict our search space. A little thought will convince you that Jeremy needn't consider any strategy not of the form "call just when $y > q$" for some fixed threshold q. The reason is that no matter what Louise does, a higher value of y cannot increase Jeremy's incentive to fold.

So let's assume that Jeremy has picked such a threshold q and consider Louise's best response when holding the value x. If she passes, she wins $1 when $y < x$ (probability x) and otherwise loses $1, so her expectation is

$$x \cdot \$1 + (1-x) \cdot (-\$1) = \$(2x - 1).$$

Suppose $x > q$. Then when $y < q$ Louise wins $1 whatever she does;

when $y > q$ the game is in Louise's favor just when x is more than halfway from q to 1, that is, when $x > (q+1)/2$, so that's when she should raise.

What about when $x < q$? Should Louise ever raise? If she doesn't, we know her expectation is $\$(2x - 1)$, which is dismally low when x is small. If she does raise, she wins $1 when her bluff works, that is, when $y < q$, otherwise she loses $2, for a net expectation of

$$q \cdot \$1 + (1-q) \cdot (-\$2) = \$(3q - 2).$$

Setting $2x - 1 = 3q - 2$ gives us $x = (3q-1)/2$ and tells us that Louise should raise whenever $x < (3q - 1)/2$, as well as when $x > (q+1)/2$.

We could at this point compute (as a function of q) what this strategy brings, on average, to Louise's coffers, and then minimize that (using calculus—ugh) to find Bob's choice for his threshold q. But we know that at equilibrium, if Bob actually drew the number q, he should be indifferent to the choice of folding or calling. Of course if he folds he loses $1. If he calls, he wins $2 when Louise was bluffing and loses $2 otherwise. For this to cost Bob $1 on average, Louise's probability of bluffing *given that she raised* has to be 1/4; that means

$$\frac{3q - 1}{2} = \frac{(1 - q)/2}{3},$$

giving $9q - 3 = 1 - q$, $q = 0.4$.

We conclude that Louise raises when $x < 0.1$ or $x > 0.7$, and Bob calls when $y > 0.4$. The value of the game is computed as follows:

- When $x < 0.1$ Louise bluffs, winning $1 when $y < 0.4$ and losing $2 otherwise for a net gain of $-\$0.80$.

- When $0.1 < x < 0.7$ Louise passes and, with her average holding of $x = 0.4$, earns $\$0.40 - \$0.60 = -\$0.20$.

- Finally, when $x > 0.7$ Louise breaks even on average when $y > 0.7$ as well, but she picks up $1 when $y < 0.4$, and $2 when $0.4 < y < 0.7$ and Bob calls her raise. This gives her a net average profit of $1 even.

Bottom line: Louise makes $0.1 \cdot (-\$0.80) + 0.6 \cdot (-\$0.20) + 0.3 \cdot \$1 = -\$0.08 - \$0.12 + \$0.30 = \$0.10$. So the game is worth exactly a dime to Louise.

The next puzzle, also involving equilibrium strategies, is equally hard to fully analyze—but we're being asked for much less than that.

Swedish Lottery

In a proposed mechanism for the Swedish National Lottery, each participant chooses a positive integer. The person who submits the lowest number not chosen by anyone else is the winner. (If no number is chosen by exactly one person, there is no winner.)

If just three people participate, but each employs an optimal, equilibrium, randomized strategy, what is the largest number that has positive probability of being submitted?

Solution: Suppose k is the highest number any player is willing to play. If a player chooses k, he wins anytime the other two players agree, except if they agree on k. But if he chooses $k+1$, he wins anytime they agree, *period*. Hence, $k+1$ is a better play than k, and we cannot be in equilibrium. The contradiction shows that arbitrarily high submissions must be considered—sometimes one should choose 1,487,564.

The actual equilibrium strategy calls for each player to submit the number j with probability $(1-r)r^{j-1}$, where

$$r = -\frac{1}{3} - \frac{2}{\sqrt[3]{17+3\sqrt{33}}} + \frac{\sqrt[3]{17+3\sqrt{33}}}{3},$$

which is about 0.543689. The probabilities for choosing 1, 2, 3, and 4 are, respectively, about 0.456311, 0.248091, 0.134884, and 0.073335.

I do not know if our "Swedish lottery" was ever implemented or even seriously considered for any official lottery, but don't you think it should have been?

Suppose a puzzle presents a process, that is, one that moves in time from one state to another—deterministically, randomly, or under your control. It's often helpful to determine whether the process is *reversible*, in the sense that you can tell from the state you're in at the moment what the *previous* state must have been.

If there are only a finite number of possible states, then reversibility tells you that no matter what state you start in, you must return to that state. Why? You must return to some state, surely, but why the state you started with?

The proof is by contradiction. Let S be the first repeated state, and say that this repetition—that is, the second occurrence of state S—takes place at time t. Reversibility says that the previous state must be some particular state R. But if S was not the starting state, then S's first occurrence was also preceded by R—so R was repeated before S was, contradicting our choice of S.

To apply this "reversibility theorem" the trickiest part is often figuring out what information should go into the "state." Here's a typical example.

Pegs on the Corners

Four pegs begin on the plane at the corners of a square. At any time, you may cause one peg to jump over a second, placing the first on the opposite side of the second, but at the same distance as before. The jumped peg remains in place. Can you maneuver the pegs to the corners of a larger square?

Solution: Note first that if the pegs begin on the points of a grid (i.e., points on the plane with integer coordinates), then they will remain on grid points.

In particular, if they sit initially at the corners of a unit grid square, then they certainly cannot later find themselves at the corners of a *smaller* square since no smaller square is available on the grid points. But why not a larger one?

Here's the key observation: The jump step is reversible! If you could get to a larger square, you could reverse the process and end up at a smaller square, which we now know is impossible.

Slightly trickier:

Touring an Island

Aloysius is lost while driving his Porsche on an island in which every intersection is a meeting of three (two-way) streets. He decides to adopt the following algorithm: Starting in an arbitrary direction from his current intersection, he turns right at the next intersection, then left at the next, then right, then left, and so forth.

Prove that Aloysius must return eventually to the intersection at which he began this procedure.

Solution: Between intersections Aloysius' current state can be characterized by a triple consisting of the edge he's on, the direction he's going on the edge, and the type of his last turn (right or left). From this information you can determine Aloysius' last move, and an application of our reversibility theorem does the rest.

Trickier still:

Fibonacci Multiples

Show that every positive integer has a multiple that's a Fibonacci number.

Solution: We need to show that for any n, as we generate the Fibonacci numbers (1, 1, 2, 3, 5, 8, 13, 21, 34, etc.) we eventually find one that is equal to zero modulo n. We normally define the Fibonacci numbers by specifying that $F_1 = F_2 = 1$ and for $k > 2$, $F_k = F_{k-1} + F_{k-2}$.

As it stands, this does not seem like a reversible process; given the value (modulo n) of, say, F_{k+1}, we can't immediately determine the value (again, modulo n) of F_k. But if we keep track of the value modulo n of *two consecutive* Fibonacci numbers, then we can go backwards by just subtracting.

For example, if $n = 9$ and $F_8 \equiv 3 \mod 9$ and $F_7 \equiv 4 \mod 9$ then we know $F_7 \equiv 3 - 4 \equiv 8 \mod 9$. Putting it another way, if we define $D_k = (F_k \mod n, F_{k+1} \mod n)$, then the D_k's constitute a reversible process.

But so what? We'd like to find a k such that one of the coordinates of D_k is zero, but how do we know it doesn't cycle among pairs that don't contain a zero?

Aha, a second trick: Start the Fibonacci numbers with $F_0 = 0$, that is, start one step ahead with 0, 1, 1, 2, 3, etc. Then $D_0 = (0,1)$ and therefore eventually (after at most n^2 steps, in fact) D_k will cycle back to (0,1), which means that F_k is a multiple of n.

If we take $n = 9$ as above, for instance, our D_k's, beginning with D_0, are (0,1), (1,1), (1,2), (2,3), (3,5), (5,8), (8,4), (4,3), (3,7), (7,1), (1,8), (8,0) and we can stop there with $F_{12} = 144 \equiv 0 \mod 9$. The D sequence continues (0,8), (8,8), (8,7), (7,6), (6,4), (4,1), (1,5), (5,6), (6,2), (2,8), (8,1), (1,0), (0,1) so it cycles back in this case after only 24 steps.

Light Bulbs in a Circle

In a circle are light bulbs numbered clockwise 1 through n, $n > 1$, all initially on. At time t, you examine bulb number t (modulo n), and if it's on, you change the state of bulb $t+1$ (modulo n); that is, you turn off the clockwise-next bulb if it's on, and on if it's off. If bulb t is off, you do nothing.

Prove that if you continue around and around the ring in this manner, eventually all the bulbs will again be on.

Solution: We observe first that there is no danger of turning all the lights off; if a change is made at time t, bulb t (modulo n) is still on. Moreover, if we look at the circle just *after* time t, we can deduce the state of the bulbs before t (by changing the state of bulb $t+1$ if bulb t is on). Thus the process is reversible, *provided* we take care to include in the state information not only which bulbs are on and which off, but which bulb is the one whose state determined the last action.

The number of possible states is less than $n \times 2^n$, thus is finite, and we can apply the above argument to conclude that we will eventually cycle back to the initial all-on state—moreover, we can even insist that that we reach such a point at a moment when we are slated to again examine bulb #1.

In the next puzzle, reversibility is just a part of the solution.

Emptying a Bucket

You are presented with three large buckets, each containing an integral number of ounces of some non-evaporating fluid. At any time, you may double the contents of one bucket by pouring into it from a fuller one; in other words, you may pour from a bucket containing x ounces into one containing $y \leq x$ ounces until the latter contains $2y$ ounces (and the former, $x-y$).

Prove that no matter what the initial contents, you can, eventually, empty one of the buckets.

Solution: There is more than one way to solve this puzzle, but our strategy here will be to show that the contents of one of the buckets can always be *increased* until one of the *other* buckets is empty.

To do this, we first note that we can assume there is exactly one bucket containing an odd number of ounces of fluid. This is true because if there are no odd buckets, we can scale down by a power of 2; if there are two or more odd buckets, one step with two of them will reduce their number to one or none.

Second, note that with an odd and an even bucket we can always do a reverse step, that is, get half the contents of the even bucket into the odd one. This is because each state of this pair of buckets can be reached from at most one state, thus if you take enough steps, you must cycle back to your original state; the state *just before* you return is the result of your "reverse step."

Finally, we argue that as long as there is no empty bucket, the odd bucket's contents can always be increased. If there is a bucket whose contents are divisible by 4, we can empty half of it into the odd bucket; if not, one forward operation between the even buckets will create such a bucket. ♡

Our last "reversibility" puzzle is—there's no other word for it—unbelievable.

Icᴇ Crᴇam Cakᴇ

On the table before you is a cylindrical ice-cream cake with chocolate icing on top. From it you cut successive wedges of the same angle θ. Each time a wedge is cut, it is turned upside-down and reinserted into the cake. Prove that, regardless of the value of θ, after a finite number of such operations all the icing is back on top of the cake!

Solution: If you think you have a proof that no finite number of operations can restore all the icing when x is an irrational angle, you have justification. When θ is irrational, that is, is not a rational multiple of 2π radians, every cut will be at a different angle. But the very first inverted wedge will create a boundary-line on top (and bottom) of the cake between iced and un-iced areas. If we never again cut at that point, how can we erase that boundary-line?

In fact, we *can* cut again at the boundary-line, because when a wedge is inverted, its iced/un-iced pattern is not only complemented but reversed. Thus, boundary-lines move.

In analyzing this puzzle, and indeed many serious algorithmic problems as well, it helps to redefine the operation so that it is only the "state space"—here, the icing pattern on the cake—and not the operation itself, that changes from step to step. In this case, that means rotating the cake after each operation so that you are always cutting in the same place.

We will use standard mathematical notation, reading angles counterclockwise around the cake counting "east" as 0 radians. One *operation* will consist of cutting the cake at $-\theta$ and 0, flipping that piece over, then rotating the whole cake clockwise by angle θ.

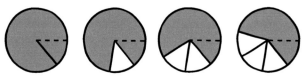

Suppose that k operations is the minimum needed to get all the way around the cake; that is, $k\theta \geq 2\pi$ but $(k-1)\theta < 2\pi$. Putting it another way, k is the greatest integer in $2\pi/\theta$. Let δ be the amount by which you overshoot the first cut, that is, $\delta = k\theta - 2\pi$; then $0 \leq \delta < \theta$.

At this point we advise the reader to try some reasonable angle, say a bit more than $\pi/4$ (making $k = 4$). You will find that after only seven cuts ($2k-1$ in general), all new cuts are in the same places as old ones!

Let S be the following set of $2k-1$ angles, listed in counterclockwise order around the cake:

$$S = \{0,\ \theta-\delta,\ \theta,\ 2\theta-\delta,\ 2\theta,\ 3\theta-\delta,\ 3\theta,\ \ldots,\ (k-1)\theta-\delta,\ (k-1)\theta\}.$$

The figure below shows the cuts in the case $\theta = 93.5°$, where $k = 4$ and $\delta = 4 \cdot 93.5° - 360° = 14°$.

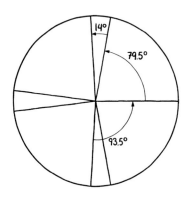

We claim that these are the angles of all cuts that you will ever make. In fact, it is not hard to see that the set S is closed under our operation. The angle 0 maps to itself, while the angle $-\theta = 2\pi - \theta = (k-1)\theta - \delta$ maps to θ. The only line on the piece thus cut is at angle $(k-1)\theta$, which then moves to $\theta - \delta$. The rest of the angles are shifted up by θ as the cake rotates, so $m\theta$ moves to $(m+1)\theta$ and $m\theta - \delta$ moves to $(m+1)\theta - \delta$ for $1 \leq m \leq k-2$.

All the angles in S are indeed cut, in just two passes around the cake, so S represents our set of cuts exactly. And here's where we use reversibility: Since there are only $2k-1$ potential cake slices and each can only have either all its icing on top or all on bottom, there are only 2^{2k-1} possible states of the cake. Our operation is completely reversible, so we must return to the all-icing-on-top state that we started with.

In fact, we return after many fewer than 2^{2k-1} steps. To see this, notice that the portions into which the cuts in S divide the cake come in only two sizes, δ and $\theta - \delta$, with $k-1$ of the former and k of the latter. (These sizes could be equal, but the types are still distinct.)

A θ-wedge of cake consists of two of these portions, one of each type. One operation flips the next two portions, one of each type, reading clockwise around the cake. Notice that the portions of a given type remain in order after a flip, since the flip involves only two portions of different type.

To get all the icing back on top, each portion must be flipped an even number of times; thus the number of operations required must be an even multiple of $k-1$ (to get the δ portions right) and simultaneously an even multiple of k (to get the $\theta - \delta$ portions right). The smallest number of operations that fits the bill is $2k(k-1)$, and that is the precise answer—except when $\delta = 0$, that is, when $\theta = 2\pi/k$ for some integer k, in which case only $2k$ operations are required.

Notice that except in the $\delta = 0$ case, to get all the icing on the *bottom* you'd need the number of flips to be simultaneously an odd multiple of $k-1$ and an odd multiple of k, which is impossible since either $k-1$ or k must be an even number. So it never happens that all the icing is on the bottom.

A certain very well known mathematician's reaction, upon hearing the ice cream cake puzzle, was: "I find it hard to believe that the icing ever returns to the top. But, one thing I'm sure of: If it does, there must also be a time when it's all on the bottom!"

Moth's Tour

A moth alights on the "12" of a clock face, and begins randomly walking around the dial. Each time it hits a number, it proceeds to the next clockwise number, or the next counterclockwise number, with equal probability. It continues until it has been at every number.

What is the probability that the moth finishes at the number "6"?

Solution: No surprise: It pays to think about how the process ends.

Let's generalize slightly and think about the probability that the moth's tour ends at i, where i is any number on the dial except 12. Consider the first time that the moth gets within one number of i. Suppose this happens when the moth reaches $i - 1$ (the argument is similar if it happens at $i + 1$). Then i will be the last number visited if and only if the moth gets all the way around to $i+1$ before it gets to i.

But the probability of this event doesn't depend on i. So the probability of ending at 6 is the same as the probability of ending at any other number, other than 12 itself. Therefore the probability that the moth's tour ends at 6 is $1/11$.

Boardroom Reduction

The Board of Trustees of the National Museum of Mathematics has grown too large—50 members, now—and its members have agreed to the following reduction protocol. The board will vote on whether to (further) reduce its size. A majority of ayes results in the immediate ejection of the newest board member; then another vote is taken, and so on. If at any point half or more of the surviving members vote nay, the session is terminated and the board remains as it currently is.

Suppose that each member's highest priority is to remain on the board, but given that, agrees that the smaller the board, the better.

To what size will this protocol reduce the board?

Solution: We start from the end, of course. If the board gets down to two members they will certainly both survive, as member 2 (numbered from most senior to newest) will vote to retain herself. Thus, member 3 will not be happy if the board comes down to members 1, 2, and 3; members 1 and 2 would then vote "aye" and eject him.

It follows that at size 4, the board would be stable; member 3 would vote nay to prevent reduction to 1, 2, and 3, and member 4 would vote nay to save him/herself immediately.

The pattern suggests that perhaps the stable board sizes are exactly the powers of 2; if so, the board will reduce to 32 members and then stop. Is this right? Suppose not. Let n be the least number for which the claim fails, and let k be the greatest power of 2 strictly below n.

Since n is our least counterexample, the board *is* stable when it gets down to members 1 through k. Thus those k will vote "aye" until then, outvoting the remaining $n-k$ unless $n = 2k$. Thus when $n = 2k$ the k newer members must vote "nay" to save themselves, stopping the process when n is itself a power of 2. Thus n is not a counterexample after all, and our claim is correct.

Note we have used not just retrograde analysis, but also induction and contradiction to solve this puzzle—not to mention consideration of small numbers. So this puzzle could easily have gone into at least three other chapters of the book.

For our theorem we will look at a surprising (to some) fact about numbers, which we can formulate in terms of two steadfast blinkers.

Suppose that two regular blinkers begin with synchronized blinks at time 0, and afterwards there is an average of one blink per minute from the two blinkers together. However, they never blink simultaneously again (equivalently, the ratio of their frequencies is irrational).

The theorem then tells us that for every positive integer t, there will be exactly one blink between time t and time $t+1$.

Theorem. *Let p and q be irrational numbers between 0 and 1 that sum to 1. Let P be the set of real numbers of the form n/p, where n is a positive integer, and Q the set of reals of the form n/q. Then for every positive integer t, there is exactly one element of $P \cup Q$ in the interval $[t, t+1)$.*

Let's first see why the theorem applies. Let p be the rate of the first blinker; that is, it blinks p times per second. Since the first blink will be at time 0, the next will be at time $1/p$, then $2/p$, and so forth; that is, its blinking times after time 0 will be the set P.

Similarly, the second blinker will blink at rate q and its blinking times will be 0 together with times in the set Q.

Saying that together they blink an average of once a second is the same as saying $p+q = 1$, and if p/q (which we could write as $p/(1-p)$ or $(1-q)/q$) is irrational than so are p and q.

The conclusion of the theorem is then that there is exactly one blink in each interval between any positive integer and its successor. Why did I think that might be surprising? We know that we'll see one blink *on average* in such intervals, but exactly one every time? Since the two rates are not rationally related, we know that there will be some moment when there are two blinks less than (say) one millionth of a minute apart. Yet, if you believe the theorem, wedged between those two blinks there must be an integer!

But the theorem is true; let's prove it.

Proof. Of course, we start from the end. The key is to notice that the theorem's conclusion is equivalent to the following statement: For any positive integer t, the number of blinks (by both blinkers together) after time 0 and before time t is exactly $t-1$. In symbols, $|(P \cup Q) \cap (0, t)| = t-1$.

Why? Because then the number of blinks between time t and time $t+1$ has to be $(t + 1 - 1) - (t - 1) = 1$.

The number of times the first blinker blinks after time 0 and before time t is $\lfloor pt \rfloor$, the greatest integer less than or equal to pt. Similarly, the second blinker blinks $\lfloor qt \rfloor$ times in that period. The total number of blinks in the open interval $(0, t)$ is thus $\lfloor pt \rfloor + \lfloor qt \rfloor$. We know $pt + qt = (p + q)t = t$ is an integer, but pt and qt aren't integers; so what can $\lfloor rt \rfloor + \lfloor st \rfloor$ be? It's strictly less than the integer $pt + qt$ but can't be less by as much as 2, because we only cut off two fractions between 0 and 1. Since it's an integer itself, it must be exactly $rt + st - 1 = t - 1$, and we are done! ♡

22. Seeing Is Believing

Your mind's eye is a powerful tool. It's not a coincidence that when you suddenly understand something, you say "I see it now." Drawing a picture—paper is allowed, too, or laptop or electronic pad—can be the magic ingredient that helps you solve a puzzle, even one that doesn't seem at first to call for it.

Meeting the Ferry

Every day at noon GMT a ferry leaves New York and simultaneously another leaves Le Havre. Each trip takes seven days and seven nights, arriving before noon on the eighth day. How many of these cross-Atlantic ferries does one of them pass on its way across the pond?

Solution: Draw it! Put New York on the left of your page and Le Havre on the right; imagine time running down the page. Then each ferry trip is a slanted line across the page, and you can verify that each line meets 13 others on the way.

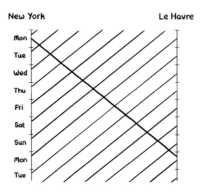

Mathematical Bookworm

The three volumes of Jacobson's *Lectures in Abstract Algebra* sit in order on your shelf. Each has $2''$ of pages and a front and back cover each $\frac{1}{4}''$, thus a total width of $2\frac{1}{2}''$.

A tiny bookworm bores its way straight through from page 1, Vol I to the last page of Vol III. How far does it travel?

Solution: Sounds like the answer should be $3 \times 2\frac{1}{2}'' - 2 \times \frac{1}{4}'' = 7''$, that is, the total width of the three volumes less the front cover of Volume I and the back cover of Volume III.

But *visualize* the three volumes lined up on your shelf. Where is the first page of Volume I? And the last page of Volume III? That's right, the worm passes only through the middle volume and two additional covers, a total of $2\frac{1}{2}'' + 2 \times \frac{1}{4}'' = 3''$.

Yes, this means that in a sense the correct way to order multi-volume books on your shelf is right-to-left, not left-to-right. If you pull the books off the shelf together with the intent of reading them in one sitting (not recommended), they are then stacked correctly.

Rolling Pencil

A pencil whose cross-section is a regular pentagon has the maker's logo imprinted on one of its five faces. If the pencil is rolled on the table, what is the probability that it stops with the logo facing up?

Solution: A mental picture is all you need for this one. The pencil will end up with one of the five faces down, therefore none facing straight up—so the answer is zero. If you count facing *partially* up, make that $\frac{2}{5}$.

Splitting a Hexagon

Is there a hexagon that can be cut into four congruent triangles by a single line?

Solution: Drawing random hexagons and trying to split them this way will give you a headache. Better: Start with four copies of some triangle, and try to abut them all to a single line segment, to make a hexagon. The triangles have a total of 12 sides; uniting three pairs of sides will get you down to nine sides, so you'll need to get some additional pairs of triangle-sides to line up.

This is most easily done if you use right triangles, in which case you want each side of the segment to support one long leg and one short leg, while the remaining legs join up perpendicular to the segment, as pictured below.

Not exactly the kind of hexagon that first springs to mind, though!

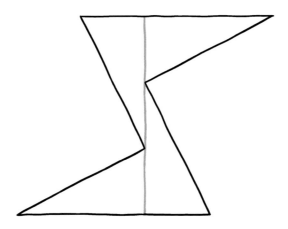

Circular Shadows I

Suppose all three coordinate-plane projections of a convex solid are disks. Must the solid be a perfect ball?

Solution: No, in fact you can take a ball and slice off a piece from it without affecting its coordinate-plane projections. Just pick a point on its surface that's far from any coordinate plane through the ball's center—for example, the point $(\frac{1}{\sqrt{3}}, \frac{1}{\sqrt{3}}, \frac{1}{\sqrt{3}})$ on the surface of the unit-radius ball centered at the origin.

Mark a small circle on the ball's surface around your point and shave off the cap bounded by the circle. That creates a little flat spot on the sphere that will go undetected by the projections.

If you really want to avoid wasting material when you 3D-print your solid, you could do this with eight points, one centered in each octant, making the circles just touch the coordinate planes.

But you can also go the other way and construct a big object that contains every other object, convex or not, with the same disk-shaped coordinate projections: Intersect three long cylinders of the same radius, each with its axis along a different coordinate axis.

Trapped in Thickland

The inhabitants of Thickland, a world somewhere between Edwin Abbott's *Flatland* and our three-dimensional universe, are an infinite set of congruent convex polyhedra that live between two parallel planes. Up until now, they have been free to escape from their slab, but haven't wanted to. Now, however, they have been reproducing rapidly and thinking about colonizing other slabs. Their high priest is worried that conditions are so crowded, no inhabitant of Thickland can escape the slab unless others move first.

Is that even possible?

Solution: Yes, even if the inhabitants are regular tetrahedra. First, note that if you view a regular tetrahedron edge-on, as shown on the left of the figure below, you see a square. Arrange light and dark tetrahedra as shown, with the near edge of the dark ones running SW to NE, and of the light ones NW to SE. One of the two parallel planes enclosing Thickland contains the near edge of each tetrahedron, the other the far edge.

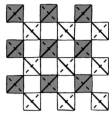

At the moment the tetrahedra have lots of freedom; each can pop straight out of Thickland by just traveling perpendicular to the planes. But now let's squash the tetrahedra together as shown. Now every dark tetrahedron is locked in by two light ones (its NW and SE neighbors) from above and two other light ones (its SW and NE neighbors) from below, and similarly for light tetrahedra. So no one can escape unless at least two other tetrahedra move over. In fact, if you jam them in so that their faces touch, infinitely many tetrahedra will need to move just to get one out.

Polygon Midpoints

Let n be an odd integer, and let a sequence of n distinct points be given in the plane. Find the vertices of a (possibly self-intersecting) n-gon that has the given points, in the given order, as midpoints of its sides.

Solution: Let the midpoints given by $M_1, \ldots, M_n$, and suppose that indeed there are points $V_1, \ldots, V_n$ which are the vertices of a polygon such that M_1 is the midpoint of the side V_1, V_2, M_2 the midpoint of A_2, A_3, and so forth, ending with M_n being the midpoint of A_n, A_1.

It's clear that if we know any of the A_i (say, A_1), the rest of the vertices are determined; we could reflect A_1 180° around M_1 to get A_2, and continue. That will work find if, when we finally reflect A_n around M_n, we find ourselves back at A_1. But why should we?

Let's try it: Take an arbitrary point P in the plane, reflect it around M_1 to get P_2, then M_2 to get P_3, etc., ending, say, at the point Q. You won't find that $Q = P_1$ unless you're fantastically lucky. But what happens if you do the procedure again, starting at Q? Then the vectors QP, P_1Q_1, Q_2P_2, P_3Q_3 etc., are all the same, and since n is odd, you end back at P.

Aha! In that case, let's do the procedure one more time, starting at the midpoint R between P and Q. Now you finish at R, and the polygon has been constructed!

It turns out that this solution is unique. When n is even, generally there is no solution; but if some starting point P works (by getting back to P after reflections) then *any* point will work, giving infinitely many solutions.

Not Burning Brownies

When you bake a pan of brownies, those that share an edge with the edge of the pan will often burn. For example, if you bake 16 square brownies in a square pan, 12 are subject to burning. Design a pan and divide it into brownie shapes to make 16 identically shaped brownies of which as few as possible will burn. Can you get it down to four burned brownies? Great! How about three?

Solution: You can beat the squares with sixteen 14×1 rectangular brownies, arranged as in the figure below, only four of which lie on the boundary.

It seems that you ought to be able to get down to three burnt brownies, but ordinary triangles won't do the trick. What if you curve them?

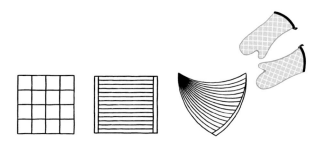

The 3-burn shape in the figure is constructed using circular arcs with the property that the straight line distance between the end points is equal to the radius of the circle.

Protecting the Statue

Michelangelo's *David*, in Florence, is protected (on the plane) by laser beams in such a way that no-one can approach the statue or any of the laser-beam sources without crossing a beam.

What's the minimum number of lasers needed to accomplish this? (Beams reach 100 meters, say.)

Solution: You'll need a non-convex protected area around David, in order to protect the lasers themselves as well as David; in fact you'll want every edge of your polygon to be incident to a concave corner.

You can get this with a hexagon shaped like a three-pointed star. Arrange the lasers at the vertices of a regular hexagon around David, beamed outward to cross at three corners, as shown below.

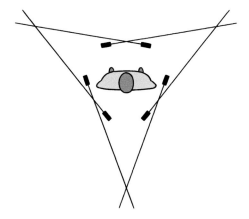

To see that six laser-beam sources are needed, let C be the convex closure of the protected area; it has at least three sides. At each vertex of C, at least two beams must cross, but no beam can contribute to two vertices because then the laser source for that beam would lie outside the protected area. So there must be at least $3 \times 2 = 6$ lasers.

Gluing Pyramids

A solid square-base pyramid, with all edges of unit length, and a solid triangle-base pyramid (tetrahedron), also with all edges of unit length, are glued together by matching two triangular faces.

How many faces does the resulting solid have?

Solution: This problem showed up in 1980 on the Preliminary Scholastic Aptitude Exam (PSAT), but, to the embarrassment of the Educational Testing Service (ETS), the answer they marked as correct was wrong. A confident student called the ETS to task when his exam was returned to him. Luckily for us, the *correct* answer boasts a marvelous, intuitive proof.

The square-base pyramid has five faces and the tetrahedron four. Since the two glued triangular faces disappear, the resulting solid has $5 + 4 - 2 = 7$ faces, right? This, apparently, was the intended line of reasoning. It may have occurred to the composer that in theory, some pair of faces, one from each pyramid, could in the gluing process become adjacent and coplanar. They would thus become a single face and further reduce the count. But, *surely*, such a coincidence can be ruled out. After all, the two solids are not even the same shape.

In fact, this does happen (twice): The glued polyhedron has only five faces.

You can see this in your own head. Imagine *two* square-based pyramids, sitting side-by-side on a table with their square faces down and abutting. Now, draw a mental line between the two apices; observe that its length is one unit, the same as the lengths of all the pyramid edges.

Thus, between the two square-based pyramids, we have in effect constructed a regular tetrahedron. The two planes, each of which contains a triangular face from each square-based pyramid, also contain a side of the tetrahedron; the result follows. ♡ (Check the figure below if you find this hard to visualize.)

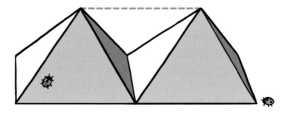

Precarious Picture

Suppose that you wish to hang a picture with a string attached at two points on the frame. If you hang it by looping the string over two nails in the ordinary way, as shown below, and one of the nails comes out, the picture will still hang (albeit lopsidedly) on the other nail.

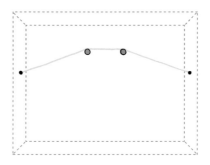

334

Can you hang it so that the picture falls if *either* nail comes out?

Solution: One of several ways to hang the picture is illustrated below, with slack so you can see better how it works. This solution requires passing the string over the first nail, looping it over the second, sending it back over the first nail, then looping it again over the second nail but counter-clockwise.

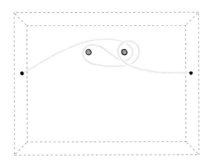

There are also some non-topological solutions: For example, you can pinch a loop of the string between two closely spaced nails, assuming the nail-head width is not much larger than the string diameter. But why rely on friction when you can use mathematics?

Finding the Rectangles

Prove that any tiling of a regular 400-gon by parallelograms must contain at least 100 rectangles.

Solution: Walk across the tiling from any edge to the opposite edge. That path must cross a similar path from the 90-degree-away edges, at a rectangle. Since the sides of that rectangle are parallel or perpendicular to only four edges of the polygon, there must be at least 100 such rectangles.

Tiling a Polygon

A "rhombus" is a quadrilateral with four equal sides; we consider two rhombi to be different if you can't translate (move without rotation) one to coincide with the other. Given a regular polygon with 100 sides, you can take any two non-parallel sides, make two copies of each and translate them to form a rhombus. You get $\binom{50}{2}$ different rhombi that way. You can use translated copies of these to tile your 100-gon; show that if you do, you will use each different rhombus exactly once!

Solution: Let $\vec{u}$ be one of the sides of the $2n$-gon; a $\vec{u}$-rhombus is any of the $n-1$ rhombi using $\vec{u}$ as one of its two vectors. In a tiling, the tile next to a $\vec{u}$-side must be a $\vec{u}$-rhombus, as must the tile on the other side of that one, and so forth until we reach the opposite side of the $2n$-gon. Notice that each step of this path proceeds in the same direction (i.e., right or left) with respect to the vector $\vec{u}$, as must any other path of $\vec{u}$-rhombi; but then there can be no other $\vec{u}$-rhombi, since they would generate paths with no way to close and nowhere to go.

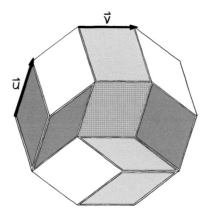

The similarly defined path for a different side $\vec{v}$ must cross the $\vec{u}$, and the shared tile is of course made up of $\vec{u}$ and $\vec{v}$. Can they cross twice? No, because a second crossing would have $\vec{u}$ and $\vec{v}$ meeting at an angle greater than π inside the common rhombus. ♡

Tiling with Crosses

Can you tile the plane with 5-square crosses? Can you tile 3-space with 7-cube crosses?

Solution: Tiling the plane with 5-square crosses is a snap if you organize them into diagonals, as shown on the left side of the figure below.

To tile 3-space with 7-cube crosses, use five of the seven cubes of each of a lot of crosses to tile a unit-thickness flat slab, much as the flat crosses were used to tile the plane. However, between each pair of diagonal strips, leave blank a strip of cattycorner dominoes.

Each of these is filled in one cube from below the slab, and one from above. ♡

Cube Magic

Can you pass a cube through a hole in a smaller cube?

Solution: Yes. To pass a unit cube through a hole in a second unit cube, it suffices to identify a projection of the (second) cube which contains a unit square in its interior. A square cylindrical hole of side slightly more than 1 can then be made in the second cube, leaving room through which to pass the first cube.

You can then do the same, with even smaller tolerances, if the second cube is just a bit smaller than a unit cube.

The easiest (but not the only) projection you can try this with is the regular hexagon you get by placing three vertices and the centroid on a plane perpendicular to your line of sight. You can see this hexagon by viewing the cube so that one of its vertices is centered.

Letting A be the projection of one of the visible faces onto the plane, we observe that its long diagonal is the same length ($\sqrt{2}$) as a unit square's, since that line has not been foreshortened. If we slide a copy of A over to the center of the hexagon, then widen it to form a unit square B, B's widened corners will not reach the vertices of the hexagon (since the distance between opposing vertices of the hexagon exceeds the distance between opposing sides).

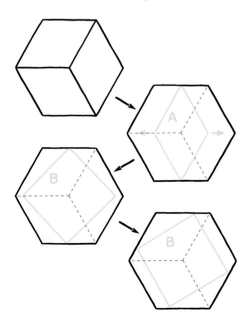

It follows that if we now tilt B slightly, all four of its corners will lie strictly inside the hexagon. ♡

Circles in Space

Can you partition all of 3-dimensional space into circles?

Solution: It's an odd thing to ask—partitioning space into objects of lower dimension. But there's nothing to prevent you from trying to do that, and in fact, in this case it is indeed possible.

But how? We can certainly partition 3-space minus a point into spheres, namely all spheres having the missing point as their center. Can we partition a sphere into circles? If we leave out the poles, you can certainly partition the rest into circles of lattitude. In fact we can omit any two points P and Q, and partition the rest into circles whose centers lie on the plane through P, Q and the sphere's center; for example, circles whose ratio of distance to P and Q is the same along the short route from P to Q as it is on the long route.

With these observations we can construct a partition of all of 3-space into circles as follows. We start with a line of unit-radius circles on the XY-plane, one centered at $(4n+1, 0)$ for each integer n. The key is that every sphere centered at the origin hits the union of those circles in exactly two points. And, very importantly, the origin is covered (by the unit circle centered at (1,0)).

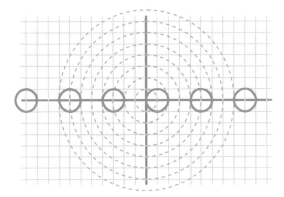

Thus, if we partition every sphere centered at the origin, with its two intersection points omitted, into circles, then all those circles plus our line of circles on the plane partition space perfectly! ♡

Invisible Corners

Can it be that you are standing outside a polyhedron and can't see any of its vertices?

Solution: Yes. Imagine six long planks arranged so that they meet in their middles to form the walls of a cubical room, but do not quite touch. From the middle of this room you won't be able to see any vertices. These planks can easily be hooked up far from the room to form a polygon, with the necessary additional vertices well hidden.

[If there is a plane separating you from the polyhedron, this can't happen.]

For our theorem, I can't resist using one of the most famous of all theorems with no-word proofs. A picture is all you need!

A *lozenge* is a rhombus formed from a pair of unit-edge triangles glued together along an edge. Suppose a regular hexagon with integer sides is tiled with lozenges. The tiles come in three varieties, depending on orientation.

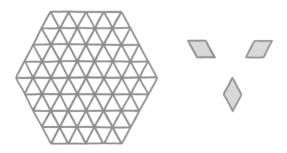

Theorem. *Any tiling of a regular hexagon with integer sides by unit lozenges uses precisely the same number of lozenges of each of the three orientations.*

Proof.

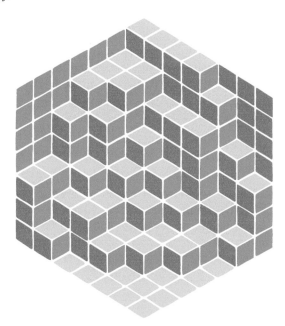

23. Infinite Choice

Need an axiom? As often as not, the Axiom of Choice is the axiom of choice.

The Axiom of Choice (abbreviated AC here and in logic books) says that given any collection of nonempty sets, you can choose an element from each set. Seems pretty reasonable, right? What's to stop you?

Indeed, if the collection is finite, there's no problem (you can prove by induction that one item can be selected from each set.) If the sets have distinguished objects, you can use them; for instance, among infinitely many pair of shoes, you could choose the left shoe from each pair. But suppose you have infinitely many pairs of socks?

There is a name, actually, for the set of all ways to choose an element from each set: It's called the *product* of the sets. The size of the product, if both the sets and the collection are finite, is the ordinary product of the sizes of the sets. It seems unlikely that the product of an infinite collection of nonempty sets would suddenly decide to be empty, but that can happen if you don't have AC. And if you're a mathematician, you might find it handy to have tools like Zorn's Lemma, the well-ordering theorem, or the Hausdorff maximal principle (each equivalent to AC), as well as myriad other theorems that require AC without being obvious about it.

On the other hand, AC has some pretty weird consequences. The most famous of these is the Banach–Tarski paradox: A unit ball in 3-space can be partitioned into five subsets which can then be reassembled (using only rotations and translations) to form two unit balls! Mathematicians are used to having to get around problems like this by restricting attention to "measurable" sets. But, as you will see later, there are other bizarre consequences of AC that mathematicians are less familiar with.

So is AC true or false? I like to tell my students that although neither the axiom of choice nor its negation can be disproved, either can be made to look ridiculous. Both AC and "not AC" are consistent with the usual axioms of set theory, so you have a choice; as a puzzle solver you will mostly want to choose AC. (Assumptions like the Axiom of Determinacy that are not consistent with AC can occasionally come in handy, but we will not pursue such contingencies here.)

You might reasonably ask: How do collections of sets that don't have obvious "choice functions" arise? One way is in the consideration of equivalence relations.

A (binary) relation is technically a set of pairs, but you can think of it as a property that may or may not be true of a given pair of items. For example, among people, "x knows y" is a relation; so is "x and y are siblings." The latter of these is a *symmetric* relation, one that is satisfied by the pair (x, y) if it is satisfied by (y, x). It is also a *transitive* relation: If (x, y) and (y, z) satisfy the relation, so must (x, z).

Suppose we change the relation to "x and y are either full siblings, or are the same person." Then the relation is also *reflexive*, meaning that (x, x) satisfies the relation for any x. Relations that are symmetric, transitive, and reflexive are said to be *equivalence* relations. If a set has an equivalence relation, the set partitions neatly into *equivalence classes*: subsets whose elements are all related to one another but not to anyone outside the subset. For example, for the above relation, the equivalence classes are the sets of people that share a given pair of parents.

Suppose humans are forced to settle a new planet and it's decided that exactly one person from each set of siblings will go. Such a collection is called a *set of representatives* of the equivalence relation. In this case we don't need to apply AC to get a set of representatives; for example, we could always send the eldest of the siblings (and anyway there are only finitely many sibling sets that we know of).

But, generally, we may indeed need AC to get a set of representatives. Consider, for example, the set of all (countably) infinite binary sequences. Suppose $(a_1, a_2, \dots)$ is one such sequence and $(b_1, b_2, \dots)$ is another. Say that these two sequences are related if for all but finitely many indices i, $a_i = b_i$. You can easily verify that this relation is symmetric, transitive, and reflexive, therefore is an equivalence relation. Thus the set of all infinite binary sequences breaks up into (a lot of) equivalence classes. One of these, for example, is the set of all binary sequences that have only finitely many 1's.

According to AC, it is possible to pick one sequence from each equivalence class. From the aforementioned class of sequences with only finitely many 1's, you might want to pick the all-zero sequence. But for most classes there won't be an obvious choice, so it's hard to see how you can avoid using AC. Let's go ahead and invoke AC to get our set of representatives. What good is it? Well, if you're a prisoner. . .

Hats and Infinity

Each of an infinite collection of prisoners, numbered $1, 2, \ldots,$ is to be fitted with a red or black hat. At a prearranged signal, all the prisoners are revealed to one another, so that everyone gets to see all his fellow prisoners' hat colors—but no communication is permitted. Each prisoner is then taken aside and asked to guess the color of his own hat.

All the prisoners will be freed provided *only finitely many* guess wrongly. The prisoners have a chance to conspire beforehand; is there a strategy that will ensure freedom?

Solution: If we associate "red" with the digit 1 and "black" with 0, assignments of hats to prisoners correspond precisely to infinite binary sequences. Suppose that the prisoners believe in AC and have the wherewithal to agree in advance on a set of representatives for the equivalence relation described above. Thus, for each set S of related sequences, they have identified a particular member.

When the curtain goes up, prisoner i sees the whole binary sequence except for its ith entry. That's enough for him to identify the equivalence class S to which that sequence belongs. He then looks up (or remembers) the representative that everyone has agreed upon for the set S, and guesses that his own hat color corresponds to the ith entry of that representing sequence.

Consequently, everyone's guess will correspond to this same representing sequence, and since that sequence is itself in S, only finitely many prisoners will have guessed wrong. ♡

It can be shown that under certain assumptions that contradict AC, the prisoners have essentially no chance, even if the warden is obligated to assign colors uniformly at random.

Try this similar puzzle.

All Right or All Wrong

This time the circumstances are the same but the objective is different: The guesses must either be *all right* or *all wrong*. Is there a winning strategy?

Solution: This version perhaps sounds even tougher, as only two successful outcomes are possible. But it does admit a cute solution if there are only finitely many prisoners (stop here and see if you can do the finite version before reading further).

Yes, if there are only finitely many prisoners, they can simply decide to assume that the number of red hats is even. In other words, a prisoner who sees an odd number of red hats guesses that his own hat is red, otherwise that it is black. Then if, indeed, the number of red hats is even, the prisoners are all correct; if odd, all wrong.

In the infinite case, this won't work unless the warden assigns only finitely many hats of one color. But the prisoners' agreed-upon set of representatives makes things easy. Each prisoner again identifies the equivalence class of the hat assignment and finds its representative; he then assumes that (say) the number of discrepancies between the actual sequence and the representative is even, and guesses his own color accordingly.

However, there's a simpler solution: Everyone guesses "green"! ♡

The Axiom of Choice has the powerful consequence that any set can be well-ordered. (A set X is said to be well-ordered if it has an order relation on it with the property that all non-empty subsets of X have a lowest element.) In addition, it's possible to ensure that the set of elements below any element of X has "cardinality" (i.e., size) less than the cardinality of X itself. We won't prove these consequences here, but we'll use them to solve the following innocent-sounding problem.

Double Cover by Lines

Let $\mathcal{L}_\theta$ be the set of all lines on the plane at angle θ to the horizontal. If θ and θ' are two different angles, the union of the sets $\mathcal{L}_\theta$ and $\mathcal{L}_{\theta'}$ constitutes a *double cover* of the plane, that is, every point belongs to exactly two lines.

Can this be done in any other way? That is, can you cover each point of the plane exactly twice using a set of lines that contains lines in more than two different directions?

Solution: Since this problem appears here, you have a big clue that the answer is yes if you have AC. You can prove it by well-ordering all the points on the plane in such a way that the set of points below any point has cardinality less than $2^{\aleph_0}$. ($2^{\aleph_0}$ stands for the cardinality of the set of all real numbers, also the set of angles, the set of points on the plane, and many other sets; it's strictly greater than $\aleph_0$, the cardinality of the set of whole numbers. The symbol $\aleph$, pronounced "aleph," is the first letter of the Hebrew alphabet.) We start by picking three lines that cross one another, so that we already have our three directions. Let P be the lowest point in our well-ordering that isn't already double-covered, and pick a line through P that misses the three points that are currently double-covered.

Now repeat the process with a new point Q, this time avoiding a larger, but still finite, set of points that are already double-covered. We can do this until we have a countably infinite number of lines in our set. But why stop there? Until we're done, there's always a lowest point of the plane that hasn't been double-covered yet. And because the number of points so far considered is less than $2^{\aleph_0}$, there's always an angle available that misses all points that are currently double-covered.

This construction is perhaps disappointing in that it does not leave you with any geometry you can wrap your head around. No better construction is known.

Let us return to direct applications of AC. If you were unimpressed by AC's power to free prisoners, perhaps the next puzzle will get your attention.

Wild Guess

David and Carolyn are mathematicians who are unafraid of the infinite and cheerfully invoke the Axiom of Choice when needed. They elect to play the following two-move game. For her move, Carolyn chooses an infinite sequence of real numbers, and puts each number in an opaque box. David gets to open as many boxes as he wants—even infinitely many—but must leave one box unopened. To win, he must guess exactly the real number in that box.

On whom will you bet in this game, Carolyn or David?

Solution: I can hear you thinking: "I'm betting on Carolyn! She can just put, say, a random real number between 0 and 1 in each box. David has no clue what's in the unopened box, hence his probability of winning is zero."

In fact David has an algorithm that will guarantee him at least a 99% chance of winning, regardless of Carolyn's strategy. You don't believe me? That shows good judgment on your part, but it's nonetheless true given AC.

Here's how it works. Before the game even starts, David proceeds much as the prisoners in *Hats and Infinity* did, but this time with sequences of real numbers instead of bits. (If you state the problem using bits instead of reals, David can use the prisoners' set of representatives.) So, two sequences of reals $x_1, x_2, \ldots$ and $y_1, y_2, \ldots$ are related if $x_i \neq y_i$ for only finitely many indices i. Again this is an equivalence relation, and David invokes AC to choose from each equivalence class of real sequences one representative sequence. He can even show his set of representatives to Carolyn; it won't help her!

Now Carolyn picks her sequence, let's call it $c_1, c_2, \ldots$, and boxes up the numbers. David takes the boxes and breaks them up into 100 infinite rows. We may as well assume the boxes in row 1 contain c_1, c_{101}, c_{201}, etc., while row 2 begins with c_2 and c_{102}; finally row 100 contains c_{100}, c_{200} and so forth. Each row is itself a sequence of real numbers.

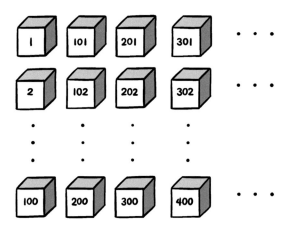

In a crucial step, David now chooses, uniformly at random, an integer between 1 and 100. Say it's 44.

David first opens all the boxes in row 1. This sequence belongs to some equivalence class, say S_1, whose representative he now looks up. The representative, say $d_1, d_2, \ldots$, will differ from row 1 in finitely many positions, the last of which is, say, the 289th entry in each. David writes on a piece of paper "$n_1 = 289$." (If the sequences happen to be identical, he writes "$n_1 = 0$.")

He repeats the procedure with row 2, this time getting a (probably) different class S_2 with its own representative, and ends up writing down perhaps "$n_2 = 4{,}183{,}206$."

This continues with every row except the 44th. When he's done with row 100, David has written down values for the 99 numbers $n_1, n_2, \ldots, n_{43}$ and $n_{45}, \ldots, n_{100}$. He sets m equal to one more than the maximum of these numbers.

Now he finally turns to row 44, opening not all the boxes, but only those after the mth. This is enough for him to determine the equivalence class S_{44} of that row, and to retrieve its representative sequence, which we'll denote by $f_1, f_2, \ldots$. David now guesses that the mth box in row 44 contains the real number f_m.

So why does this work? Well, ultimately there is some largest index (n_{44}) which represents the last position where row 44 and $f_1, f_2, \ldots$ disagree. Unless $n_{44} \geq m$, David's guess will be correct.

But m was greater than the maximum of $n_1, n_2, \ldots, n_{43}$ and $n_{45}, \ldots, n_{100}$, so in order for n_{44} to equal or exceed m, n_{44} would have to be the unique largest of the 100 numbers $n_1, \ldots, n_{100}$. What's the probability of that? Since 44 was random, and only one of these numbers can be the unique largest, it's at most $1/100$. Bingo!

If you find that you have to read this proof more than once to appreciate it, you are not alone. But it's legitimate. The Axiom of Choice really does imply that David can win this game with any desired probability $p < 1$. Is there something wrong with your intuition, or is the Axiom of Choice suspect? You be the judge. If it's any help, you can also use AC to predict the future (see *Notes & Sources*).

I hear the following objection. Suppose Carolyn chooses her numbers independently, each uniformly random between 0 and 1. Then the content of any box, in particular the mth box of row 44, is random and there's no way David can guess it with probability 99% or indeed any probability greater than zero.

The counter to this objection is that when we say infinitely many random numbers are independent, we usually mean that each value is independent of any *finite* number of the others. If you want each number to be independent of *all* of the others, and you believe in AC, you're stuck! It's interesting, in my view, that AC—which we usually think of as an *aid* to constructing infinite things—prevents you from constructing an infinite sequence of random reals, or even random bits, that are independent in this strong sense.

We now move on to some other puzzles involving infinity (but we'll be assuming the axiom of choice holds, in everything that follows). We start with a an infinite expression.

Exponent upon Exponent

Part I: If $x^{x^{x^{x^{\cdot^{\cdot^{\cdot}}}}}} = 2$, what is x?

Part II: If $x^{x^{x^{x^{\cdot^{\cdot^{\cdot}}}}}} = 4$, what is x?

Part III: How do you explain getting the same answer to Parts I and II?

Solution: If it means anything at all, $x^{x^{x^{\cdot^{\cdot^{\cdot}}}}}$ must be the limit of the sequence

$$x, \ x^x, \ x^{x^x}, \ x^{x^{x^x}}, \ \ldots,$$

assuming that limit exists. Note that this expression is not the same as

$$(\ldots(((x^x)^x)^x)\ldots).$$

The exponent of the bottom x in the expression $x^{x^{x^{\cdot^{\cdot^{\cdot}}}}}$ is the same as the expression itself; thus, if $x^{x^{x^{\cdot^{\cdot^{\cdot}}}}} = 2$ then $x^2 = 2$, $x = \sqrt{2}$.

That's one part of the puzzle out of the way.

For the second part, similar reasoning tells you that if $x^{x^{x^{\cdot^{\cdot^{\cdot}}}}} = 4$, then $x^4 = 4$, thus $x = \sqrt[4]{2} = \sqrt{2}$.

Aha. Something's wrong here. What exactly is $\sqrt{2}^{\sqrt{2}^{\sqrt{2}^{\cdot^{\cdot^{\cdot}}}}}$? It can't be both 2 and 4. Which is it, if either? Or is it something else, or nothing at all?

We need to look at the sequence $\sqrt{2}, \ \sqrt{2}^{\sqrt{2}}, \ \sqrt{2}^{\sqrt{2}^{\sqrt{2}}}, \ \ldots$, and determine whether it has a limit. In fact, it does; the sequence is increasing and bounded above.

To show the former, we name the sequence $s_1, s_2, \ldots$ and prove by induction that $1 < s_i < s_{i+1}$ for each $i \geq 1$. This is easy: Since $\sqrt{2} > 1$, $s_2 = \sqrt{2}^{\sqrt{2}} > \sqrt{2} = s_1$; and $s_{i+1} = \sqrt{2}^{s_i} > \sqrt{2}^{s_{i-1}} = s_i$.

To get the bound, observe that if we replace the top $\sqrt{2}$ in any s_i by the larger value 2, the whole expression collapses to 2. (We could similarly show the sequence is bounded by 4, but of course if it's bounded by 2 it's automatically bounded by anything bigger than 2.)

Now that we know the limit exists, let us call it y; it must indeed satisfy $\sqrt{2}^y = y$. Looking at the equation $x = y^{1/y}$, we observe (perhaps using elementary calculus—sorry) that x is strictly increasing in y up to its maximum at $y = e$ and strictly decreasing thereafter. Thus, there are at most two values of y corresponding to any given value of x, and for $x = \sqrt{2}$, we know the values: $y = 2$ and $y = 4$.

Since our sequence is bounded by 2, we rule out 4 and conclude that $y = 2$. ♡

Generalizing the above argument, we see that $x^{x^{x^{\cdot^{\cdot^{\cdot}}}}}$ is meaningful and equal to the lower root of $x = y^{1/y}$, as long as $x \leq e^{1/e}$. For $x = e^{1/e}$, the expression is equal to e, but as soon as x exceeds $e^{1/e}$, the sequence diverges to infinity. This is why Part II of the puzzle fails:

There is no x such that $x^{x^{x^{\cdot^{\cdot^{\cdot}}}}} = 4$.

We have already noted that there are different infinities; the smallest is *countable* infinity, which we called $\aleph_0$. A set is said to be countable if you can put its members into one-to-one correspondence with the positive integers. The set of rational numbers (fractions) is countable; reference to a particularly nice way of putting them into correspondence with the positive integers is given in *Notes & Sources*. The set of *real* numbers, however, is not countable, as observed by the brilliant Georg Cantor in 1878.

Here's a simple puzzle for which you can make use of countability.

Find the Robot

At time $t = 0$ a robot is placed at some unknown grid point in 3-dimensional space. Every minute, the robot moves a fixed, unknown distance in a fixed, unknown direction, to a new grid point. Each minute, you are allowed to probe any single point in space. Devise an algorithm that is guaranteed to find the robot in finite time.

Solution: There are only countably many integer vectors (a, b, c, x, y, z), each coding a starting position (a, b, c) and displacement (x, y, z). List them as v_1, v_2, etc., and at time t, probe the point $(a, b, c) + t(x, y, z)$ where $(a, b, c, x, y, z) = v_t$.

Figure Eights in the Plane

How many disjoint topological "figure 8s" can be drawn on the plane?

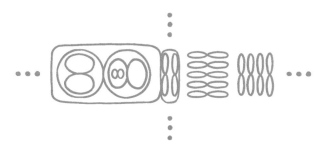

Solution: We could draw concentric circles on the plane with all possible positive real diameters, hence if the puzzle asked for circles instead of figure eights, the answer would be "uncountably many," or more precisely, "the cardinality of the reals."

However, we can only draw countably many eights. Associate with each 8 a pair of rational points (points of the plane with both coordinates rational numbers), one in each loop; no two figure eights can share a pair of points. Hence, the cardinality of our set of 8s is no greater than the set of pairs of pairs of rational numbers, which is countable. ♡

Was that too easy? Try the next one!

Y's in the Plane

Prove that only countably many disjoint **Y**'s can be drawn in the plane.

Solution: Here is a particularly neat proof (there is more than one way to do this). Associate with each **Y** three rational circles (rational center and radius) containing the endpoints, and small enough so that none contains or intersects any other arm of the **Y**. We claim that no three **Y**s can all have the same triple of circles; for, if that were so, you could connect the hub of each **Y** to the center of each circle by following the appropriate arm until you hit the circle, then following a radius to the circle's center. This would give a planar embedding of the graph $K_{3,3}$, sometimes known as the "gas-water-electricity network."

In other words, we have created six points in the plane, divided into two sets of three each, with each point of one set connected by a curve to each point of the other set, and no two curves crossing. This is impossible; in fact, readers who know Kuratowski's Theorem will recognize this graph as one of the two basic nonplanar graphs.

To see for yourself that $K_{3,3}$ cannot be embedded in the plane without crossings, let the two vertex sets be $\{u, v, w\}$ and $\{x, y, z\}$. If we could embed it without crossings the sequence u, x, v, y, w, z would represent consecutive vertices of a (topological) hexagon. The edge uy would have to lie inside or outside the hexagon (let us say inside); then vz would have to lie outside to avoid crossing uy, and wx has no place to go. $\heartsuit$

There's no shortage of serious mathematical theorems that use some form of AC in their proofs; they're rampant in algebra, analysis and topology. Here's one that concerns only basic graph theory, which we make use of elsewhere in this volume.

Recall that a graph is a set of points, called *vertices*, together with a symmetric relation between pairs of distinct vertices. Two related vertices are said to be *adjacent* and to constitute an *edge* of the graph, usually illustrated by connecting the vertices by a line segment or curve. A graph is *two-colorable* if its vertices can be partitioned into two sets X and Y so that no two vertices in the same one of these sets are adjacent. Below is an example of a graph with its vertices two-colored (by black and white).

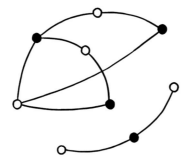

The graph whose vertices are the digits 0 through 9 with 0 adjacent to 1, 1 to 2, 2 to 3, ..., 8 to 9 and 9 to 0, can be colored with two colors by putting the even digits in X and the odd digits in Y. But if we try two-coloring the same construction but with an odd number of vertices—an "odd cycle"—it won't work.

In fact, no graph containing an odd cycle can be two-colorable. A famous, elementary theorem of graph theory says the converse.

Theorem. *If a graph contains no odd subset of vertices whose adjacencies include those of an odd cycle, then the graph is two-colorable.*

The idea of the proof is this: Pick some vertex v and put it in X, then put all the vertices that are adjacent to v in Y, then all the vertices adjacent to *them* in X again, and so forth. The absence of odd cycles assures you that you will never find yourself trying to put the same vertex in both X and Y. If at any stage you find that you are finding no new adjacencies but haven't used up all the vertices of the graph, pick an unassigned vertex v', put it in X, and continue as before.

That's all very fine, but if the graph is infinite and has infinitely many connected components, the part of the proof asking you to "pick an unassigned vertex" requires the Axiom of Choice. Here's an example of such a graph: Let the vertices be all real numbers, with x adjacent to y if either $|x - y| = 1$ or $|x - y| = \sqrt{2}$. This graph has no odd cycles, because if you start a cycle at x and return to x, you will necessarily use an even number of 1-sized steps (some number up, and an equal number down) and likewise an even number of $\sqrt{2}$-sized steps; this is because no multiple of $\sqrt{2}$ is an integer.

Nonetheless, under some axioms that contradict AC, this graph has no two-coloring! Indeed, if you try to *construct* a set of vertices that could serve as X, you will encounter only frustration. *Three*-colorings are easily found, though; it's really quite reasonable, in graph-theoretic terms, to postulate that this graph has chromatic number 3, that is, can be properly colored with three colors but no fewer.

24. Startling Transformation

Often a puzzle that looks impenetrable becomes suddenly transparent if you just look at it a different way. Yes, that may require some creativity and imagination on your part; but great ideas, no matter how brilliant they appear at first, don't arise out of nothing. A little experience may go a long way to sparking your next bit of genius.

Sinking 15

Carol and Desmond are playing pool with billiard balls number 1 through 9. They take turns sinking balls into pockets. The first to sink three numbers that sum to 15 wins. Does Carol (the first to play) have a winning strategy?

Solution: If Carol and Desmond record their sunk balls on the magic square below, then their objective is to fill all three squares in a row, column, or diagonal. Thus, they are playing Tic-Tac-Toe!

2	7	6
9	5	1
4	3	8

As a consequence, Carol does not have a winning strategy; best play leads to a draw.

Slicing the Cube

Before you is a circular saw and a $3 \times 3 \times 3$ wooden cube that you must cut into twenty-seven $1 \times 1 \times 1$ cubelets. What's the smallest number of slices you must make in order to do this? You are allowed to stack pieces prior to running them through the saw.

Solution: If you hold the cube together (carefully!) throughout, you can make two horizontal cuts all the way through the cube, then two vertical North–South cuts and two vertical East–West cuts—six cuts in all—to reduce the cube to 27 cubelets. Can you do it with fewer than six cuts?

   No, because the center cubelet has to be cut out on all six sides, and no cut can do more than one.

Expecting the Worst

Choose n numbers uniformly at random from the unit interval $[0,1]$. What is the expected value of their minimum?

Solution: The answer is $1/(n+1)$. (And the average value of the next-smallest is $2/(n+1)$, and so forth.) This kind of problem can be solved using calculus, but when the solution is so nice, you might suspect there is a simpler way—and there is. We make use of the interval's more symmetrical cousin, the circle.

   Choose $n+1$ numbers $x_0, \ldots, x_n$ uniformly and independently from a circle of circumference one. By symmetry, the expected distance between neighboring numbers is $1/(n+1)$. Cut the circle at x_0 and open it up to form a line segment of length one. The remaining numbers $x_1, \ldots, x_n$ will be uniformly distributed on this line, and the least of these is the next x_i to the right of x_0 (assuming you opened up the circle clockwise). Thus its expected value is the same as the expected value of the distance between x_0 and the next selected point to its right, which is $1/(n+1)$.

   The idea of bending a line segment around to make a circle, then exploiting the symmetry of the circle, will be useful later for proving our theorem. In the meantime, let's look at a betting game.

Next Card Red

Paula shuffles a deck of cards thoroughly, then plays cards face up one at a time, from the top of the deck. At any time, Victor can interrupt Paula and bet $1 that the next card will be red. He bets once and only once; if he never interrupts, he's automatically betting on the last card.

What's Victor's best strategy? How much better than even can he do? (Assume there are 26 red and 26 black cards in the deck.)

Solution: We know that Victor's expectation in this game, if he plays it well, is at least 0—because he can just bet on the first card, which is equally likely to be one of the 26 red cards or to be one of the 26 black cards. Or, he could wait for the last card, with the same result.

But Victor is a smart guy and knows that if he waits for a point when he's seen more black cards go by than red, he can bet at that moment and enjoy odds in his favor. Of course, he could wait until he's seen *all* the black cards go by and then bet knowing he's going to win, but that might not ever happen. Perhaps best is the following conservative strategy: He waits until the first time that the number of black cards he's seen exceeds the number of red, and takes the resulting (probably small) odds in his favor. For example, if the very first card is black, he bets on the next one, accepting a positive expectation of $(26/51) \cdot \$1 + (25/51) \cdot (-\$1)$ which is about 2 cents.

Of course, though unlikely, it possible that the black cards seen never get ahead of the red ones, in which case Victor is stuck betting on the last card, which will be black. It seems like it might be difficult to compute Victor's expectation from this or any moderately complex strategy.

But there is a way—because it's a fair game! Not only has Victor no way to earn an advantage, he has no way to lose one either: All strategies are equally ineffective.

This fact is a consequence of the martingale stopping time theorem, and can also be established by induction on the number of cards of each color in the deck. But there is another proof, which I will describe below, and which must surely be in "the book." (As readers may know or recall from Chapter 9, the late, great mathematician Paul Erdős often spoke of a book owned by God in which is written the best proof of each theorem. I imagine Erdős is reading the book now with great enjoyment, but the rest of us will have to wait.)

Suppose Victor has elected a strategy S, and let us apply S to a slightly modified variation of the game. In the new variation, Victor interrupts Paula as before, but this time he is betting not on the *next* card in the deck, but instead on the *last* card of the deck.

In any given position, the last card has precisely the same probability of being red as the next card. Thus, the strategy S has the same expected value in the new game as it did before.

But, of course, the astute reader will already have observed that the new variation is a pretty uninteresting game; Victor wins if and only if the last card is red, regardless of his strategy. ♡

Magnetic Dollars

One million magnetic "susans" (Susan B. Anthony dollar coins) are tossed into two urns in the following fashion: The urns begin with one coin in each, then the remaining 999,998 coins are thrown in the air one by one. If there are x coins in one urn and y in the other, magnetic attraction will cause the next coin to land in the first urn with probability $x/(x+y)$, and in the second with probability $y/(x+y)$.

How much should you be willing to pay, in advance, for the contents of the urn that ends up with fewer susans?

Solution: You might very reasonably worry that one urn will "take over" and leave very little for the other. In my experience, faced with this problem, most people are unwilling to offer $100 in advance for the contents of the lesser urn.

In fact the lesser urn is worth, on average, a quarter of a million dollars. This is because the final content of (say) Urn A is precisely equally likely to be any whole number of dollars from $1 all the way to $999,999. Thus, the average amount in the lesser urn is

$$\frac{1}{999,999} \left(2 \cdot \$1 + 2 \cdot \$2 + \ldots 2 \cdot \$499,999 + 1 \cdot \$500,000 \right)$$

which is $250,000 plus change.

How can we see this? There are several proofs but here's my favorite. Imagine that we shuffle a fresh-from-the-box deck of cards in the following careful manner. We place the first card, say the Ace of Spades, face down on the table. Then we take the second card, maybe the Deuce of Spades, and randomly place it over or under the ace. Now there are three possible slots for the Three of Spades; choose one at random, putting it on top of the current pile, under the current pile, or between the ace and deuce, each with probability $\frac{1}{3}$.

We continue in this manner; the nth card is inserted in one of n possible slots, each with probability $1/n$. The result eventually is a perfectly random permutation of the cards.

Now think of the cards that go in above the Ace of Spades as the coins (not counting the first) that go into Urn A, and the ones that go under as coins going into Urn B. The magnetic rule above describes the process perfectly, because if there are currently $x-1$ cards above the Ace of Spades (thus x slots) and $y-1$ cards under the Ace of Spades (thus y slots), the probability that the next card goes above the Ace of Spades is $x/(x+y)$.

Since in the final shuffled deck the Ace of Spades is equally likely to be in any position, the number of slots ending up above it is equally likely to be anything from 1 to the number of cards in our deck, which for our puzzle is 999,999.

Integral Rectangles

A rectangle in the plane is partitioned into smaller rectangles, each of which has either integer height or integer width (or both). Prove that the large rectangle itself has this property.

Solution: This problem is famous for its many proofs, each a kind of transformation. Here's one proof that turns the picture into a graph.

Place the big rectangle with its lower left corner at the origin of a Cartesian grid, and sides parallel to the coordinate axes. Let G be the graph whose vertices are all the points with integer coordinates that are also corners of small rectangles (indicated by shaded circles), plus the small rectangles themselves (indicated by white circles at their centers).

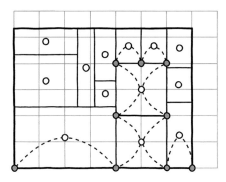

Put an edge between a shaded vertex and a white vertex if the shaded vertex is a corner of the white vertex's rectangle, as shown in the example.

Every white vertex has 0, 2, or 4 neighbors (since one of its dimensions is an integer), and every shaded vertex has two or four neighbors, except that shaded vertices at the corners of the big rectangle have just one neighbor each.

In any graph there must be an even number of vertices that have an odd number of neighbors, since every edge contributes two to the total number of neighbors of all vertices. Thus the origin can't be the only vertex of G with an odd number of neighbors, and it follows that at least one of the other corners of the big rectangle must be an integer point as well, and we are done. $\heartsuit$

Laser Gun

You find yourself standing in a large rectangular room with mirrored walls. At another point in the room is your enemy, brandishing a laser gun. You and she are fixed points in the room; your only defense is that you may summon bodyguards (also points) and position them in the room to absorb the laser rays for you. How many bodyguards do you need to block all possible shots by the enemy?

Solution: You can certainly protect yourself if you have *continuum* many bodyguards, for example, by arranging them in a circle around yourself with, of course, your enemy outside the circle.

That number can be reduced to a countably infinite number of bodyguards, because there are actually only countably many directions your enemy can fire in that will hit you. One way to see this is to view the room as a rectangle in the plane, with you at P and your enemy at Q. You can now tile the plane with copies of the room, by repeatedly reflecting the room about its walls, each copy containing a new copy of your enemy (see figure below).

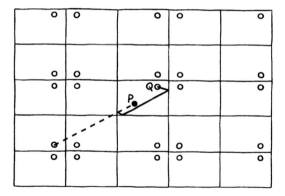

Every possible accurate shot by the enemy can be represented on this picture by a straight line from some copy of Q to P; every time such a line crosses a boundary between rectangles, the real laser beam would be bouncing off a wall. In the figure, one such (dotted) line is indicated; a solid line shows the corresponding path of the laser beam back in the original room.

Thus, you can pick a point on every shot line and station a guard there and so protect yourself with countably many guards. But back in the original room the shot-lines cross each other often; maybe if you place guards at the intersections, you can get away with a finite number.

Would you believe 16?

Your object is going to be to intercept every shot at its halfway point. To do this, you first trace a copy of the above plane tiling, pin it to the plane at your position P, and shrink the copy by a factor of two vertically and horizontally. The many copies of Q on the shrunk copy will be our bodyguard positions; they serve our purpose because each copy of Q on the original tiling appears halfway between it and you on the shrunk copy.

In the second figure, the shrunk copy is in gray, and some virtual laser paths are indicated; you can see that they pass, at their half-way points, through the corresponding smaller dots in the gray grid.

Of course, there are infinitely many such points, but we claim that they are all reflections of the right set of 16 points in the original room. Four of the points will already be in the original room; the four points in the room to the left of the original room can be reflected back to give four new points, and similarly for the room above the original one. Finally, the four points in the room above *and* to the left of the

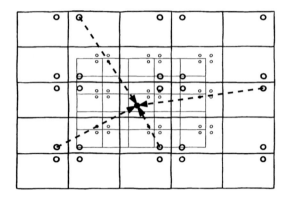

original room can be reflected twice to provide the last four points in the original room. In the third figure, the 12 new points (centers filled in gray) have been added in black to the original rectangle. A virtual laser path is added, with its corresponding real path crossing one of the new points.

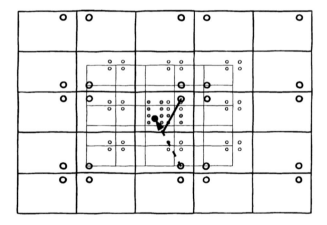

Since every room looks exactly like the original or one of the other three we just examined, all the bodyguard points in the plane are reflections of the 16 points we have now identified in the original room. Since every line from a copy of Q passed through a reflected bodyguard, the actual shot hits a "real" bodyguard at its halfway point (if not sooner) and is absorbed.

If the locations of P and Q are carefully chosen, some of the 16 bodyguard locations will coincide; but in general the full 16 are needed.

Random Intervals

The points $1, 2, \ldots, 1000$ on the number line are paired up at random, to form the endpoints of 500 intervals. What is the probability that among these intervals is one which intersects all the others?

Solution: Amazingly, the answer is exactly $\frac{2}{3}$, regardless of the number of intervals (as long as there are at least two).

Suppose the interval endpoints are chosen from $\{1, 2, \ldots, 2n\}$. We will label the points $A(1)$, $B(1)$, $A(2)$, $B(2)$, $\ldots$, $A(n-2)$, $B(n-2)$ recursively as follows. Referring to points $\{n+1, \ldots, 2n\}$ as the *right side* and $\{1, \ldots, n\}$ as the *left side*, we begin by setting $A(1) = n$ and letting $B(1)$ be its mate. Suppose we have assigned labels up to $A(j)$ and $B(j)$, where $B(j)$ is on the left side; then $A(j+1)$ is taken as the left-most point on the right side not yet labeled, and $B(j+1)$ as its mate. If $B(j)$ is on the right side, $A(j+1)$ is the right-most unlabeled point on the left side and again $B(j+1)$ is its mate.

If $A(j) < B(j)$, we say that the jth interval "went right," otherwise it "went left." Points labeled $A(\cdot)$ are said to be *inner* endpoints, the others *outer*.

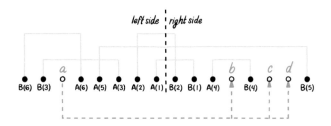

It is easily checked by induction that after the labels $A(j)$ and $B(j)$ have been assigned, either an equal number of points have been labeled on each side (in case $A(j) < B(j)$) or two more points have been labeled on the left (in case $A(j) > B(j)$).

When the labels $A(n-2)$ and $B(n-2)$ have been assigned, four unlabeled endpoints remain, say $a < b < c < d$. Of the three equiprobable ways of pairing them up, we claim two of them result in a "big" interval which intersects all others, and the third does not.

In case $A(n-2) < B(n-2)$, we have a and b on the left and c and d on the right, else only a is on the left. In either case, all inner endpoints lie between a and c, else one of them would have been labeled. It follows that the interval $[a, c]$ meets all others, and likewise $[a, d]$, so unless a is paired with b, we get a big interval.

Suppose, on the other hand, that the pairing is indeed $[a, b]$ and $[c, d]$. Neither of these can qualify as a big interval since they do not intersect each other; suppose some other interval qualifies, say $[e, f]$, labeled by $A(j)$ and $B(j)$.

When a and b are on the left, the inner endpoint $A(j)$ lies between b and c, thus $[e, f]$ cannot intersect both $[a, b]$ and $[c, d]$, contradicting our assumption.

In the opposite case, since $[e, f]$ meets $[c, d]$, f is an outer endpoint (so $f = B(j)$) and $[e, f]$ went right; since the last labeled pair went left, there is some $k > j$ for which $[A(k), B(k)]$ went left, but $[A(k-1), B(k-1)]$ went right. Then $A(k) < n$, but $A(k) < A(j)$ since $A(k)$ is a later-labeled, left-side inner point. But then $[A(j), B(j)]$ does not, after all, intersect $[B(k), A(k)]$, and this final contradiction proves the claim.

Infected Cubes

An infection spreads among the n^3 unit cubes of an $n \times n \times n$ cube, in the following manner: If a unit cube has three or more infected neighbors, then it becomes infected itself. (Neighbors are orthogonal only, so each little cube has at most six neighbors.)

Prove that you *can* infect the whole big cube starting with just n^2 sick unit cubes.

Solution: You might recall the *Infected Checkerboard* puzzle where you were asked to show that you couldn't infect all of an $n \times n$ checkerboard if you began with fewer than n sick squares. Essentially the same proof works in higher dimension to show that you need at least n^{d-1} initially sick squares to infect all of a d-dimensional hypercube of side n (where d infected neighbors infect a unit hypercube).

But this time, it's not obvious how to choose the initial n^{d-1} unit hypercubes ("cells") to infect.

In what follows we will consider only the 3-dimensional case, but the construction easily generalizes. Label the cells by vectors (x, y, z), with $x, y, z \in \{1, 2, \ldots, n\}$, so that two cells will be neighbors if all their coordinates are the same except in one position, where their values differ by 1.

Start by infecting all cells such that $x + y + z \equiv 0 \mod n$. These cells form a "diagonal subspace" that is cut up into many pieces. From these initially sick cells, the infection fills up the big cube in a weird way. It barely succeeds, relying on many apparent coincidences that allow the growth to continue. Very different from the 2-dimensional case! It seems remarkable that the process manages to infect the whole big cube.

To prove it really works, we transform the puzzle to a game. The players are you and your adversary—a doctor, say—who is trying to trap you, the infector. Let us choose a k and start with the infection as above.

In the game, the doctor begins by putting you at some cell $C = (x, y, z)$. Now repeat: The doctor chooses a coordinate (first, second, or third) and you move either forward or backward in that dimension (if the corresponding coordinate is 1 or n, you won't have a choice). You win if you can reach some cell the sum of whose coordinates is a multiple of n; the doctor wins if she can keep you wandering forever.

We now claim that if you have a winning strategy, then the big cube will indeed be fully infected.

To see this, we first refine the claim to state that if you can win starting from cell C, then C itself will become infected. Since from C, the doctor has three choices of coordinate, a winning strategy for you must work for all three possibilities. This implies that your strategy also wins if started at any of the three neighbors of C that you were preparing to move to. By induction (over the number of steps to a win), all three of these neighbors of C will be infected, thus C will as well. The base case of the induction is when the starting point C has coordinates summing to k modulo n, in which case it is of course already sick.

Now all we need is to provide you with a winning strategy; For any cell $C = (x, y, z)$, let c be the number $x + y + z + \frac{1}{2} \mod n$. After the doctor chooses a coordinate, say the first, you increment x if it's less than c (thus c goes down as well, though possibly around the corner from $\frac{1}{2}$ to $n - \frac{1}{2}$) and decrement x if $x > c$.

If you ever get to a cell where $c = \frac{1}{2}$, you've won the game. So if $x = 1$ the algorithm will not ask you to decrement it. If $x = n$ then x will always be greater than c, so you will not be asked to increment it. It follows that the move prescribed by the algorithm is always legal, unless you have already won the game.

We now assert that the doctor can't force you to cycle. Suppose to the contrary that starting from C the algorithm cycles forever. Let I

be the set of coordinates (among the first, second, and third) chosen infinitely often by the doctor. We may as well assume that you are past the point where any index not in I will ever be chosen. Let m be the biggest value ever encountered at a coordinate in I. Let J be the set of indices in I which are at the moment equal to m.

If it ever happens that $c > m$, then you will be incrementing at every step, pushing c up until it snaps around the corner to $\frac{1}{2}$ and you win. Therefore it must be that c is always below m. But then, whenever the doctor chooses a coordinate $j \in J$, the jth coordinate must decrease to $m-1$. It follows that eventually J will disappear, leaving you forever with a smaller maximum value m. This can't go on forever, so we have our contradiction.

We conclude that the above algorithm will win the game for you, no matter where the doctor starts you off or how she chases you around. The existence of a winning strategy means that the infection really does capture the whole big cube, and we are done. ♡

Gladiators, Version I

Paula and Victor each manage a team of gladiators. Paula's gladiators have strengths $p_1, p_2, \ldots, p_m$ and Victor's, $v_1, v_2, \ldots, v_n$. Gladiators fight one-on-one to the death, and when a gladiator of strength x meets a gladiator of strength y, the former wins with probability $x/(x+y)$, and the latter with probability $y/(x+y)$. Moreover, if the gladiator of strength x wins, he gains in confidence and inherits his opponent's strength, so that his own strength improves to $x+y$; similarly, if the other gladiator wins, his strength improves from y to $x+y$.

After each match, Paula puts forward a gladiator (from those on her team who are still alive), and Victor must choose one of his to face Paula's. The winning team is the one which remains with at least one live player.

What's Victor's best strategy? For example, if Paula begins with her best gladiator, should Victor respond from strength or weakness?

Solution: All strategies for Victor are equally good. To see this, imagine that strength is money. Paula begins with $P = p_1 + \cdots + p_m$ dollars and Victor with $V = v_1 + \cdots + v_n$ dollars. When a gladiator of strength x beats a gladiator of strength y, the former's team gains \$$y$ while the latter's loses \$$y$; the total amount of money always remains the same. Eventually, either Paula will finish with \$$P$+\$$V$ and Victor with zero, or the other way 'round.

The key observation is that every match is a fair game. If Victor puts up a gladiator of strength x against one of strength y, then his expected financial gain is

$$\frac{x}{x+y} \cdot \$y + \frac{y}{x+y} \cdot (-\$x) = \$0.$$

Thus, the whole tournament is a fair game, and it follows that Victor's expected worth at the conclusion is the same as his starting stake, $\$V$. We then have

$$q \cdot (\$P + \$V) + (1 - q) \cdot \$0 = \$V,$$

where q is the probability that Victor wins. Thus, $q = V/(P+V)$, independent of anyone's strategy in the tournament. ♡

Here's another, more combinatorial proof. Using approximation by rationals and clearing of denominators, we may assume that all the strengths are integers. Each gladiator is assigned x balls if his initial strength is x, and all the balls are put into a uniformly random vertical order. When two gladiators battle, the one whose topmost ball is highest wins (this happens with the required $x/(x+y)$ probability) and the loser's balls accrue to the winner.

The surviving gladiator's new set of balls is still uniformly randomly distributed in the original vertical order, just as if he had started with the full set; hence, the outcome of each match is independent of previous events, as required. But regardless of strategy, Victor will win if and only if the top ball in the whole order is one of his; this happens with probability $V/(P+V)$.

Gladiators, Version II

Again Paula and Victor must face off in the Colosseum, but this time, confidence is not a factor, and when a gladiator wins, he keeps the same strength he had before.

As before, prior to each match, Paula chooses her entry first. What is Victor's best strategy? Whom should he play if Paula opens with her best man?

Solution: Obviously, the change in rules makes strategy considerations in this game completely different from the previous one—or does it? No, again the strategy makes no difference!

For this game, we take away each gladiator's money (and balls), and turn him into a light bulb.

The mathematician's ideal light bulb has the following property: Its burnout time is completely memoryless. That means that knowing how long the bulb has been burning tells us absolutely nothing about how long it will continue to burn.

You may know that the unique probability distribution with this property is the exponential; if the expected (average) lifetime of the bulb is x, then the probability that it is still burning at time t is $e^{-t/x}$. However, no formula is necessary for this puzzle. You only need to know that a memoryless probability distribution exists.

Given two bulbs of expected lifetimes x and y, respectively, the probability that the first outlasts the second is $x/(x+y)$. To see this without calculus, consider a light fixture that uses one "x-bulb" (with average lifetime x) and one "y-bulb"; every time a bulb burns out, we replace it with another of the same type. When a bulb does burn out, the probability that it is the y-bulb is a constant independent of the past. But that constant must be $x/(x+y)$, because over a long period of time, we will use y-bulbs and x-bulbs in proportion $x{:}y$.

Back in the Colosseum, we imagine that the matching of two gladiators corresponds to turning on their corresponding light bulbs until one (the loser) burns out, then turning off the winner until its next match; since the distribution is memoryless, the winner's strength in its next match is unchanged. Substituting light bulbs for the gladiators may be less than satisfactory for the spectators, but it's a valid model for the fighting.

During the tournament, Paula and Victor each have exactly one light bulb lit at any given time; the winner is the one whose total lighting time (of all the bulbs/gladiators on her/his team) is the larger. Since this has nothing to do with the order in which the bulbs are lit, the probability that Victor wins is independent of strategy. (Note: That probability is a more complex function of the gladiator strengths than in the previous game.)

More Magnetic Dollars

We return to *Magnetic Dollars*, but we strengthen their attractive power just a bit.

This time, an infinite sequence of coins will be tossed into the two urns. When one urn contains x coins and the other y, the next coin will fall into the first urn with probability $x^{1.01}/(x^{1.01}+y^{1.01})$, otherwise into the second urn.

Prove that after some point, one of the urns will never get another coin!

Solution: The really neat way to show that one urn gets all but finitely many coins is to employ those continuous memoryless waiting times that proved so useful in the previous puzzle.

Look just at the first urn and suppose that it acquires coins by waiting an average of $1/n^{1.01}$ hours between its nth coin and its $(n+1)$st coin, where the waiting time is memoryless. Coins will arrive slowly and sporadically at first, then faster and faster; since the series $\sum_{n=1}^{\infty} 1/n^{1.01}$ converges, the urn will explode with infinitely many coins at some random moment (typically about 4 days after the process begins).

Now suppose we start two such processes simultaneously, one with each urn. If at some time t, there are x coins in the first urn and y in the second, then (as we saw with the gladiator-light bulbs) the probability that the next coin goes to the first urn is

$$\frac{1/y^{1.01}}{1/x^{1.01} + 1/y^{1.01}} = \frac{x^{1.01}}{x^{1.01} + y^{1.01}},$$

exactly what it should be. Nor does it matter how long it's been since the xth coin in the first urn (or yth in the second) arrived, since the process is memoryless. It follows that this accelerated experiment is faithful to the puzzle.

However, you can see what happens now; with probability 1, the two explosion times are different. (For this, you only need to know that our memoryless waiting time has a continuous distribution.) But the experiment ends at the first explosion, at which time the other urn is stuck with whatever finite number of coins it had. ♡

Seems like kind of a scary experiment, doesn't it? The slow urn never got to finish because, in effect, time ended.

For our theorem, we look at a combinatorial object that arises with surprising frequency called a "parking function."

Imagine that there are n parking spaces $1, 2, \ldots, n$ (in that order) marked on a one-way street. Cars numbered 1 through n enter the street in some order, each with a preferred parking spot (a random number between 1 and n). Each car drives to its preferred spot and parks there if the spot is available; otherwise, it takes the next available spot. If there are no empty spots beyond the filled preferred spot, the car must leave the street without parking; tough luck!

We want to know: What is the probability that all n cars will be able to park?

Theorem. *The order in which the cars arrive makes no difference, that is, the issue of whether all cars find a spot depends only on what the cars' preferred parking spots are. If each car's preferred spot is independently chosen and uniformly random, the probability of success is $(n+1)^{n-1}/n^n$.*

The total number of ways to assign preferred spots to the n cars is n^n; we need to figure out how many of these will lead to everyone finding a spot. For example, if everyone prefers a *different* spot, all will be well. Or if everyone prefers slot #1. But not if everyone prefers slot #2. In general, we're in trouble if there's some k for which fewer than k cars want slots from 1 through k; with some thought, you'll be able to see that if there is no such k, everyone will get parked, no matter how the cars are ordered.

Another way to say the same thing: If we order the preferred spots x_i so that $x_1 \leq x_2 \leq \cdots \leq x_n$, then everyone gets parked (in which case the mapping from i to x_i is called a parking function) provided $x_i \leq i$ for each i. (If the spot i ends up left open, then there must have been fewer than i cars wanting a spot between 1 and i.) This is a pretty natural mathematical condition; it would be nice to know the probability that it holds.

But it turns out we don't need to know the condition to prove the theorem.

We transform the situation by adding an $n+1$st slot and imagining that the slots now close up to form a circle. We'll let each car chose a preferred slot from 1 all the way to $n+1$; each car in turn goes to its preferred slot. Then, as before, if the slot is occupied it continues (clockwise, say) around the circle and parks in the first available slot.

In this scenario every car will find a slot, and there will be one empty slot left over. If that empty slot happens to be slot $n+1$, the list of preferences would have worked in the original problem; otherwise, it would have failed.

Now we take advantage of the circle's symmetry: We notice that if all the car's preferences were advanced by k, modulo $n+1$, then the whole processes would simply take place k slots clockwise from where it was; consequently, the final empty slot would also advance by k. But that implies that for every assignment of preferred slots, exactly 1 of its $n+1$ rotations works.

We conclude that the probability that the final empty slot is slot $n+1$ is $1/(n+1)$. The total number of ways to assign preferences in our circular case is $(n+1)^n$, hence the total number of *good* assignments is $(n+1)^n/(n+1) = (n+1)^{n-1}$. Note that, necessarily, none of those good assignments has any car preferring slot $n+1$.

We conclude that the probability of success in the original problem is $(n+1)^{n-1}/n^n$. ♡

How big is this number? For large n, we can write

$$(n+1)^{n-1}/n^n = (1 + \frac{1}{n})^n/n \sim e/n.$$

For example, if there are 100 cars, the theorem pegs the probability of success at about 2.678033494%, close to $e/100 \sim 2.718281828\%$. One might say that it's a pleasant surprise, but not shockingly lucky, if all the cars find parking spots.

Notes and Sources

1. Out for the Count

Half Grown: Sent by Jeff Steif, of Chalmers University in Sweden.

Powers of Two: Classic.

Watermelons: Adapted from Gary Antonick's *New York Times* "Number Play" blog, 11/7/12; but the puzzle appeared earlier in *Kvant*, June/August 1998, contributed by I. F. Sharygin.

Bags of Marbles: Contributed by pathologist Dick Plotz, of Providence RI.

Salaries and Raises: Idea from Rouse Ball [9].

Efficient Pizza-Cutting: Subject of a MoMath (National Museum of Mathematics) event of 10/30/19, led by Paul Zeitz.

Attention Paraskevidekatriaphobes: As far as I can tell this remarkable fact was first observed by Bancroft Brown (a Dartmouth math professor, like your author), who published his calculation in the *American Mathematical Monthly* Vol. 40 (1933), p. 607. My present-day colleague Dana Williams is the one who brought all this to my attention.

The origin of superstition concerning Friday the 13th is usually traced to the date of an order given by King Philip IV of France (Philip the Fair), dismantling the Knights Templar.

Meet the Williams Sisters: See also Hess [62].

Rating the Horses: Suggested by Saul Rosenthal of TwoSigma.

Shoelaces at the Airport: Told to me by Dick Hess at IPP35 (the 35th Annual International Puzzle Party).

Rows and Columns: This is a classical theorem, simple and surprising; I was reminded of it by Dan Romik, a math professor at the University of California, Davis. Donald Knuth, in Vol. III of *The Art of Computer Programming*, traces the result to a footnote in a 1955 book by Hermann Boerner. Bridget Tenner, a student of famed combinatorialist Richard Stanley, wrote a paper called "A Non-Messing-Up Phenomenon for Posets" generalizing this theorem.

Three-way Election: Dreamed up by Ehud Friedgut of the Weizmann Institute, Israel.

Sequencing the Digits: Remembered from the Putnam Exam, many years ago.

Faulty Combination Lock: Suggested to me by Amit Chakrabarti of Dartmouth; it had been proposed for the International Mathematical Olympiad in 1988 by East Germany. The optimality proof given (supplied by Amit) doesn't quite work for 8 replaced by 10, and fails increasingly thereafter. Thus the general puzzle, when the lock has n numbers per dial, is unsolved as far as I know.

Losing at Dice: I discovered this curious fact accidently, 40 or so years ago, while constructing homework problems for an elementary probability course at Emory University.

Splitting the Stacks: From Einar Steingrimsson, University of Strathclyde, Glasgow.

North by Northwest: Original.

Early Commuter: From the one and only Martin Gardner.

Alternate Connection: From the First Iberoamerican Mathematical Olympiad [7].

Bindweed and Honeysuckle: From Karthik Tadinada, but recast by me in recollection of a delightful tragicomic song by Michael Flanders & Donald Swann.

theorem: The sequence we associated with each tree is known as the *Prüfer code* [92] for the tree. Besides helping us prove the theorem, the Prüfer code gives us additional information. For example, you can easily check that the number of neighbors of a vertex of the tree is always one more than the number of times it appears in the code; leaves do not appear at all. From this we could deduce that when n is large, the number of leaves in a random tree is about n/e, that is, about 37% of the vertices are leaves.

2. Achieving Parity

Bacterial Reproduction: Based on Supplementary Problem C/4, [68], p. 14.

Fourth Corner: Brought to my attention by Paul Zeitz.

Unanimous Hats: My version of a classic.

Half-right Hats: My version of a classic.

Red and Blue Hats in a Line: Passed on to me by Girija Narlikar of Bell Labs, who heard it at a party. Note that an analogue of the given solution works with any number k of hat colors. The colors are assigned numbers from 0 to $k-1$, say, and the last prisoner in line guesses the color corresponding to the sum, modulo k, of the color-numbers of the hats in front of him. As before, all other prisoners are saved, if they're careful with their modular arithmetic.

Prisoners and Gloves: Suggested by game theorist Sergiu Hart of the Hebrew University.

Even-sum Billiards: Inspired by a problem from John Urschel's column,
https://www.theplayerstribune.com/author/jurschel.

Chameleons: Boris Schein, an algebraist at the University of Arkansas, sent me this puzzle; it may be quite old. On one occasion it was given to an 8th-grader in Kharkov, on another to a young Harvard grad interviewing at a big finance firm; both solved it.

Missing Digit: Lifted from Elwyn Berlekamp and Joe Buhler's Puzzles Column in *Emissary*, Spring/Fall 2006; they heard it from the number theorist Hendrik Lenstra.

Subtracting Around the Corner: When I was in high school, a substitute math teacher told my class that some WWII prisoner of war entertained himself by trying various sequences of four numbers to see how long he could get them to survive under the above operations. The process (with four numbers) is sometimes known as a difference box or diffy box. Among real numbers, there is an (essentially) unique quad that never terminates [12]. That quad is $(0, 1, q(q-1), q)$, where q is the unique real root of $q^3 - q^2 - q - 1 = 0$, namely,

$$q = \frac{1}{3} \left(1 + \sqrt[3]{19 + 3\sqrt{33}} + \sqrt[3]{19 - 3\sqrt{33}} \right) \sim 1.8393,$$

so $q(q-1) \sim 1.5437$. Remarkably, this number arises also in the solution to another, seemingly unrelated, puzzle in this volume.

Uniting the Loops: This is one of a number of ingenious puzzles devised by writer/mathematician Barry Cipra, of Northfield Minnesota.

theorem: Conjectured in [67] and proved at Danny Kleitman's 65th Birthday conference in August 1999 by Noga Alon, Tom Bohman, Ron Holzman and Danny himself [2].

3. Intermediate Math

Squaring the Mountain State: My version of a classic.

Monk on a Mountain: Ancient.

Cutting the Necklace: Suggested by Pablo Soberón of Baruch College, City University of New York. For more information see [4].

Three Sticks: Moscow Mathematical Olympiad 2000, Problem D4, contributed by A. V. Shapovalov [40].

Hazards of Electronic Coinflipping: See also *Fair Play* in Chapter 19, Hammer and Tongs. The actual value of p works out to $\frac{1}{2} \pm \sqrt{\sqrt{\frac{2}{3}} - \frac{3}{4}}$. But there's no need to limit yourself to three people; you can choose uniformly among n people, for any positive integer n, in a bounded number of flips. Johan Wästlund suggests the following method. Suppose you have enough power for f flips. You allot as many of the 2^f outcomes as can be divided into $n-1$ equivalent sets; left over are at most $n-2$ of each level, that is, for each k at most $n-2$ outcomes involving exactly k heads. Now the IVT can be applied as in the $n = 3$ case as long as the total probability of the left-out outcomes, at $p = \frac{1}{2}$, is at most $\frac{1}{n}$. The latter is achieved when f is just over $2\log_2(n)$ flips—only a factor of two away from the information-theoretic lower bound.

Bugs on a Pyramid: Devised by Dr. Kasra Alishahi, of Sharif University; sent to me by Mahdi Saffari.

Skipping a Number: Based on the Putnam exam problem (#A1, from 2004) that introduced the now-famous (in certain circles) Shanille O'Keal.

Splitting a Polygon: From a Moscow Mathematical Olympiad in the 1990's.

Garnering Fruit: Sent to me by Arseniy Akopyan, author of the unique and delightful book *Geometry in Figures*. The first of the two solutions to the two-fruit version came from Pablo Soberón of Baruch College, City University of New York.

theorem: Known since ancient times.

4. Graphography

Air Routes in Aerostan: Moscow Mathematical Olympiad 2003, Problem A5, contributed by R. M. Fedorov [40].

Spiders on a Cube: Passed to me by Matt Baker, Georgia Tech.

Handshakes at a Party: Classic.

Snake Game: From the 12th All Soviet Union Mathematical Competition, Tashkent, 1978.

Bracing the Grid: Passed to me by geometer Bob Connelly of Cornell, based on [29].

Competing for Programmers: Moscow Mathematical Olympiad 2011.

Wires under the Hudson: This is a variation of a puzzle publicized by Martin Gardner, and sometimes called the Graham-Knowlton Problem. To electricians it is the "WIP" (wire identification problem). In Gardner's version, you can tie any number of wires together at either end, and test at either end, as well. Our solution, found in [99, 57], satisfies our additional constraints and involves only two operations at each end (thus three river crossings, not counting additional crossings to untie and perhaps actually *use* the wires). The solution is not unique, however, so even if your three-crossing solution is different, it may be just as good.

Two Monks on a Mountain: Brought to my attention by Yuliy Baryshnikov of the University of Illinois at Urbana-Champaign.

Worst Route: Adapted from a puzzle found in [84]. If the number of houses were odd, distances would matter *slightly*: The postman would then begin or end at the middle house, zigzag between high-numbered and low-numbered houses, and end or begin at the closest house to the middle house. Example: If the addresses were 1,2,4,8,16 one of the maximum-distance routes would be 4,8,1,16,2.

theorem: The idea of switching odd and even edges used in the proof is known as the "alternating path method" and is an important tool in graph theory.

5. Algebra Too

Bat and Ball: An oldie, as you can see from the prices!

Two Runners: From [62].

Belt Around the Earth: Origin unknown, but Enrique Treviño of Lake Forest College has informed me that a version appears in the notes to a 1702 translation of Euclid by William Whiston.

Odd Run of Heads: A variation of this puzzle was suggested, but not used, for an IMO in the early '80s (see [68]).

Hopping and Skipping: Devised by mathematician James B. Shearer of IBM, this puzzle appeared in the April 2007 edition of IBM's puzzle site "Ponder This," http://domino.research.ibm.com/Comm/wwwr_ponder.nsf/challenges/April2007.html.

Area–Perimeter Match: Problem 176, p. 71 of [75].

Three Negatives: Due to Mark Kantrowitz of Carnegie Mellon University.

Red and Black Sides: Moscow Mathematical Olympiad 1998, Problem C2 [39], contributed by V. V. Proizvolov. A nice non-algebraic approach is instead to project the red lines onto, say, the left edge of the square. If the projection covers the whole left edge, we're done; suppose it misses an interval I. Then every rectangle that intersects the horizontal band through I has a red horizontal side, so projecting these downward covers the bottom edge of the square.

Recovering the Polynomial: Sent to me by Joe Buhler of Reed College, who believes it may be quite old (but, maybe not as old as Delphi). Note that without the restriction that the Oracle must be fed integers, the polynomial can be determined in one step with (say) $x = \pi$. Of course, the Oracle would have to find a way to convey $p(\pi)$ in a finite amount of time; if she gives a decimal expansion digit by digit, there'd be no way to know when to cut her off.

Gaming the Quilt: Sent to me by Howard Karloff of Goldman Sachs.

Strength of Schedule: Moscow Mathematical Olympiad 1997, Problem C5 [39], contributed by B. R. Frenkin.

Two Round-robins: Moscow Mathematical Olympiad 1997, Problem B5 [39], contributed by B. R. Frenkin.

Alternative Dice: This problem is famous enough so that its solution has a name: "Sicherman dice." Martin Gardner wrote in *Scientific American* magazine, in 1978, about their discovery by one Colonel George Sicherman, of Buffalo NY.

theorem: The model described, often now called the "Galton–Watson tree," had already been studied independently by I. J. Bienaymé.

6. Safety in Numbers

Broken ATM: Moscow Mathematical Olympiad 1999, Problem A4 [39]. Hidden in this puzzle is a form of the Chinese Remainder Theorem.

Subsets with Constraints: The first part of the puzzle was presented by the late, long-time puzzle maven Sol Golomb (University of Southern California) at the Seventh

Gathering for Gardner; the second part was suggested by Prasad Tetali of Georgia Tech; the third, an obvious variation.

Cards from Their Sum: The trick is a version of Colm Mulcahy's "Little Fibs." The two solutions are described in the Coda of Mulcahy's book [87], and were supplied by combinatorialist Neil Calkin of Clemson University.

Divisibility Game: Moscow Mathematical Olympiad 2003, Problem A6 [40], contributed by A. S. Chebotarev.

Prime Test: Moscow Mathematical Olympiad 1998, Problem B1 [39], contributed by A. K. Kovaldzhi.

Numbers on Foreheads: Sent to me by Noga Alon of Princeton University.

Rectangles Tall and Wide: There are other proofs; see, for example,
http://www-math.mit.edu/~rstan/transparencies/tilings3.pdf.

Locker Doors: Classic.

Factorial Coincidence: From [96], sent to me by Christof Schmalenbach of IBM.

Even Split: This puzzle, with n replaced by 100, appeared in the 4th All Soviet Union Mathematical Competition, Simferopol, 1970. It is elegant enough to be called a theorem and in fact it is a theorem in [37].

Factorials and Squares: Both the puzzle and the solution given are due to Jeremy Kun, author of "A Programmer's Introduction to Mathematics." Passed to me by Diana White of the National Association of Math Circles. (There are other equally valid ways to get the answer.)

theorem: This probably goes back to ancient Greece, but proofs of this fundamental fact are still of interest in the 21st century; see, for example, [15].

7. The Law of Small Numbers

Domino Task: Moscow Mathematical Olympiad 2004, Problem A6 [40], contributed by A. V. Shapovalov.

Spinning Switches: This puzzle reached me via Sasha Barg of the University of Maryland, but seems to be known in many places. Although no fixed number of steps can guarantee turning the bulb on in the three-switch version, a smart randomized algorithm can get the bulb on in at most $5\frac{5}{7}$ steps *on average*, against any strategy by an adversary who sets the initial configuration and turns the platform.

Candles on a Cake: Moscow Mathematical Olympiad 2003, Problem D4 [40], contributed by G. R. Chelnokov. For a somewhat similar idea, see *Gasoline Crisis* below.

Lost Boarding Pass: I first heard the puzzle at the Fifth Gathering for Gardner; the version here, from probabilist Ander Holroyd. Later the puzzle's popularity zoomed when it appeared on *Car Talk*.

Flying Saucers: My version of a puzzle suggested by Sergiu Hart.

Gasoline Crisis: This puzzle has been around for a long time and can be found, for example, in [79].

Coins on the Table: This nice puzzle came to me by way of computer scientist Guy Kindler, during a marvelous visiting year by each of us at the Institute for Advanced Study in Princeton.

Coins in a Row: Passed to me by mathematician Ehud Friedgut, this puzzle was alleged to have been used by a high-tech company in Israel to test job candidates.

Powers of Roots: Idea from 29th Annual Virginia Tech Mathematics Contest, 2007, and from *Using Your Head is Permitted*, brand.site.co.il/riddles, October 2008.

Coconut Classic: This is the "Williams version" of an ancient classic, where the monkey gets a sixth coconut in the morning. With the original version, you need the ending number to be 1 mod 5, achieved by beginning with $5^6 - 4$ coconuts instead of $5^5 - 4$. That ends with $5 \times 4^5 - 4 = 5116$ coconuts, giving each man 1023 and the monkey 1.

theorem: Known now as a "1-factorization of the complete graph on an even number of vertices," such constructions have been around at least since 1859 [94].

8. Weighs and Means

Tipping the Scales: The Second All Soviet Union Mathematical Competition, Leningrad 1968.

Men with Sisters: Classic.

Watches on the Table: The 10th All Soviet Union Mathematical Competition, Dushanbe, 1976.

Raising Art Value: Original.

Waiting for Heads: Original, more or less.

Finding a Jack: Idea from Dexter Senft of Lehman Bros; appears as Problem 40 in Fred Mosteller's delightful book [86].

Sums of Two Squares: Ross Honsberger [64].

Increasing Routes: This puzzle, and its elegant solution, were passed to me by Ehud Friedgut, of Hebrew University. With his help the puzzle was traced back to the fourth problem in the second round of the "Bundeswettbewerb Mathematik 1994." The Bundeswettbewerb is one of two big old German mathematical competitions.

theorem: Shockingly little-known, this result is recommended for inclusion in the next new graph theory text!

9. The Power of Negative Thinking

Chomp: Due to the late David Gale of U.C. Berkeley (although a roughly equivalent number-theoretic game had been studied earlier by the Dutch mathematician Frederik Schuh).

Dots and Boxes Variation: From Dartmouth professor Sergi Elizalde.

Big Pairs in a Matrix: 2011 Moscow Olympiad; from MIT puzzle maven Tanya Khovanova's blog, http://www.tanyakhovanova.com.

Bugging a Disk: This puzzle was published by CMU math professor Alan Frieze on CMU's puzzle website http://www.cs.cmu.edu/puzzle/solution26.pdf; the proof above was suggested by George Wang and Leo Zhang.

Pairs at Maximum Distance: From the Putnam Exam of 1957.

Lemming on a Chessboard: Devised by Kevin Purbhoo when he was a high school student at Northern Secondary School in Toronto. Purbhoo is now on the faculty at the University of Waterloo.

Curve on a Sphere: Passed to me by physicist Senya Shlosman, who heard it from Alex Krasnoshel'skii. The solution given is Senya's. Omer Angel, of the University of British Columbia, has a different proof that is less elementary, but still elegant and educational.

Let C be our closed curve and $\hat{C}$ its convex closure in 3-space, that is, the smallest convex set containing C. If C is not contained in a hemisphere, then $\hat{C}$ contains the origin 0; otherwise, there would be a plane through 0 cutting off $\hat{C}$ from 0. Thus, by Carathéodory's Theorem, there is a set of four points on C some convex combination of which is 0. Putting it another way, the tetrahedron whose vertices are those four points contains the origin.

Now, move the points continuously toward one another along the curve. When the points merge their tetrahedron will no longer contain the origin, so somewhere during the process, there was a time when the origin lay on a *face* of the tetrahedron. The three points that determine that face lie on a great circle, and each pair has a shortest route along that circle not containing the third point. Hence, the sum of the pairwise distances of the three points is 2π, impossible since they all lie on C.

Soldiers in a Field: From the 6th All Soviet Union Mathematical Competition in Voronezh, 1966.

Alternating Powers: Spotted in *Emissary*, Fall 2004; see [61]. The proof is from Noam Elkies, Harvard mathematician and (musical) composer.

Halfway Points: Moscow Mathematical Olympiad 1998, Problem B6 [39], contributed by V. V. Proizvolov.

theorem: The statement was originally conjectured by James Joseph Sylvester (arguably, the first great mathematician to leave Europe for America) back in 1893. It was first proved by the Hungarian mathematician Tibor Gallai, but the elegant proof given was found by L. M. Kelly and published in 1948 [28].

10. In All Probability

Winning at Wimbledon: Suggested by Dick Hess.

Birthday Match: Problem 32 of [86]. You may have wondered, since the answer 253 just barely works, whether accounting for the possibility that one or more of your shipmates might have been born on February 29 could push the answer up to 254; in fact, it does. (The answer, 23, to the better-known version of the puzzle is unaffected.) If you yourself were born on February 29, you'd need to query a daunting 1013 people to be a favorite to find a birthmate.

Other Side of the Coin: A similar calculation shows that you should switch doors in the notorious "Monty Hall" problem, provided you assume Monty chooses randomly which door to open when two are available.

Boy Born on Tuesday: Created by Thomas Starbird of JPL and unleashed by puzzle designer Gary Foshee at the Ninth Gathering for Gardner.

Whose Bullet?: Original.

Second Ace: Suggested by a puzzle in Littlewood's *Miscellany* [82].

Stopping After the Boy: Classic, promulgated by Martin Gardner.

Points on a Circle: Suggested by combinatorics legend Richard Stanley, of the University of Miami. For k points instead of 3, similar reasoning will lead to probability $2k/2^k$ of being containable in a semicircle. One can also do this in higher dimension; for example, the probability that four random points on a sphere can be contained in a hemisphere turns out to be $1/4$.

Comparing Numbers, Version I: To the best of my knowledge, this problem originated the late, supremely creative Tom Cover of Stanford University [26].

Comparing Numbers, Version II: Source unknown.

Biased Betting: Original.

Home-field Advantage: Original. For a faster way to do some of the computations, see [78].

Service Options: Adapted from independent suggestions by authors Hirokazu Iwasawa (Iwahiro) and Dick Hess, and Cambridge University's Geoffrey Grimmett.

Who Won the Series?: Heard from Pradeep Mutalik, mathematics writer for the online magazine QUANTA, at the Eleventh Gathering for Gardner. The solution given is my own.

Dishwashing Game: Source unknown.

Random Judge: Problem 3 of [86].

Wins in a Row: Problem 2 of [86], and popularized by Martin Gardner in his "Mathematical Games" column in *Scientific American*. Gardner gave a (correct) algebraic proof of the answer ("no") but acknowledging the value of a proof by reasoning, he also provided two arguments that you'd be better off playing black first: (1) You must win the crucial middle game, thus you want to be playing white second; and (2) You must win a game as black, so you're better off with two chances to do so. In fact, neither argument is convincing, and even together they are not a proof.

Split Games: Original (although my Dartmouth colleague Peter Doyle came up with something similar, independently).

Angry Baseball: Suggested, at the author's Charles River Probability Lecture entitled "Probability in Your Head," by Po-Shen Loh. Po-Shen is on the faculty at Carnegie Mellon and is the current coach of the U.S. Mathematical Olympic Team.

Two-point Conversion: Adapted from a problem in John Urschel's column, https://www.theplayerstribune.com/author/jurschel. John is a former starting guard with the NFL's Baltimore Ravens, now a PhD student in mathematics at MIT.

Random Chord: A classic paradox that I learned as a teenager from Martin Gardner's column.

Random Bias: Brought to me by Paul Zeitz of the University of San Francisco. Paul is the founder of San Francisco's Proof School, and a frequent contributor at the National Museum of Mathematics.

Coin Testing: Original.

Coin Game: This game is often called "Penney Ante," a pun on the name of its inventor [90]. Conway's formula appears in [52]; see also [41, 51]. The quantity A·B is sometimes called the "correlation" of A and B, but I avoided that terminology as it suggests that A·B = B·A which is not generally the case.

Sleeping Beauty: For Arntzenius and Dorr's argument see [8, 33]; more arguments and many references can be found in my article [105].

Leading All the Way and *theorem*: Bertrand's original paper [16] was followed by a nice proof by Désiré André [8].

11. Working for the System

No Twins Today: Classic.

Natives in a Circle: Moscow Mathematical Olympiad 1998, Problem B3 [39], contributed by B. R. Frenkin. Note that in fact anytime the proportion of liars is $1/2$, the anthropologist will be able to deduce this from their answers.

Poker Quickie: Sent to me by puzzle maven Stan Wagon of Macalester College, who found it in [47].

Fewest Slopes: Moscow Mathematical Olympiad 1993, Problem D3 [39], contributed by A. V. Andjans.

Two Different Distances: Goes back at least to Einhorn and Schoenberg [35].

First Odd Number: The prime version was brought to me by the late Herb Wilf, of the University of Pennsylvania. I've heard it attributed to Don Knuth.

Measuring with Fuses: This and other fuse puzzles seem to have spread like wildfire some years ago. Recreational mathematics expert Dick Hess has put together a miniature volume called *Shoelace Clock Puzzles* devoted to them; he first heard the one above from Carl Morris of Harvard University. (Hess considers multiple fuses—shoelaces, for him—of various lengths, but lights them only at ends.)

King's Salary: This puzzle was devised by Johan Wästlund of Chalmers University, (loosely!) inspired by historical events in Sweden.

Packing Slashes: Heard from Vladimir Chernov of Dartmouth. Lyle Ramshaw, a researcher at Hewlett-Packard, has devised schemes for the general $n \times n$ case, but as of this writing the optimal solutions for odd n are not known to me. See [18].

Unbroken Lines: Devised by mathematics writer Barry Cipra, this one is directly inspired by a work of the late Sol LeWitt.

Unbroken Curves: Again by Cipra, but with less direct connection to LeWitt.

Conway's Immobilizer: I don't know the composer of this puzzle but the name came from the late John Horton Conway's claim that it immobilized one solver in his chair for six hours. The first solution given here was suggested by my former PhD student Ewa Infeld; the second was devised by Takashi Chiba in response to a puzzle column in Japan, and sent to me by Ko Sakai of the Graduate School of Pure and Applied Sciences, University of Tsukuba. See [24] for yet more solutions.

Seven Cities of Gold: Adapted from a suggestion by Frank Morgan of Williams College.

theorem: The given uniqueness proof is the author's, but there are many others. The Moser Spindle (discovered by brothers William and Leo Moser in 1961) cannot be vertex-colored with three colors without having adjacent vertices of the same color. As a consequence, the "unit-distance graph," whose vertices are all points on the plane with two adjacent when they are distance 1 apart, has chromatic number at least 4. This lower bound stood until 2018 when computer scientist and biologist Aubrey de Grey found a 1581-vertex graph that raised the lower bound to 5. The upper bound remains at 7.

12. The Pigeonhole Principle

Shoes, Socks & Gloves: Original.

Polyhedron Faces: 1973 Moscow Mathematical Olympiad, but with a solution that used Euler's formula.

Lines Through a Grid: Moscow Mathematical Olympiad 1996, Problem C2 [39], contributed by A. V. Shapovalov.

Same Sum Subsets: Based on a problem from the 1972 International Mathematical Olympiad. There doesn't seem to be any easy way to actually *find* two disjoint subsets of Brad's numbers with the same sum; the number of pairs of disjoint nonempty subsets of a 10-element set is a daunting $(3^{10} - 2 \cdot 2^{10} + 1)/2 = 28{,}501$. Indeed, given n numbers (not necessarily distinct), even the problem of determining whether you can divide *all* of them into two disjoint sets with the same sum is notoriously difficult; it's one of the "original" NP-complete problems from around 1970. That means there's probably no way to decide this efficiently. Yet, the pigeonhole principle tells us that if the n numbers are distinct integers between 1 and $2^n/n$, and disjoint subsets (not necessarily a partition) with the same sum are required, then the answer to the existence question is easy—it's always "yes"!

Lattice Points and Line Segments: Source unknown; best guess, some Moscow Mathematical Olympiad.

Adding, Multiplying, and Grouping: Moscow Mathematical Olympiad 2000, Problem B2 [40], contributed by S. A. Shestakov.

Line Up by Height: The theorem first appeared in [38].

Ascending and Descending: Moscow Mathematical Olympiad 1998, Problem C6 [39], contributed by A. Ya. Kanel-Belov and V. N. Latyshev.

Zeroes and Ones: The final idea was passed to me by the late David Gale.

Same Sum Dice: Brought to me by David Kempe of the University of Southern California. Similar results can be found in a paper by the very notable mathematicians Persi Diaconis, Ron Graham and Bernd Sturmfels [31]. Pointed out by Greg Warrington of the University of Vermont: The puzzle (and proof) still works with m n-sided dice versus n m-sided dice.

Zero-sum Vectors: Sent to me by Noga Alon of Princeton University. The problem appeared as D6 in the 1996 Moscow Mathematical Olympiad [39]. The puzzle's statement is "tight" in at least two senses. First, if any of the 2^n $\{-1, +1\}$-vectors is left off the original list, the result fails—even if the others are represented arbitrarily often! Suppose, for example, that $n = 5$ and that the missing vector is $y = \langle +1, +1, +1, -1, -1 \rangle$. In all other vectors, change all $+1$'s among the first three coordinates to zeroes, and all -1's among the last two coordinates to zero. Since y is missing, the zero vector will not appear among the altered vectors; and any sum of altered vectors will have a negative number among its first three coordinates, or a positive number among its last two, or both.

Second, as noted by mathematician and hacker Bill Gosper, for any n there is a way to alter the vectors in such a way that there is no way to get a zero sum short of adding up the entire list. We leave it to the reader to verify this fact.

theorem: In fact the numbers $\{nr\}$, for r irrational, are not only dense but remarkably (and usefully) evenly spaced: Roughly speaking, if you run n through the integers from 1 to any large number, the fraction of the values $\{nr\}$ that appear in some subinterval of $[0, 1)$ will be close to the length of that subinterval. This observation is fundamental in a branch of mathematics sometimes called "discrepancy theory," for which a wonderful source is [11].

13. Information, Please

Finding the Counterfeit: Classical; suggested by Robert DeDomenico. David Gontier found a solution in a 1946 article [34] by the late, great physicist Freeman Dyson.
When I distributed the puzzle for the National Museum of Mathematics' "Mindbenders for the Quarantined," subscriber Markus Schmidmeier pointed out that there are non-adaptive solutions such as (with numbered coins) 1,2,3,4 versus 5,6,7,8; 1,2,9,10 versus 3,4,8,11; and 1,3,5,9 versus 2,6,10,12. Another subscriber, Bob Henderson, pointed out that this solution allows you to diagnose also the case of no counterfeit—in which case all weighings balance.

Three Natives at the Crossroads: Puzzles of this sort were studied and popularized by Martin Gardner and Raymond Smullyan, among others; this particular version came to me from two mathematical physicists, Vladas Sidoravicius and Senya Shlosman.

Attic Lamp: Classical, but losing ground as incandescent bulbs disappear even from memory. By the way, with enough patience you can extend the solution to *five* switches: Flip switch E on a year or two in advance, to acquire a fifth possible state, "burnt out"!

Players and Winners: Devised and communicated to me by Alon Orlitsky, of the University of California, San Diego.

Missing Card: The trick first appeared in Wallace Lee's book *Math Miracles*, in which he credits its invention to William Fitch Cheney, Jr., a.k.a. "Fitch," around 1950. See also [69, 87].

Peek Advantage: Brought to my attention by professional gambler Jeff Norman, who was offered the chance to bet a lot more than $100 on the "color of the flop" prior to a deal of Texas hold 'em, with the extra inducement of being allowed to peek at one of his own hole cards before deciding which color to bet on.

Bias Test: Original. The theorem mentioned in the text was proved by Gheorghe Zbăganu [22, 109].

Dot-Town Exodus: Classical. Steve Babbage, a manager and cryptographer with Vodafone, points out that if the residents of Dot-town begin to worry that a suicide was not caused by knowing one's dot color—but perhaps because some Dot-towner "has finally cracked under the strain of living in such a ludicrous environment"—then under certain circumstances the rest of the town may yet survive the stranger's incursion.

Conversation on a Bus: A creation of John H. Conway's. Variations can be found in a paper of Tanya Khovanova's posted at https://arxiv.org/pdf/1210.5460.pdf.

Matching Coins: Brought to my attention by Oded Regev of the Courant Institute, NYU. In [56] the authors show that with more sophisticated versions of this scheme, Sonny and Cher can get as close as they want to a fraction x of success, where x is the unique solution to the equation

$$-x \log_2 x - (1-x) \log_2 (1-x) + (1-x) \log_2 3 = 1,$$

about 0.8016, but they can do no better. Moreover, this applies whether the coinflips are random or adversarial.

Two Sheriffs: This puzzle was devised and presented in [10] to give a toy example of how two parties that share information, but have no common secret, can establish a common secret over an open channel, and then use it to communicate in secret.

theorem: Properly speaking, the study of secret codes is *cryptology*, which divides into designing codes (cryptography) and breaking them (cryptanalysis). In practice "cryptography" is often used to cover the whole subject. A particularly entertaining source is [65].

14. Great Expectation

Bidding in the Dark: Sent by Maya Bar Hillel, University of Jerusalem.

Rolling All the Numbers: Classic.

Spaghetti Loops: Classic.

Ping-Pong Progression: Adapted from a problem posed on 2/28/17 at the MoMath Masters, an annual puzzle-solving contest put on by the National Museum of Mathematics.

Drawing Socks: Suggested by Yan Zhang of San Jose State University.

Random Intersection: Observed and proved by Jim Propp, of the University of Massachusetts, Lowell. He cannot be blamed for the calculus-avoiding "proof" given here, however.

Roulette for the Unwary: My variation on Problem 7 from [86].

An Attractive Game: Problem 6 from [86]. Our proof shows that the game becomes fair if the payoff for two of your number is $3 instead of $2, and for three $5 instead of $3.

Next Card Bet: I heard this problem from probabilist Russ Lyons, of Indiana University, who heard it from Yuval Peres, who heard it from Sergiu Hart; but the problem goes back apparently to a paper of Tom Cover [25]. If the original $1 stake is *not* divisible, but is composed of 100 indivisible cents, things become more complicated and it turns out that Victor does a dollar worse. A dynamic program (written by Ioana Dumitriu, now a professor at UCSD) shows that optimal play guarantees ending with at least $8.08. Warning: you need to be a bit more conservative in the "100 cents" game than in the continuous version; if instead you always bet the

nearest number of cents to the fraction $(b-r)/(b+r)$ of your current worth, you
could go bankrupt before half the deck is gone.

Serious Candidates: Devised by probabilist David Aldous of U.C. Berkeley, who was
inspired by the prediction market and the Republican party's presidential nomination process in 2012.

Rolling a Six: Devised and told to me by MIT probabilist Elchanon Mossel, who
dreamed it up as an easy problem for his undergraduate probability students and
then realized the answer was not 3. The solution given is your author's.

Napkins in a Random Setting: Posed on the spot by the late, great John Horton Conway at a math conference banquet where the circular table, coffee cups and napkins
were as described.

Roulette for Parking Money: Due to Dick Hess, who was inspired by a note written
by Bill Cutler. It's called "Bus Ticket Roulette" in [63]. Here we have assumed that
the table has both a "0" and a "00" but the same scheme is optimal in Monte Carlo
where there's only the single zero.

Buffon's Needle: Ramaley's paper [93].

Covering the Stains: Devised by Naoki Inaba and sent to me by Iwasawa Hirokazu,
known also as Iwahiro; both are prolific puzzle composers. The actual maximum
number of points in the plane that can always be covered by disjoint open disks
is not known (see [5]); embarrassingly, the best results currently are that 12 points
are always coverable and 42 are not. My guess? 25.

Colors and Distances: Sent by probabilist Ander Holroyd, who heard it from Russ
Lyons. My setting is a real town which, the last time I looked, was pretty evenly
split.

Painting the Fence: From the Spring '17 *Emissary* Puzzles Column by Elwyn
Berlekamp and Joe P. Buhler, to which it was contributed by Paul Cuff who recalls
it from one of Tom Cover's seminars at Stanford.

Filling the Cup: Classic.

theorem: For (much) more on the probabilistic method, the reader is enthusiastically
referred to [3].

15. Brilliant Induction

IHOP: Yes, I know, "IHOP" doesn't really work as an abbreviation for "induction
hypothesis." Lizz Moseman, a former PhD student of mine, had an earlier math
teacher who used that term and Lizz and I both found it irresistible.

Uniform Unit Distances: Posed in the (now defunct) journal *Mathematical Spectrum*
and solved by David Seal of Winchester College (UK) in Vol. 6, Issue 2, 1973–4.
As you may have noticed, the sets S_n we constructed look like projections onto
the plane of an n-dimensional hypercube.

Swapping Executives: Moscow Mathematical Olympiad 2002, Problem C4 [40], contributed by A. V. Shapovalov.

Odd Light Flips: Moscow Mathematical Olympiad 1995, Problem C6 [39], contributed by A. Ya. Kanel-Belov. A proof using linear algebra is also possible.

Truly Even Split: This puzzle was suggested by Muthu Muthukrishnan who heard it
from eminent computer theorist Bob Tarjan.

Non-repeating String: Moscow Mathematical Olympiad 1993, Problem A5 [39], contributed by A. V. Spivak.

Baby Frog: Moscow Mathematical Olympiad 1999, Problem D4 [39], contributed by
A. I. Bufetov. Note that the given proof provides a pretty efficient scheme. Since a
circular clearing of radius $\sqrt{2}$ times a cell side (thus area $2\pi/2^{2n}$, relative to the

big square) must contain a grid cell, the baby frog can be guided to a circle of area A with only $\lceil \frac{1}{2} \log_2(2\pi/A) \rceil$ croaks.

Guarding the Gallery: The question (with 11 replaced by n) was asked by the late Victor Klee, a brilliant geometer who spent most of his career at the University of Washington, and answered by combinatorialist Václav Chvátal. The lovely proof in this volume was found by Steve Fisk [45].

Path Through the Cells: Moscow Mathematical Olympiad 1999, Problem D5 [39], contributed by N. L. Chernyatyev. Notice that the fact that the cells of a telephone network most likely constitute a *planar* graph is not needed for this result.

Profit and Loss: This puzzle is adapted from one which appeared on the 1977 International Mathematical Olympiad, submitted by a Vietnamese composer. Thanks to Titu Andreescu for telling me about it. The solution given is my own; the industrious reader will not find it difficult to generalize it to the case where x and y have a greatest common divisor $\gcd(x, y)$ other than 1. The result is that $f(x, y) = x + y - 1 - \gcd(x, y)$.

Uniformity at the Bakery: This lovely puzzle is from a Russian competition and appears in [97]. The given argument also works if all the weights are rational numbers, since we can just change units so that the weights are integers. But what if the weights are irrational? Regard the real numbers $\mathbb{R}$ as a vector space over the rationals $\mathbb{Q}$, and let V be the (finite-dimensional) subspace generated by the weights of the bagels. Let α be any member of a basis for V, and let q_i be the rational coefficient of α when the weight of the ith bagel is expressed in this basis. Now the same argument used in the rational case shows that all the q_i's must be 0, but this is a contradiction since then α was not in V to begin with.

Summing Fractions: From the 3rd All Soviet Union Mathematical Competition, Kiev, 1969.

Tiling with L's: Told to me by Rick Kenyon of Yale University, an expert on *random* tilings.

Traveling Salesmen: From the 11th All Soviet Union Mathematical Competition, Tallinn, 1977; thanks to Barukh Ziv of Intel for sending me the intended solution. The solution given was devised by me and Bruce Shepherd of the University of British Columbia; another nice solution was sent to me by Emmanuel Boussard of Boussard & Gavaudan Asset Management.

Lame Rook: Composed by Rustam Sadykov and Alexander Shapovalov for a 1998 Olympics, and told to me by Rustam. I love that unexpected question at the end of the puzzle statement! Two additional remarks: (1) there's a non-inductive proof using Pick's Theorem, and (2) in fairy chess, the lame rook is called the "wazir."

theorem: A good source on Ramsey theory is [58].

16. Journey Into Space

Easy Cake Division: Martin Gardner, in [53], attributes the puzzle to Coxeter. The solution given works for any regular polygon and also for any triangle, provided you cut to the incenter.

Painting the Cubes: Puzzle and solution both told to me by Paul Zeitz.

Curves on Potatoes: Communicated to me by Dick Hess, who heard it from Dieter Gebhardt; appears in [14].

Painting the Polyhedron: Told to me by Emina Soljanin (formerly a Distinguished Member of Technical Staff at Bell Labs, now at Rutgers).

Boarding the Manhole: Classic.

Slabs in 3-Space: From an early Putnam Exam.

Bugs on Four Lines: This puzzle was passed to me by Matt Baker, of Georgia Tech. It is sometimes called "the four travellers' problem" and appears on the website "Interactive Mathematics Miscellany and Puzzles" at http://www.cut-the-knot.org/.

Circular Shadows II: From the 5th All Soviet Union Mathematical Competition, Riga, 1971.

Box in a Box: Told to me by Anthony Quas (University of Victoria) who heard it, and the solution given, from Isaac Kornfeld, a professor at Northwestern University; Kornfeld had heard the puzzle many years ago in Moscow. Additional very nice solutions were sent to me by Mike Todd of Cornell University and by Marc Massar. Mike's uses vectors and the triangle inequality, while Marc's is based on the observation that for an $a \times b \times c$ box, $(a + b + c)^2$ is equal to the area of the box plus the square of its diagonal length.

Angles in Space: I was tested on this puzzle during a visit to MIT, and was stumped. It turns out that the question was for some time (since the late 1940's) an open problem of Paul Erdős and Victor Klee, then was solved by George Danzig and Branko Grünbaum in 1962.

The question of the maximum number of points in n-space that determine only angles that are *strictly* acute remained open much longer, for a long time known only to be between $2n-1$ and $2^n - 1$. Only recently [54] has it been shown that you can get halfway to the upper bound: There's a set of $2^{n-1} + 1$ points with all angles strictly acute.

Curve and Three Shadows: The question was raised by number theorist Hendrik Lenstra, after hearing about Oskar van Deventer's puzzle "Oskar's Cube" in which three orthogonal sticks poke through the sides of a cube, each of which has a maze cut out of it. Since the mazes can't have loops without a piece falling out, it wasn't clear whether the puzzle could be designed so that the stick intersection can follow a closed curve in space.

theorem: The **Monge Circles** theorem, with its famous sphere proof, was first brought to my attention by computer theorist Dana Randall of Georgia Tech. The flaw was pointed out to me by Jerome Lewis, Professor of Computer Science at the University of South Carolina Upstate.

17. Nimble Nimbers

Life Is a Bowl of Cherries: The game is ancient; the analysis goes back at least as far as 1935 [98].

Life Isn't a Bowl of Cherries?: For more on this and the previous puzzle, [13] is the fun place to go.

Whim-Nim: Devised by John H. Conway, who called it simply "Whim."

Option Hats: This puzzle dates back to the U.C. Santa Barbara 1998 PhD thesis of Todd Ebert (now at Cal State Long Beach). It was the subject of a *New York Times* article entitled "Why Mathematicians Now Care About Their Hat Color," by Sara Robinson, April 10, 2001 (see http://web.csulb.edu/~tebert/hatProblem/nyTimes.htm). The radius-1 covering code of size 7, for the $n = 5$ case, was taken from [21].

Chessboard Guess: From [81] and sent to me by one of that book's authors, Anany Levitin.

Majority Hats: Posted by Thane Plambeck of Counterwave, Inc. Many insights, plus the dynamic program and its results, were contributed by Johan Wästlund of Chalmers University. Wästlund points out that there is an application of the result to gambling at a "responsible" casino where you buy chips once in advance, and are paid in cash, so that you can't gamble with your winnings and perhaps get addicted. The dynamic program maximizes your probability of coming out ahead given that you bought 100 chips that can be bet at even odds on fair coinflips. (Yes, you'd've been better off had you bought only 99 chips.) But one can sometimes do better in *Majority Hats* than in the responsible casino; with computer help, Wästlund now has a protocol that has pushed the majority probability for 100 prisoners to 1156660500373338319469/1180591620717411303424, or approximately 97.9729552603863%.

Fifteen Bits and a Spy: Told to me by László Lovász, then of Microsoft Research, now president of the Hungarian Academy of Sciences. Laci is uncertain of its origin.

theorem: A nice source for error-correcting codes is [91].

18. Unlimited Potentials

Signs in an Array: Classic. We were not asked how long the process takes, but using the facts that (1) every line needs to be flipped at most once, and (b) flipping every line that some scheme *doesn't* flip gives the same result, it can be deduced that you can always get all the sums non-negative in at most $\lfloor (m+n)/2 \rfloor$ flips.

Righting the Pancakes: Moscow Mathematical Olympiad 2000, Problem A5 [40], contributed by A. V. Shapovalov.

Breaking a Chocolate Bar: I'd love to know who first came up with this one.

Red Points and Blue: From a Putnam Exam of the 1960's. Another nice solution was sent to me by tax expert Carl Giffels. The claim is that any group of n red and n blue points, with no three collinear, can be divided by a line so that each half-plane contains the same number of red points as blue. To see this, pick any point P that doesn't lie on any line containing more than one of the $2n$ points. Draw a line through P and rotate it slowly about P. The line may start with (say) more red points than blue on its right-hand side, but $180°$ later the reverse will be true; thus, at some point, the numbers of red and blue points on the "right" side of the line (and therefore also on the left) must be equal. Now, each half-plane can be further sub-divided the same way (this time by a half-line ending at the previous line), and the sub-divisions divided similarly, until every region contains only a single red and a single blue point. The line segments connecting those pairs cannot cross.

The puzzle's conclusion is true in higher dimensions as well, as pointed out to me by Pablo Soberón of Baruch College, City University of NY. In three dimensions, you have n points each of colors red, blue, and green, no three collinear. You want disjoint triangles connecting a 3-way matching, and you get it by induction via the Ham Sandwich Theorem (a.k.a. the Stone–Tukey Theorem). If n is even, use the theorem to find an appropriate plane, missing all the points, then apply the theorem to each half. If n is odd, it will cut a point of each color and you can use that triangle along with the ones obtained by induction. And if the plane cuts through 3 or more red (say) points? No problem, split them and include the right number in each induction piece.

Bacteria on the Plane: Variation of a puzzle posed by Fields Medal winner Maxim Kontsevich (see [39], p. 110).

Pegs on the Half-plane: Variation of a problem described in [13], Vol II; I am told the problem was invented originally by the second author, Conway. In his problem, diagonal jumps were not permitted; one can nonetheless get a peg to the line $y = 4$ without much difficulty, but an argument like the one below shows that no higher position can be reached.

Returning to my variation, Dieter Rautenbach wrote to me that a summer school student, Niko Klewinghaus, came up with a proof that one can get a peg up eight units and that this is the highest possible value.

Pegs in a Square: There is more than one way to solve this puzzle, which is *part* of a problem which appeared at the 1993 International Mathematical Olympiad. The proof given was communicated to me by combinatorialist Benny Sudakov of ETH Zurich. At the Olympiad, contestants were asked to determine precisely for which n the squares were reducible—pretty tough to do on the spot!

First-grade Division: Passed to the third author by Ori Gurel-Gurevich, a mathematician at the Hebrew University of Jerusalem. Gurel-Gurevich heard it from his army friend Alon Amit, but the puzzle may go back much further in time. For generalizations, see [46].

Infected Checkerboard: The puzzle seems to have started life somewhere in the old Soviet Union, then migrated to Hungary where Spencer heard it. Its generalization contributed to a whole new field of study called "bootstrap percolation," initiated by Béla Bollobás [17] in 1968.

Impressionable Thinkers: Suggested to me by Sasha Razborov, of the University of Chicago; he tells me that it was considered for an International Mathematics Olympiad, but rejected as too hard. It was posed and solved in [55].

The puzzle can be generalized considerably, for example, by adding weights to vertices (meaning that some citizens' opinions are more highly prized than others'), allowing loops (citizens who consider their own current opinions as well), allowing tie-breaking mechanisms, and even allowing different thresholds for opinion changes.

Frames on a Chessboard: Sent by Ehud Friedgut of Hebrew University, it is a variation of a problem which appeared on an Israeli youth mathematics contest. In the contest, the frames were 3×3 and 4×4, and a counting argument says that you cannot reach every color configuration. The point is that the order in which the frames are laid down is irrelevant; all you need to know is which of the 5^2 ways to put the 4×4 frame down is utilized, and which of the 6^2 ways to put the 3×3 frame. Altogether there are thus $2^{25} \times 2^{36} = 2^{61}$ ways to *try* to get color configurations; not enough. With Friedgut's modification, however, we need some more subtle argument, such as the one given.

Bugs on a Polyhedron: Presented in a paper by Anton Klyachko [70].

Bugs on a Line: The analysis given was done by Ander Holroyd (then at the University of British Columbia) and Jim Propp, at a meeting of the Institute for Elementary Studies in Banff, Alberta, 2003. The bug was proposed by Propp as a way to simulate deterministically a random walk on the non-negative integers in which steps are made, independently, to the left with probability $1/3$ and to the right with probability $2/3$. In such a walk, a given bug drops off to the left or proceeds to infinity on the right with equal probability; as we saw, the deterministic model gives a strict alternation instead, after the first couple of trips. The argument can be extended to other random walks.

Flipping the Pentagon: First brought to my attention by Andy Liu, this now-notorious puzzle was Problem #3 at the 1986 International Mathematical Olympiad, held in Poland. Joseph Keane, a US contestant, received a special prize for his sets-of-consecutive-vertices solution.

Picking the Athletic Committee: Brought to me by my Dartmouth computer science colleague Deeparnab Chakrabarty, who needed the general result (with the number 3 replaced by an arbitrary integer k) for a research paper. Recently Iwasawa Hirokazu has generalized the result even further.

Bulgarian Solitaire: See [1]. For non-triangular numbers of chips, you end up cycling through triangular formations with extra chips shifting along the new diagonal. *theorem*: See [85] for more on the infinite case.

19. Hammer and Tongs

Phone Call: I cannot recall from whom I first heard this, many years ago.

Crossing the River: Appears in the medieval Latin manuscript *Propositiones ad Acuendos Juvenes* (Problems to Sharpen the Young), attributed to the polymath Alcuin of York, 735–804.

Sprinklers in a Field: Moscow Mathematical Olympiad 1996, Problem A3 [39], contributed by I. F. Sharygin.

Fair Play: I was reminded of this puzzle by Tamás Lengyel, of Occidental College. More improvements are possible and [88] shows how to get the last drop of blood out of the process, minimizing the expected number of flips to get a decision, regardless of the coin's probability of landing heads up. The problem of extracting unbiased, random bits from various tainted random sources is of major importance in the theory of computing, and the subject of many research papers and significant breakthroughs in recent years.

Finding the Missing Number: An alternative for those who are weak on addition but strong on dexterity is to use fingers and toes to keep track of which ones digits, and which tens digits (respectively), have appeared odd numbers of times.

Identifying the Majority: The algorithm given is described in [44].

Poorly Placed Dominoes: Original, but inspired by Problem 429 in [83].

Unbreakable Domino Cover: Sent by aeronautical engineer Bob Henderson.

Filling the Bucket: Passed to me by Dartmouth PhD student Grant Molnar. The puzzle was composed by Caleb Stanford and appeared on the 2017 Utah Math Olympiad. Note that if the buckets contain k gallons instead of just two, you can still win—but you'll need a lot of buckets and a lot of time, since the series $1 + 1/3 + 1/5 + \ldots$ takes exponentially many terms to get to k. (The two-gallon case can be done with 7 buckets, but for $k = 3$ you already need 43 buckets and lots of time.)

Polygon on the Grid: Moscow Mathematical Olympiad 2000, Problem C3 [40], contributed by Gregory Galperin, now of Eastern Illinois University.

One-bulb Room: I heard this puzzle from Adam Chalcraft, who has the distinction of having represented Great Britain internationally in unicycle hockey. The puzzle has also appeared on IBM's puzzle site and was reprinted in *Emissary*, the newsletter of the Mathematical Sciences Research Institute in Berkeley, California. A version even appeared on the justly famous public radio program *Car Talk* in 2003.

Closely related, but much more challenging, is the puzzle of the *Two-Bulb Room* where the conditions are similar but there is a second bulb; the catch is that all prisoners must use the same protocol. An elegant solution known as the "see-saw protocol" is presented in [43].

Sums and Differences: From the 5th All Soviet Union Mathematical Competition, Riga, 1971. The argument given works for any odd number of numbers greater than 3; for even numbers, the contradiction is reached even more quickly. There is, however, a set of three numbers with no pair satisfying the requirements of the puzzle: $\{1, 2, 3\}$!

Prisoner and Dog: Brought to my attention by Giulio Genovese, a mathematician at Harvard's McCarroll Lab. The prisoner's strategy described, though good enough for speed ratio 1:4, can be improved upon.

Love in Kleptopia: This puzzle was passed to me by Caroline Calderbank, daughter of mathematicians Ingrid Daubechies and Rob Calderbank.

Badly Designed Clock: Posed by Andy Latto, a Boston-area software engineer, at the Gathering for Gardner IV. It can be solved algebraically or geometrically, with sufficient care and patience; the proof here was supplied to Andy by Michael Larsen, a mathematics professor at Indiana University. The idea of a third hand (instead of a second clock) came to me from David Gale.

Worms and Water: Told to me by Balint Virag, a probabilist at the University of Toronto.

Generating the Rationals: From The 13th All Soviet Union Mathematical Competition, Tbilisi, 1979.

Funny Dice: Non-transitive (or "intransitive") sets of dice have been around for a long time; a recent article [23] showed that in fact randomly designed dice often have this property.

Sharing a Pizza: Devised by Daniel E. Brown in 1996, this puzzle has attracted a lot of attention, in part because of your author's conjecture that Alice could always get at least $4/9$ of the pizza—proved independently by two groups [20, 72].

Names in Boxes: This puzzle has a short but fascinating history. Devised by Danish computer scientist Peter Bro Miltersen, a version appeared in a prize-winning paper of his and Anna Gal's [48]. But Miltersen didn't think there was a solution until one was pointed out to him over lunch by colleague Sven Skyum. The puzzle reached me via quantum computing expert Dorit Aharonov.

Eugene Curtin and Max Warshauer [30] have recently proved that the solution given cannot be improved upon.

Lambert Bright and Rory Larson, and independently Richard Stanley of MIT, proposed the following variation. Suppose each prisoner *must* look in 50 boxes, and the requirement for survival is that every prisoner *not* find his own name. Despite having the diametrically opposite objective from before, it seems that the prisoners can do no better than to follow exactly the same strategy. Here, though, they survive only if every cycle has *more* than 50 boxes in it, which can only happen if there is just one big cycle—for which their chances are precisely 1 in 100. Not great, but a lot better than 1 in 2^{100}. They do just as well even if every prisoner is required to look in 99 boxes—again, they follow the strategy and win just when the random permutation has just one big cycle. In this case it is immediately obvious that no better strategy is available, because the very first prisoner, no matter what he does, has only a 1% probability of avoiding his own name. The amazing thing is that following the strategy, if that first prisoner succeeds, then automatically every other prisoner will succeed as well.

Life-Saving Transposition: Sent to *Emissary* via Kiran Kedlaya by Piotr Krason.

Self-referential Number: Devised and told to me by probabilist Ander Holroyd.

theorem: Due to Georg Alexander Pick in 1899. There are many proofs; the one presented is an amalgam of several.

20. Let's Get Physical

Rotating Coin: Another classic popularized by Martin Gardner.

Pie in the Sky: Original (but of course, I may be the millionth person to have thought of this question).

Returning Pool Shot: Moscow Mathematical Olympiad 2004, Problem B3 [40], contributed by A. Ya. Kanel-Belov.

Falling Ants: As far as I know the first publication of the puzzle was in Francis Su's "Math Fun Facts" web-column at Harvey Mudd College; Francis recalls hearing it in Europe from someone he can't trace named Felix Vardy. The puzzle then showed up in the Spring/Fall 2003 issue of *Emmissary*. Dan Amir, a former Rector of Tel Aviv University, read the puzzle in *Emmissary* and posed it to Noga Alon, who brought it to the Institute for Advanced Study; I first heard it from Avi Wigderson of the I.A.S., in late 2003.

Ants on the Circle: Elwyn Berlekamp and I came up with this one together (but of course others may have thought of it as well—turning a line into a circle is, as we have seen, often a useful thing to do).

Sphere and Quadrilateral: Told to me by Tanya Khovanova, who lists it among the "coffin puzzles" on her blog. These are puzzles with simple solutions that are difficult to find, especially for someone taking a timed exam. According to Tanya and others, such puzzles were used in the Soviet Union to keep "undesirables," for example, Jewish students, out of the best schools.

Two Balls and a Wall: Disseminated by Dick Hess and Gary Antonick at the Eleventh Gathering for Gardner, based on a discovery [49] by Gregory Galperin.

theorem: Brought to my attention by Yuval Peres.

21. Back from the Future

Portrait: Goes back at least to a puzzle submitted by G. H. Knight to an Oxford journal in 1872 [71], p. 240.

Three-way Duel: Classical, and Problem 20 of [86]. The argument that Bob should also shoot in the air was advanced, convincingly, to me by Gerry Myerson of Macquarie U., Sydney.

Testing Ostrich Eggs: Sent to me by Tamás Lengyel, it appears in [73] as Problem #166.

Pancake Stacks: Brought to my attention by Bill Gasarch of the University of Maryland, this puzzle appeared in the 12th All Soviet Union Mathematical Olympiad, Tashkent, 1978.

Chinese Nim: Also known as Wythoff's Game, introduced in 1907 [106].

Turning the Die: I was challenged by a superior officer in the U.S. Navy to play this game for drinks, at a stateside bar during the Vietnam War. I declined, but of course analyzed the game so I would be prepared on the next occasion—which, perhaps luckily, never materialized.

Game of Desperation: Brought to my attention by my PhD student, Rachel Esselstein, the puzzle appeared at the 28th Annual USA Mathematical Olympiad in 1999. See also [42].

Deterministic Poker: From an early Martin Gardner "Mathematical Games" column in *Scientific American*.

Bluffing with Reals: Origin unknown.

Swedish Lottery: This nice lottery idea was brought to my attention by Olle Häggström of Chalmers University (Göteborg, Sweden). I've tried it a few times with groups of twenty or so students; twice, the winning number was 6.

Pegs on the Corners: Told to me by computer theorist Mikkel Thorup of the University of Copenhagen, who heard it from Assaf Naor (now at Princeton), who heard it from graduate students at the Hebrew University of Jerusalem.

Touring an Island: Seen on Carnegie Mellon University's "Puzzle Toad" web page, and found in [50].

Fibonacci Multiples: Told to me by Richard Stanley.

Light Bulbs in a Circle: From the International Mathematical Olympiad of 1993. Puzzle maven Tom Verhoeff tells me that by applying the theory of linear feedback shift registers, he has determined that the bulbs first return to all-on at time $t =$ 181,080,508,308,501,851,221,811,810,889, much greater than the age of the universe in seconds. Check out his demo [102].

Emptying a Bucket: From the 5th All Soviet Union Mathematical Olympiad, Riga, 1971. The puzzle showed up again, minus the hardware, on the Putnam Exam in 1993, and reached me via Christian Borgs, then at Microsoft Research. The solution given is mine, but there is also an elegant number-theoretic solution found independently by Svante Janson of Uppsala University, Sweden, and Garth Payne of Penn State.

Ice Cream Cake: Told to me by French graduate student Thierry Mora, who heard it from his prep-school teacher Thomas Lafforgue; Stan Wagon tells me that it comes from a Moscow Mathematical Olympiad (which, if you've read this far in *Notes & Sources*, should not be a surprise.) The puzzle originally involved a second angle as well, indicating the amount of cake passed over between wedge-cuttings; it *still* requires only finitely many operations to get all the icing back on top, as ambitious readers will verify. But, the puzzle as stated here (where the second angle is 0) is already surprising and challenging.

Moth's Tour: Suggested to me by Richard Stanley. Thirty years ago, at a Chinese restaurant in Atlanta, Laci Lovász and I [80] managed to prove that the cycle is the only graph, apart from the complete graph, with this nice property (that the last new vertex visited by a random walk is equally likely to be any vertex other than the starting point).

Boardroom Reduction: Original. Many complex versions abound of a similar puzzle involving pirates and gold coins, but it seemed to me that the essence of that puzzle could be achieved without the coins.

theorem: This marvelous fact (in another context) was observed by Lord Rayleigh, and probably many others since—and maybe before. A nice modern reference is [95]. Sometimes called "Beatty's problem" (after Samuel Beatty 1881–1970), the puzzle appeared as Problem 3117 in *The American Mathematical Monthly* **34** (1927), pp. 159–159, and again on the 20th Putnam Exam, November 21, 1959.

22. Seeing Is Believing

Meeting the Ferry: Due to Edouard Lucas, 19th Century French mathematician.
Mathematical Bookworm: From Martin Gardner.
Rolling Pencil: My late colleague Laurie Snell caught me on this one, which appears in [104].
Splitting a Hexagon: From Tanya Khovanova's blog.
Circular Shadows I: Moscow Mathematical Olympiad 1997, Problem C1 [39], contributed by A. Ya. Kanel-Belov.
Trapped in Thickland: Moscow Mathematical Olympiad 2000, Problem D6 [40], contributed by A. Ya. Kanel-Belov.
Polygon Midpoints: Sent by Barukh Ziv who found it in a book by I. M. Yaglom [107]. The solution given is Ziv's. Incidentally, if you don't care for self-intersecting polygons, note that any finite set of points in the plane can be taken to be the vertices of a non-self-intersecting polygon: Just pick a point C inside the convex closure of your set of points, then connect your points in clockwise order around C.
Not Burning Brownies: Told to me by Brown University's Michael Littman, devised by him and fellow computer scientist David McAllester.
Protecting the Statue: Moscow Mathematical Olympiad 2001, Problem C2 [40], contributed by V. A. Kleptsyn.

Gluing Pyramids: The argument given, sometimes called the "pup tent" solution, appeared in a 1982 article by Steven Young [108]. After the PSAT debacle, the Educational Testing Service created a panel, on which your author served, for reviewing the questions on their mathematics aptitude tests.

Precarious Picture: Contributed by Giulio Genovese, who heard it from more than one source in Europe.

Finding the Rectangles: Sent by Yan Zhang.

Tiling a Polygon: Sent by Dana Randall.

Tiling with Crosses: Sent by Senya Schlosman.

Cube Magic: I was reminded of this puzzle, which appeared in a Martin Gardner column, by Gregory Galperin. A polytope has the "Rupert" property if a straight tunnel can be made in it large enough to pass an identical polytope through (and, therefore, can be enlarged to pass an even bigger copy through). The cube case was originally proved, allegedly, by Prince Rupert of the Rhine in the late 17th century. As of now nine of the Archimidean polytopes are known to have the Rupert property, the last added recently in [77].

Circles in Space: Told to me by computer theorist Nick Pippenger of Harvey Mudd College; the construction given is due to Andrzej Szulkin of Stockholm University [100]. As pointed out to me by Johan Wästlund, this and some other tiling problems can also be solved using transfinite induction, as in **Double Cover by Lines** from Chapter 23.

Invisible Corners: Moscow Mathematical Olympiad 1995, Problem D7 [39], contributed by A. I. Galochkin.

theorem: This classic appears in [89] as "The Problem of the Calissons."

23. Infinite Choice

Hats and Infinity: Devised (to the best of my knowledge) by Yuval Gabay and Michael O'Connor, then graduate students at Cornell; the solution was already implicit in the work of Fred Galvin of the University of Kansas. Christopher Hardin (Smith College) and Alan D. Taylor (Union College) then included it their article [60]. Stan Wagon wrote it up as a Macalester College Problem of the Week; additional nice observations about this and the next version were made by Harvey Friedman (Ohio State), Hendrik Lenstra (Universiteit Leiden) and Joe Buhler (Reed College). It was the last of these, and (independently) Matt Baker of Georgia Tech, who communicated the puzzle to me.

All Right or All Wrong: Similar history to *Hats and Infinity* above. The "all green" solution was suggested to me by Teena Carroll of St. Norbert College.

Double Cover by Lines: Sent by Senya Shlosman, who is unaware of its origin.

Wild Guess: Sent by Sergiu Hart.

Exponent upon Exponent: Observed in 1964 by me and my college classmate Gerald Folland, now a professor at the University of Washington.

Find the Robot: Suggested by Barukh Ziv. The nice way of counting the rationals mentioned above the puzzle can be found in [19].

Figure Eights in the Plane: An oldie. I once heard it attributed to the late topologist R. L. (Robert Lee) Moore of the University of Texas.

Y's in the Plane: Supplied by Randy Dougherty of Ohio State University, a three-time winner of the U.S.A. Mathematical Olympiad.

theorem: Found in most elementary books on graph theory.

24. Startling Transformation

Sinking 15: Mentioned in Vol. II of [13], where it is attributed to E. Pericoloso Sporgersi. However, rather suspiciously, this "name" is found also on Italian railroad trains, warning passengers not to lean out of the window.

Slicing the Cube: Classic.

Expecting the Worst: Suggested by Richard Stanley.

Next Card Red: There is a discussion of this game in [27]. The modified version in our proof reminds me of a game which was described—for satiric purposes—in the *Harvard Lampoon* of March 30, 1967. The issue is dubbed "Games People Play Number" and the game in question appears to have been composed by D. C. Kenney and D. C. K. McClelland. Called "The Great Game of Absolution and Redemption," it required that the players move via dice rolls around a Monopoly-like board, until everyone has landed on the square marked "DEATH." So who wins? Well, at the beginning of the game, you were dealt a card face down from the Predestination Deck. At the conclusion, you turn your card face up, and if it says "damned," you lose.

Magnetic Dollars: Probabilists call the mechanism of this puzzle the *Pólya urn*, after the late George Pólya, a Hungarian-born professor at Stanford who famously wrote about problem-solving in mathematics.

Integral Rectangles: Our proof is #8 of fourteen in Stan Wagon's delightful article [103], and there are more!

Laser Gun: Brought to my attention by Giulio Genovese, who got it from Enrico Le Donne; they traced it to a 1990 Mathematics Olympiad in St. Petersburg. This puzzle sparked some more general research; see, for example, [76] and work by Keith Burns and Eugene Gutkin.

Random Intervals: This problem has a curious history. A colleague (Ed Scheinerman of The Johns Hopkins University) and I needed to know the answer in order to compute the diameter of a random interval graph, and we at first computed an asymptotic value of $2/3$. Later, using a lot of messy integrating, we found that the probability of finding an interval which intersects all others is *exactly* $2/3$, for any number of intervals (from 2 on up). The combinatorial proof given was found by Joyce Justicz, then taking a graduate reading course with me at Emory University.

Infected Cubes: The construction is due to Matt Cook and Erik Winfree of Caltech, the proof that it works to their colleague Len Schulman.

Gladiators, Version I: I have a theory that the "confidence" condition of Version I came about when someone tried to reconstruct Version II from the fair-game solution. The second proof of Version I is from Graham Brightwell of the London School of Economics.

Gladiators, Version II: From [66].

More Magnetic Dollars: This variation of Pólya's urn was posed and solved by Joel Spencer of New York University and his student Roberto Oliveira.

theorem: Suggested (as a puzzle) by Richard Stanley. The proof given is due to Henry Pollak of Bell Labs and now Columbia Teachers College. The theorem was originally stated and proved by Konheim and Weiss [74].

Bibliography

[1] Ethan Akin and Morton Davis, Bulgarian solitaire, *Amer. Math. Monthly* **92** #4 (1985), 310–330.

[2] Noga Alon, Tom Bohman, Ron Holzman, and Daniel J. Kleitman, On partitions of discrete boxes, *Discrete Math.* **257** #2–3 (28 November 2002), 255–258.

[3] Noga Alon and Joel H. Spencer, *The Probabilistic Method*, Fourth Edition, Wiley Series in Discrete Mathematics and Optimization, Hoboken, NJ, 2015.

[4] Noga Alon and D. B. West, The Borsuk-Ulam theorem and bisection of necklaces, *Proc. Amer. Math. Soc.* **98** #4 (December 1986), 623–628.

[5] Greg Aloupis, Robert A. Hearn, Hirokazu Iwasawa, and Ryuhei Uehara, Covering points with disjoint unit disks, *24th Canadian Conference on Computational Geometry*, Charlottetown, PEI (2012).

[6] D. André, Solution directe du problème résolu par M. Bertrand, *Comptes Rendus de l'Académie des Sciences Paris* **105** (1887), 436–437.

[7] Titu Andreescu and Zuming Feng, eds., *Mathematical Olympiads*, MAA Press, Washington DC, 2000.

[8] Frank Arntzenius, Some problems for conditionalization and reflection, *J. Philos.* **100** #7 (2003), 356–370.

[9] W. W. Rouse Ball, *Mathematical Recreations and Essays*, Macmillan & Co., London, 1892.

[10] D. Beaver, S. Haber, and P. Winkler, On the isolation of a common secret, in *The Mathematics of Paul Erdős* Vol. II, R. L. Graham and J. Nešetřil, eds., Springer-Verlag, Berlin, 1996, 121–135. (Reprinted and updated in 2013.)

[11] József Beck and William W. Chen, *Irregularities of Distribution*, Cambridge University Press, Cambridge, UK, 1987.

[12] Antonio Behn, Christopher Kribs-Zaleta, and Vadim Ponomarenko, The convergence of difference boxes, *Amer. Math. Monthly* **112** #5 (May 2005), 426–439.

[13] Elwyn R. Berlekamp, John H. Conway, and Richard K. Guy, *Winning Ways for Your Mathematical Plays*, Volumes 1–4, Taylor & Francis, Abingdon-on-Thames, UK, 2003.

[14] E. R. Berlekamp and T. Rodgers, eds., *The Mathemagician and Pied Puzzler*, AK Peters, Natick, MA, 1999.

[15] Geoffrey C. Berresford, A simpler proof of a well-known fact, *Amer. Math. Monthly* **115** #6 (June–July 2008), 524.

[16] J. Bertrand, Solution d'un problème, *Comptes Rendus de l'Académie des Sciences, Paris* **105** (1887), 369.

[17] B. Bollobás, Weakly k-saturated graphs, *Beiträge zur Graphentheorie* (Kolloquium, Manebach, May 1967), 25–31. Teubner-Verlag, Leipzig, 1968.

[18] Peter Boyland, Ivan Roth, Gabriella Pintér, István Laukó, Jon E. Schoenfield, and Stephen Wasielewski, Non-intersecting diagonals in an array, *J. Integer Seq.* **20** (2017), Article 17.2.4, 1–24.

[19] Neil Calkin and Herbert Wilf, Recounting the rationals, *Amer. Math. Monthly* **107** #4 (2000), 360–363.

[20] J. Cibulka, J. Kynčl, V. Mészáros, R. Stolař, and P. Valtr, Solution of Peter Winkler's pizza problem, in Fete of Combinatorics and Computer Science, Bolyai Society Mathematical Studies 20 (2010), Janos Bolyai Mathematical Society and Springer-Verlag, Berlin, 63–93.

[21] G. Cohen, I. Honkala, S. Litsyn, and A. Lobstein, *Covering Codes*, North-Holland, Amsterdam, 1997.

[22] Joel E. Cohen, Johannes H. B. Kemperman, and Gheorghe H. Zbăganu, Elementary inequalities that involve two nonnegative vectors or functions, *Proc. Natl. Acad. Sci. USA* **101** #42 (October 19 2004), 15018–15022.

[23] Brian Conrey, James Gabbard, Katie Grant, Andrew Liu, and Kent Morrison, Intransitive dice, *Math. Mag.* **89** #2 (April 2016), 133–143.

[24] J. H. Conway and B. Heuer, All solutions to the immobilizer problem, *Math. Intell.* **36** (2014), 78–86. https://doi.org/10.1007/s00283-014-9503-z.

[25] Thomas M. Cover, Universal Gambling Schemes and the Complexity Measures of Kolmogorov and Chaitin, Technical Report No. 12, October 14, 1974, Dept. of Statistics, Stanford University, Stanford, CA.

[26] T. M. Cover, Pick the largest number, in *Open Problems in Communication and Computation*, T. Cover and B. Gopinath, eds., Springer Verlag, Berlin, 1987, 152.

[27] T. Cover and J. Thomas, *Elements of Information Theory*, Wiley, Hoboken, NJ, 1991.

[28] H. S. M. Coxeter, A problem of collinear points, *Amer. Math. Monthly* **55** #1 (1948), 26–28.

[29] Henry Crapo and Ethan Bolker, Bracing rectangular frameworks I, *SIAM J. Applied Math.* **36** #3 (June 1979), 473–490.

[30] Eugene Curtin and Max Warshauer, The locker puzzle, *Math. Intell.* **28** #1 (2006), 28–31.

[31] P. Diaconis, R. L. Graham, and B. Sturmfels, Primitive partition identities, in Volume II of *Paul Erdős is 80*, János Bolyai Society, Budapest, 1986.

[32] D. Djukić, V. Janković, I. Matić, and N. Petrović, *The IMO Compendium*, Second Edition, Springer, New York, 2011.

[33] Cian Dorr, Sleeping beauty: In defence of Elga, *Analysis* **62** (2002), 292–296.

[34] F. J. Dyson, The problem of the pennies, *The Mathematical Gazette* **30** #291 (October 1946), 231–234.

[35] Sheldon J. Einhorn and I. J. Schoenberg, On Euclidean sets having only two distances between points II, *Indagationes Mathematicae* (Proceedings) **69** (1966), 489–504.

[36] A. Engel, *Problem-Solving Strategies*, Springer, New York, 1998.

[37] P. Erdős, A. Ginzburg, and A. Ziv, Theorem in the additive number theory, *Bull. Res. Council of Israel* **10F** (1961), 41–43.

[38] P. Erdös and G. Szekeres, A combinatorial problem in geometry, *Compos. Math.* **2** (1935), 463–470.

[39] R. Fedorov, A. Belov, A. Kovaldzhi, and I. Yashchenko, eds., *Moscow Mathematical Olympiads, 1993–1999*, MSRI/AMS 2011.

[40] R. Fedorov, A. Belov, A. Kovaldzhi, and I. Yashchenko, eds., *Moscow Mathematical Olympiads, 2000–2005*, MSRI/AMS 2011.

[41] D. Felix, Optimal Penney Ante strategy via correlation polynomial identities, *Electron. J. Comb.* **13** R35 (2006), 1–15.

[42] Thomas S. Ferguson, *A Course in Game Theory*, published by the author, 2020. Online at https://www.math.ucla.edu/~tom/Game_Theory.

[43] M. J. Fischer, S. Moran, S. Rudich, and G. Taubenfeld, The wakeup problem, *Proc. 22nd Symp. on the Theory of Computing*, Baltimore, MD, May 1990.

[44] M. J. Fischer and S. L. Salzberg, Finding a majority among n votes, *J. Algorithms* **3** 4 (December 1989), 362–380.

[45] Steve Fisk, A short proof of Chvátal's watchman theorem, *J. Comb. Theory B* **24** (1978), 374.

[46] M. L. Fredman, D. J. Kleitman, and P. Winkler, Generalization of a puzzle involving set partitions, *Barrycades and Septoku; Papers in Honor of Martin Gardner and Tom Rodgers*, Thane Plambeck and Tomas Rokicki, eds., AMA/MAA Spectrum Vol. 100, MAA Press, Providence, RI, 2020, pp. 125–129.

[47] Aaron Friedland, *Puzzles in Math and Logic*, Dover, Mineola, NY, 1971.

[48] Anna Gal and Peter Bro Miltersen, The cell probe complexity of succinct data structures, *International Colloquium on Automata, Languages and Programming (ICALP)*, Eindhoven, The Netherlands. Jos C. M. Baeten, ed., Lecture Notes in Computer Science 2719, Springer-Verlag, Berlin, 2003.

[49] G. A. Galperin, Playing pool with π: The number π from a billiard point of view, *Regul. Chaotic Dyn.* **8** (2003), 375–394.

[50] G. A. Galperin and A. K. Tolpygo, *Moscow Mathematical Olympiads*, Moscow, Prosveshchenie, 1986.

[51] Martin Gardner, *The Colossal Book of Short Puzzles and Problems*, W. W. Norton & Co., New York, NY, 2006.

[52] Martin Gardner, On the paradoxical situations that arise from nontransitive relations, *Sci. Am.* 231 4, (October 1974), 120–125.

[53] Martin Gardner, *Time Travel and Other Mathematical Bewilderments*, W. H. Freeman and Company, New York, NY, 1988. ISBN 0-7167-1925-8.

[54] Balász Gerencsér and Viktor Harangi, Too acute to be true: The story of acute sets, *Amer. Math. Monthly* **126** #10 (December 2019), 905–914.

[55] E. Goles and J. Olivos, Periodic behavior of generalized threshold functions, *Discrete Math.* **30** (1980), 187–189.

[56] Olivier Gossner, Penélope Hernández, and Abraham Neyman, Optimal use of communication resources, *Econometrica* **74** #6 (November 2006), 1603-1636.

[57] N. Goyal, S. Lodha, and S. Muthukrishnan, The Graham-Knowlton problem revisited, *Theory of Computing Systems* **39** #3 (2006), 399–412.

[58] Ronald L. Graham, Bruce L. Rothschild, and Joel H. Spencer, *Ramsey Theory*, Second Edition, Wiley Series in Discrete Mathematics and Optimization, Hoboken, NJ, 2013.

[59] L. J. Guibas and A. M. Odlyzko, String overlaps, pattern matching, and nontransitive games, *J. Comb. Theory Ser. A.* **30** (1981), 183–208.

[60] Christopher S. Hardin and Alan D. Taylor, A peculiar connection between the axiom of choice and predicting the future, *Amer. Math. Monthly* **115** #2 (2008), 91–96.

[61] G. H. Hardy, On certain oscillating series, *Q. J. Math.* **38** (1907), 269–288.

[62] Dick Hess, *Golf on the Moon*, Dover, Mineola, NY, 2014.

[63] Dick Hess, *The Population Explosion and Other Mathematical Puzzles*, World Scientific, Singapore, 2016. https://doi.org/10.1142/9886.

[64] Ross Honsberger, *Ingenuity in Mathematics*, New Mathematical Library Series **23**, Mathematical Association of America, Providence, RI, 1970.

[65] David Kahn, *The Codebreakers*, Scribner & Sons, New York, 1967 and 1996.

[66] K. S. Kaminsky, E. M. Luks, and P. I. Nelson, Strategy, nontransitive dominance and the exponential distribution, *Austral. J. Statist.* **26** #2 (1984), 111–118.

[67] K. A. Kearns and E. W. Kiss, Finite algebras of finite complexity, *Discrete Math.* **207** (1999), 89–135.

[68] Murray Klamkin, *International Mathematical Olympiads 1979–1985*, Mathematical Association of America, Providence, RI, 1986.

[69] Michael Kleber, The best card trick, *Math. Intell.* **24** #1 (Winter 2002), 9–11.

[70] Anton Klyachko, A funny property of sphere and equations over groups, *Commun. Algebra* **21** #7 (1993), 2555–2575.

[71] G. H. Knight, in *Notes and Queries, Oxford Journals (Firm)*, Oxford U. Press, Oxford, UK, 1872.

[72] K. Knauer, P. Micek, and T. Ueckerdt, How to eat 4/9 of a pizza, *Discrete Math.* **311** #16 (2011), 1635–1645.

[73] Joseph D. E. Konhauser, Dan Velleman, and Stan Wagon, *Which Way Did the Bicycle Go?*, Cambridge University Press, 1996.

[74] Alan G. Konheim and Benjamin Weiss, An occupancy discipline and applications, *SIAM J. Appl. Math.* **14** #6 (November 1966), 1266–1274.

[75] Boris Kordemsky, *The Moscow Puzzles*, Charles Scribner's Sons, New York, 1971.

[76] J.-F. Lafont and B. Schmidt, Blocking light in compact Riemannian manifolds, *Geom. Topol.* **11** (2007), 867–887.

[77] Gerárd Lavau, The truncated tetrahedron is Rupert, *Amer. Math. Monthly* **126** #10 (2019), 929–932.

[78] Tamas Lengyel, A combinatorial identity and the world series, *SIAM Rev.* **35** #2 (June 1993), 294–297.

[79] László Lovász, *Combinatorial Problems and Exercises*, North Holland, Amsterdam, 1979.

[80] L. Lovász and P. Winkler, A note on the last new vertex visited by a random walk, *J. Graph Theory* **17** #5 (November 1993), 593–596.

[81] Anany Levitin and Maria Levitin, *Algorithmic Puzzles*, Oxford U. Press, Oxford, UK, 2011.

[82] J. E. Littlewood, The dilemma of probability theory, *Littlewood's Miscellany*, Béla Bollobás, ed., Cambridge University Press, Cambridge, UK, 1986.

[83] A. Liu and B. Shawyer, eds., *Problems from Murray Klamkin, The Canadian Collection*, MAA Problem Book Series, MAA Press, Providence, RI, 2009.

[84] Chris Maslanka and Steve Tribe, *The Official Sherlock Puzzle Book*, Woodland Books, Salt Lake City, UT, 2018. ISBN 978-1-4521-7314-6.

[85] E. C. Milner and S. Shelah, Graphs with no unfriendly partitions, in *A Tribute to Paul Erdős*, A. Baker et al., eds., Cambridge University Press, Cambridge, UK, 1990, 373–384.

[86] Frederick Mosteller, *Fifty Challenging Problems in Probability*, Addison-Wesley, Boston, MA, 1965.

[87] Colm Mulcahy, *Mathematical Card Magic: Fifty-Two New Effects*, CRC press, Boca Raton, FL, 2013. ISBN: 978-14-6650-976-4.

[88] Serban Nacu and Yuval Peres, Fast simulation of new coins from old, *Ann. Appl. Probab.* **15** #1A (2005), 93–115.

[89] Roger B. Nelsen, *Proofs Without Words*, Mathematical Association of America, Providence, RI, 1993.

[90] W. Penney, Problem: Penney-ante, *J. Recreat. Math.* **2** (1969), 241.

[91] Vera Pless, *Introduction to the Theory of Error-Correcting Codes*, John Wiley & Sons, Hoboken, NJ, 1982.

[92] H. Prüfer, Neuer Beweis eines Satzes über Permutationen, *Arch. Math. Phys.* **27** (1918), 742–744.

[93] T. F. Ramaley, Buffon's noodle problem, *Amer. Math. Monthly* **76** #8 (October 1969), 916–918.

[94] M. Reiss, Über eine Steinersche combinatorische Aufgabe, welche im 45sten Bande dieses Journals, Seite 181, gestellt worden ist, *J. Reine Angew. Math.* **56** (1859), 226–244.

[95] I. J. Schoenberg, *Mathematical Time Exposures*, Mathematical Association of America, Providence, RI, 1982.

[96] Harold N. Shapiro, *Introduction to the Theory of Numbers*, Dover, Mineola, NY, 1982.

[97] D. O. Shklarsky, N. N. Chentov, and I. M. Yaglom, *The USSR Problem Book*, W. H. Freeman and Co., San Francisco, 1962.

[98] R. P. Sprague, Über mathematische Kampfspiele, *Tohoku Math. J.* **41** (1935–1936), 438–444.

[99] Roland Sprague, *Recreation in Mathematics*, Blackie & Son Ltd., London, 1963.

[100] Andrzej Szulkin, $\mathbb{R}^3$ is the union of disjoint circles, *Amer. Math. Monthly* **90** #9 (1983), 640–641.

[101] P. Vaderlind, R. Guy, and L. Larson, *The Inquisitive Problem Solver*, Mathematical Association of America, Providence, RI, 2002.

[102] Tom Verhoeff, *Circle of Lamps*, Wolfram Demonstrations Project 2020.
https://demonstrations.wolfram.com/CircleOfLamps/.

[103] Stan Wagon, Fourteen proofs of a result about tiling a rectangle, *Amer. Math. Monthly* **94** #7 (August–September 1987), 601–617.

[104] Chamont Wang, *Sense and Nonsense of Statistical Inference*, Marcel Dekker, New York, 1993.

[105] P. Winkler, The sleeping beauty controversy, *Amer. Math. Monthly* **124** #7 (August–September 2017) 579–587; also in *The Best Writing on Mathematics 2018*, M. Pitici, ed., Princeton University Press, Princeton, NJ and Oxford, UK, 2019, 117–129.

[106] W. A. Wythoff, A modification of the game of Nim, *Nieuw Arch. Wisk.* **7** (1907) 199–202.

[107] I. M. Yaglom, *Geometric Transformations I*, Translated by Allen Shields. Mathematical Association of America, Providence, RI, 1962.

[108] S. C. Young, The mental representation of geometrical knowledge, *J. Math. Behav.* **3** #2 (1982), 123–144.

[109] G. Zbăganu, A new inequality with applications in measure and information theories, *Proc. Rom. Acad. Series A* **1** #1 (2000), 15–19.

[110] Yao Zhang, *Combinatorial Problems in Mathematical Competitions* (Vol. 4, Mathematical Olympiad Series), East China Normal University Press, Shanghai, 2011.

The Author

Peter Winkler is the William Morrill Professor of Mathematics and Computer Science at Dartmouth College, and for 2019–2020, the Distinguished Visiting Professor for the Public Dissemination of Mathematics at the National Museum of Mathematics. He is the author of 160 research papers, a dozen patents, two previous puzzle books, a book on cryptographic techniques in the game of bridge, and a portfolio of compositions for ragtime piano.

Index

A

Adding, multiplying, and grouping, xxxviii, xci, 155

Air routes in Aerostan, xxii, xci, 42

Algebra, 55–65
linear, 107, 236
topology, 39

Algorithm, lxxxi, 15, 146–148
Bob's, 289
greedy, 51, 69

All right or all wrong, xli, xci, 343–344

Alternate connection, lxv, xci, 12

Alternating powers, lxvii, xci, 106–107

Alternative dice, lxvi, xci, 62–65

An attractive game, lxiv, xci, 189–190

Angles in space, lxxiii, xci, 229–230

Angry baseball, lxxiv, xci, 128

Ants on the circle, li, xci, 297–298

Area–perimeter match, xlii, 58

Ascending and descending, l, xlix, xci, 157–158

Attention paraskevidekatria-phobes, xxxii, xci, 4–5

Attic lamp, xxviii, xci, 166

Average, 91

Axiom of choice (AC), 341–352

B

Baby frog, xlvi, xci, 210

Bacterial reproduction, xxiv, xcii, 17

Bacteria on the plane, li, xci, 256–257

Badly designed clock, lxxxi, xcii, 283–284

Bags of marbles, xxiv, xcii, 2

Bat and ball, xvii, xcii, 55

Belt around the earth, xxviii, xcii, 56

Bertrand's postulate, 74

Biased betting, xlix, xcii, 118–119

Bias test, xlv, xcii, 170–171

Bidding in the dark, xxiii, xcii, 182

Big pairs in a matrix, xxv, xcii, 102–103

Bindweed and honeysuckle, lxv, xcii, 12–16

Birthday match, xx, xcii, 110–111

Bluffing with reals, lxxi, xcii, 314–315

Boarding the manhole, lv, xcii, 225–227

Boardroom reduction, lxxxvi, xcii, 323–325

Borchardt, C. W., 14

Box in a box, lxix, xcii, 228–229

Boy born on tuesday, xxi, xcii, 113–114

Bracing the grid, xlviii, xcii, 45–47

Breaking a chocolate bar, xxxiii–xxxiv, xcii, 255

Broken ATM, xxv, xcii, 67

Buffon's needle, lxxxvi, xcii, 197–198

Bugging a disk, xxix–xxx, xciii, 103–104

Bugs on a line, lxxix, xciii, 264–265

Bugs on a polyhedron, lxxv, xciii, 262–263

Bugs on a pyramid, xliv, xciii, 34

Bugs on four lines, lxii, xciii, 227–228

Bulgarian solitaire, lxxxix, xciii, 268–271

C

Cake
  candles on a, xxxvii, xciii, 79–80
  division, xxiii, xcv, 223
  ice cream, lxxxiii, xcvii, 320–322

Candles on a cake, xxxvii, xciii, 79–80

Cards from their sum, xxxiv, xciii, 68–69

Cayley, A., 14

Chameleons, lvii–lviii, xciii, 22–23

Chessboard
  frames on a, lxxiv, xcvi, 261–262
  guess, lxxiii–lxxiv, xciii, 245–246
  lemming on a, xlii, xcviii, 104–105

Chinese Nim, lii, xciii, 308–310

Chomp, xxii, xciii, 101–102

Circle
  ants on the, li, xci, 297–298
  light bulbs in a, lxxxii, xcviii, 318–319
  natives in a, xix, xcix, 138
  points on a, xxx, ci, 116
  in space, lxxx, xciii, 338–339

Circular
  shadows I, xxxii, xciii, 329–330
  shadows II, lxiv, xciii, 228

Coconut classic, lxvi–lxvii, xciii, 87–89

Coin
  game, lxxxv, xciii, 133–134
  in a row, lxi–lxii, xciii, 86–87, 286
  on the table, lix, xciii, 83–86
  testing, lxxviii, xciii, 131–132

Colors and distances, lxxxviii, xciv, 200

Comparing numbers
  version I, xxxvii–xxxviii, xciv, 116–117
  version II, xxxviii, xciv, 118

Competing for programmers, liii, xciv, 47–48

Conversation on a bus, lviii–lix, xciv 173–174

Conway, J. H., 236

Conway's immobilizer, lxviii, xciv 146–148

Corner
  fourth, xxviii, xcvi, 17–18
  invisible, lxxxv, xcviii, 339–340
  pegs on the, lxxx, c, 317

subtracting around the,
lxiii–lxiv, civ, 23–24
Coupling, 123
Covering the stains, lxxxvii, xciv,
198–200
Crossing
heterogeneous, 200
homogeneous, 200
horizontal, 72
the river, xxiii, xciv,
273–274
vertical, 72
Cryptography, 176–178
Cube
infected, lxxi, xcvii,
362–364
magic, lxxvi, xciv, 337–338
painting, xxxix, c, 224
slicing the, xxi, ciii,
353–354
spiders on a, xxxvi, civ, 43
Curve(s)
on potatoes, xliii, xciv,
224–225
on a sphere, xlix, xciv, 105
and three shadows,
lxxviii–lxxix, xciv,
230–233
unbroken, cvi, 145–146
Cutting
necklace, xxix, xciv, xcv,
31–32
pizza, 3–4

D
Deterministic poker, lxii, xciv,
313–314
Dice
alternative, lxvi, xci, 62–65
funny, lxxxiii, xcvi, 285–286

losing at, lii–liii, xcviii, 10
same sum, lxviii, cii,
159–160
set of non-transitive, 285
Discrete probability, 115
Dishwashing game, lxxii–lxxiii,
xciv, 122–124
Divisibility game, xxxvii, xciv, 70
Division
easy cake, xxiii, xcv, 223
first-grade, lxii, xcvi, 259
rule, 5
Domino task, xxv, xciv, 77–78
Dots and boxes variation, xxiv,
xciv, 102
Dot-town exodus, 1, xcv,
171–173
Double cover by lines, lii, xcv,
344–345
Drawing socks, 1, xcv, 185–186

E
Early commuter, lxiii, xcv, 12
Easy cake division, xxiii, xcv,
223
Efficient pizza-cutting, xxviii,
xcv, 3–4
Emptying a bucket,
lxxxii–lxxxiii, xcv,
319–320
Equilibrium, 313–314
Erdős–Szekeres theorem, 157
Even split, lxvi, xcv, 74–75
Even-sum billiards, liii, xcv,
21–22
Expectation, 181–204
Expecting the worst, xxviii, xcv,
354
Exponent upon exponent, lxi,
xcv, 348–349

F

Factorial(s)
  coincidence, lviii, xcv, 74
  and squares, lxxii, xcv,
    75–76
Fair play, xxiv, xcv, 275–276
Falling ants, xliii, xcv, 297
Faulty combination lock, xlvii,
  xcv, 8–10
Fewest slopes, xxvi, xcv, 139
Fibonacci multiples, lxxxi, xcv,
  318
Fifteen bits and a spy, lxxx, xcv,
  249–252
Figure eights in the plane, lxiv,
  xcv, 350
Filling
  a bucket, lx, xcvi, 279–280
  the cup, lxxxix, xcvi,
    202–204
Finding
  the counterfeit, xix, xcvi,
    163–164
  a Jack, xxxv, xcvi, 96
  the missing number, xxvii,
    xcvi, 276
  the rectangles, lxiv, xcvi,
    335
Find the robot, lxiii, xcvi,
  349–350
First-grade division, lxii, xcvi,
  259
First odd number, xxxiii, xcvi,
  140–141
Flipping the pentagon, lxxx,
  xcvi, 265–267
Flying saucers, xlviii–xlix, xcvi,
  81–83
Fourth corner, xxviii, xcvi,
  17–18
Frames on a chessboard, lxxiv,
  xcvi, 261–262
Functions
  choice, 342
  generating, 63
  parking, 367, 368
  potential, 255, 256, 260, 268
Funny dice, lxxxiii, xcvi,
  285–286

G

Galton, F., 64
Game/gaming
  attractive, lxiv, xci, 189–190
  coin, lxxxv, xciii, 133–134
  of desperation, lx–lxi, xcvi,
    312
  dishwashing, lxxii–lxxiii,
    xciv, 122–124
  divisibility, xxxvii, xciv, 70
  snake, xliv, ciii, 45
  split, lxxiv, civ, 126–128
  quilt, xlvii, xcvi, 60
Garnering fruit, lxxxiv, xcvi,
  37–40
Gasoline crisis, liv, xcvi, 83
Generating
  functions, 63
  the rationals, lxxxii, xcvi,
    285
Gladiators
  version I, lxxvi–lxxvii,
    xcvii, 364–365
  version II, lxxvii, xcvii,
    365–366
Gluing pyramids, lxi, xcvii,
  333–334
Graphography, 41–53
Greedy algorithm, 51
Guarding the gallery, xlviii,
  xcvii, 210–212

H

Half grown, xvii, xcvii, 1

Half-right hats, xxxv, xcvii, 19, 247–249

Halfway points, lxxii, xcvii, 107–108

Hammer-and-tongs approach, 273–293

Hamming code, 235–252

Ham-sandwich theorem, 39

Handshakes at a party, xlii, xcvii, 43–44

Hardy, G. H., 101, 106

Hats and infinity, xli, xcvii, 343

Hazards of electronic coinflipping, xli–xlii, xcvii, 33–34

Home-field advantage, liv, xcvii, 119–120

Hopping and skipping, xxxvi, xcvii, 57–58

I

Ice cream cake, lxxxiii, xcvii, 320–322

Identifying the majority, xl, xcvii, 276–277

Impressionable thinkers, lxx, xcvii, 260–261

Increasing routes, xlix, xcvii, 97–99

Induction, 205–221
  hypothesis, 205

Infected
    checkerboard, lxvii, xcvii, 260
    cubes, lxxi, xcvii–xcviii, 362–364

Information, 163–179

Integral rectangles, lii, xcviii, 357–358

Intermediate Value Theorem (IVT), 29–40

Invisible corners, lxxxv, xcviii, 339–340

K

King's salary, xxxviii, xcviii, 142–143

Kuratowski's theorem, 351

L

Lame rook, lxxviii, xcviii, 219–221

Laser gun, lvii, xcviii, 358–361

Lattice points, xxxiii, xcviii, 154–155

Law of large numbers, 124

Leading all the way, lxxxix, xcviii, 135–136

Lectures in Abstract Algebra (Jacobson), xx, 328

Lemming on a chessboard, xlii, xcviii, 104–105

Life is a bowl of cherries, xxvii, xcviii, 236–238

Life isn't a bowl of cherries?, xxxi, xcviii, 238–239

Life-saving transposition, lxxxvii, xcviii, 289

Light bulbs in a circle, lxxxii, xcviii, 318–319

Line(s)
    bugs on a, lxxix, xciii , 264–265
    bugs on four, lxii, xciii, 227–228
    double cover by, lii, xcv, 344–345
    through a grid, xxiii, xcviii, 153–154

up by height, xliii, xcviii,
 156–157
unbroken, xlvi, cvi,
 144–145
Line segments, xxxiii, xcviii,
 154–155
Locker doors, liv, xcviii, 73–74
Losing at dice, lii–liii, xcviii, 10
Lost boarding pass, xlii, xcviii,
 80–81
Love in Kleptopia, lxxx–lxxxi,
 xcviii, 282–283
Lozenge, 339

M

Magnetic dollars, xli, xcviii,
 356–357
Majority hats, lxxix, xcix,
 246–249
Match
 area–perimeter, xlii, xci, 58
 birthday, xx, xcii, 110–111
 coins, lxviii–lxix, xcix,
 174–176
Mathematical bookworm, xx,
 xcix, 328
Measuring with fuses, xxxv,
 xcix, 141–142
Meeting the ferry, xx, xcix, 327
Meet the Williams sisters, xxxiv,
 xcix, 5
Men with sisters, xxiv, xcix, 93
Missing
 card, xxxix, xcix, 167–169
 digit, lxi, xcix, 23
Monge point, 231, 232
Monge's theorem, 233
Monk on a mountain, xx, xcix,
 30–31
More magnetic dollars, lxxxvii,
 xcix, 366–369

Moser spindle, 149
Moth's tour, lxxxiv, xcix, 323
Multiples
 fibonacci, lxxxi, xcv, 318
 zeroes and ones, lv, cvii,
 158–159

N

Names in boxes, lxxxiv, xcix,
 288–289
Napkins in a random setting,
 lxxvii–lxxviii, xcix,
 194–195
Natives in a circle, xix, xcix, 138
Negative thinking, 101–108
Next card
 bet, lxix, xcix, 190–192
 red, xxxii, xcix, 354–356
Nimbers, 235–252
Non-repeating string, xlv, xcix,
 209–210
North by northwest, lxi, xcix,
 11–12
Not burning brownies, lii, xcix,
 332
No twins today, xvii, xcix, 137
Numbers. *See also* Comparing
 numbers
 first odd, xxxiii, xcvi,
 140–141
 on foreheads, xliv–xlv, xcix,
 71
 law of large, 124
 missing, xxvii, xcvi, 276
 rolling all the, xxvi, cii, 183
 safety, 67–76
 self-referential, lxxxix, ciii,
 289–293
 skipping a, xlv, ciii, 35
 small, 77–89

O

Odd light flips, xxxix, c, 207–208
Odd run of heads, xxix, c, 56
One-bulb room, lxx, c, 280–281
Option hats, lxix–lxx, c, 241–245
Other side of the coin, xxi, c, 111–112

P

Packing slashes, xliii, c, 143–144
Painting
    cubes, xxxix, c, 224
    fence, lxxxviii, c, 201–202
    polyhedron, li, c, 225
Pairs at maximum distance, xxxvii, c, 104
Pancake stacks, xl, c, 307–308
Parity, 17–27
Path through the cells, l, c, 212–213
Peek advantage, xliii, c, 169–170
Pegs
    on the corners, lxxx, c, 317
    on the half-plane, lvi, c, 257–258
    in a square, lx, c, 258–259
Pentagon, flipping the, lxxx, xcvi, 265–267
Permutations, 20–21
Phone call, xviii, c, 273
Picking the athletic committee, lxxxiii, c, 267–268
Pie in the sky, xxxii, ci, 295–296
Pigeonhole principle, 151–162
Ping-pong progression, xxxix, ci, 185
Players and winners, xxx, ci, 166–167
Points
    on a circle, xxx, ci, 116

halfway, lxxii, xcvii, 107–108
lattice, xxxiii, xcviii, 154–155
Monge, 231, 232
red and blue, xxxix, cii, 255–256
system, 299
two-point conversion, lxxv–lxxvi, cv, 128–129
Poker quickie, xxiii, ci, 138
Polite pizza protocol, 286
Polyhedron
    bugs on, lxxv, xciii, 262–263
    faces, xix, ci, 153
    painting, li, c, 225
Polygon
    on the grid, lx, ci, 280
    midpoints, xliv, ci, 331
Poorly placed dominoes, li, ci, 277–278
Portrait, xvii, ci, 303
Potentials, 253–271
Powers
    alternating, lxvii, xci, 106–107
    of roots, lxiv, ci, 87
    of two, xviii, ci, 1–2
Precarious picture, lxiii, ci, 334–335
Prime test, xlii, ci, 70
Prisoner(s)
    and dog, lxxix, ci, 281–282
    and gloves, xlvii, ci, 20–21
Probabilistically checkable proofs, 252
Probability, 109–136
    discrete, 115
    reverse conditional, 112
    whose bullet? xxii, cvi, 114

Profit and loss, lv, ci, 213–215
Protecting the statue, lvi–lvii, ci,
    332–333
Pyramid
    bugs on, xliv, xciii, 34
    gluing, lxi, xcvii, 333–334

R
Raising art value, xxxiii, ci,
    94–95
Ramsey's theorem, 202, 203,
    220, 221
Random
    bias, lxxvi, ci, 131
    chord, lxxvi, ci, 130–131
    intersection, lv, cii, 186–188
    intervals, lxi, cii, 361–362
    judge, lxxiii, cii, 124
    variables, 181
Rating the horses, xxxv, cii, 5–6
Rationals, generating, lxxxii,
    xcvi, 285
Recovering the polynomial,
    xlvii, cii, 59
Rectangles tall and wide, xlviii,
    cii, 71–73
Red
    and black sides, xlv–xlvi,
        cii, 58–59
    and blue hats in a line,
        xxxv–xxxvi, cii, 19–20
    points and blue, xxxix, cii,
        255–256
Returning pool shot, xl, cii,
    296–297
Reverse conditional
    probabilities, 112
Reversibility theorem, 317
Righting the pancakes, xxxi, cii,
    254–255
Rolling

all the numbers, xxvi, cii,
    183
pencil, xxi, cii, 328
a six, lxxvii, cii, 193–194
Rotating coin, xviii, cii, 295
Roulette
    for parking money, lxxxv,
        cii, 196–197
    for the unwary, lix, cii,
        188–189
Rows and columns, xliv, cii, 7

S
Salaries and raises, xxv, cii, 2–3
Same sum
    dice, lxviii, cii, 159–160
    subsets, xxx, ciii, 154
Second ace, xxvi, ciii, 114–115
Self-referential number, lxxxix,
    ciii, 289–293
Sequencing the digits, xlvi, ciii, 8
Serious candidates, lxxv, ciii,
    192–193
Service options, lxvii–lxviii, ciii,
    121
Set of non-transitive dice, 285
Seven cities of gold, lxxxvi, ciii,
    148–150
Sharing a pizza, lxxxiv, ciii,
    286–287
Shoelaces at the airport, xli, ciii,
    6
Shoes, socks, and gloves, xviii,
    ciii, 151–152
Signs in an array, xix, ciii,
    253–254
Sinking 15, xx, ciii, 353
Skipping a number, xlv, ciii, 35
Slabs in 3-space, lx, ciii, 227
Sleeping beauty, lxxxv–lxxxvi,
    ciii, 134–135

Slicing the cube, xxi, ciii, 353–354

Small numbers, 77–89

Snake game, xliv, ciii, 45
   puzzle, 52

Soldiers in a field, liv, ciii, 105–106

Space
   angles in, lxxiii, xci, 229–230
   circles in, lxxx, xciii, 338–339
   slabs in 3-space, lx, ciii, 227
   three-dimensional, 223–233

Spaghetti loops, xxx, ciii, 183–184

Sphere and quadrilateral, lxx, ciii, 298

Spiders on a cube, xxxvi, civ, 42–43

Spinning switches, xxix, civ, 78–79

Split games, lxxiv, civ, 126–128

Splitting
   hexagon, xxvii, civ, 329
   polygon, lxvi, civ, 36–37
   stacks, lvii, civ, 11

Sprague-Grundy theorem, 238

Sprinklers in a field, xxiv, civ, 275

Squaring the mountain state, xix, civ, 29–30

Stone–Tukey theorem, 39

Stopping after the boy, xxviii, civ, 115–116

Strategy-stealing argument, 102

Strength of schedule, liii, civ, 60–61

Subsets with constraints, xxix, civ, 67–68

Subtracting around the corner, lxiii–lxiv, civ, 23–24

Summing fractions, lxix, civ, 216

Sums
   cards from their, xxxiv, xciii, 68–69
   and differences, lxxv, civ, 281
   of two squares, xxxvii, civ, 96–97

Swapping executives, xxxi, civ, 206–207

Swedish lottery, lxxvi, civ, 316–317

System point, 299

T

Testing ostrich eggs, xxxii, civ, 304–306

Three-dimensional space, 223–233

Three natives at the crossroads, xxvi, civ, 165–166

Three negatives, xliv, cv, 58

Three sticks, xxxvi, cv, 32–33

Three-way
   duel, xxvii, cv, 303–304
   election, xlv, cv, 7–8

Tiling
   a polygon, lxv, cv, 335–336
   with crosses, lxxi, cv, 336–337
   with L's, lxxii, cv, 216–218

Tipping the scales, xxii, cv, 92

Touring an island, lxxxi, cv, 317

Transformation, 353–369

Trapped in thickland, xl, cv, 330–331

Traveling salesmen, lxxvii, cv, 218–219

Truly even split, xliii, cv,
208–209
Turning the die, lvi, cv, 310–312
Two balls and a wall, lxxxix, cv,
299–302
Two different distances, xxx, cv,
139–140
Two monks on a mountain, lxvi,
cv, 49–50
Two-point conversion,
lxxv–lxxvi, cv, 128–129
Two round-robins, lviii, cv,
61–62
Two runners, xxv, cv, 55
Two sheriffs, lxxxviii, cv,
176–179

U
Unanimous hats, xxxii–xxxiii,
cvi, 18–19, 248, 249
Unbreakable domino cover, lvi,
cvi, 278–279
Unbroken
curves, l, cvi, 145–146
lines, xlvi, cvi, 144–145
Unfriendly partition, 270–271
Uniformity at the bakery, lix,
cvi, 215–216
Uniform unit distances, xxvi,
cvi, 205–206
Uniting the loops, lxv, cvi,
24–27

V
Variation, dots and boxes, xxiv,
102

Vertices, 351
Von Neumann's trick, 275–276

W
Waiting for heads, xxxiv, cvi,
95–96
Watches on the table, xxix, cvi,
93–94
Watermelons, xxi, cvi, 2
Watson, H. W., 64
Weighs and means, 91–99
Whim-Nim, li, 239–241
Whose bullet?, xxii, cvi, 114
Who won the series?, lxxii, cvi,
121–122
Wild guess, lvii, cvi, 345–348
Winning at Wimbledon, xix, cvi,
110
Wins in a row, lxxiii, cvi, 125
Wires under the Hudson, lviii,
cvi, 48–49
Worms and water, lxxxii, cvi,
284–285
Worst route, lxxi–lxxii, cvi,
51–53

Y
Y's in the plane, lxxi, cvi,
350–352

Z
Zeroes and ones, lv, cvii,
158–159
Zero-sum vectors,
lxxxvii–lxxxviii, cvi,
161–162